# THE COSMIC MICROWAVE BACKGROUND: 25 YEARS LATER

PROCEEDINGS OF A MEETING ON
'THE COSMIC MICROWAVE BACKGROUND: 25 YEARS LATER',
HELD IN L'AQUILA, ITALY, JUNE 19–23, 1989

Edited by

N. MANDOLESI

*ITESRE/CNR, Bologna, Italy*

and

N. VITTORIO

*Università dell'Aquila, Italy*

KLUWER ACADEMIC PUBLISHERS

DORDRECHT / BOSTON / LONDON

Library of Congress Cataloging in Publication Data

The Cosmic microwave background, 25 years later : proceedings of a
    meeting on "The Cosmic Microwave Background, 25 Years Later", held
    in L'Aquila, Italy, June 19-23, 1989 / edited by N. Mandolesi, N.
    Vittorio.
        p.    cm. -- (Astronomical and space science library ; v. 164)
    Includes index.
    ISBN 0-7923-0849-2 (alk. paper)
    1. Cosmic background radiation--Congresses.    I. Mandolesi, N.
    II. Vittorio, N.    III. Series.
    QB991.C64C67    1990
    523.01'875344--dc20                                          90-39754

ISBN-13: 978-94-010-6779-9    e-ISBN-13: 978-94-009-0655-6
DOI: 10.1007/ 978-94-009-0655-6

Published by Kluwer Academic Publishers,
P.O. Box 17, 3300 AA Dordrecht, The Netherlands.

Kluwer Academic Publishers incorporates
the publishing programmes of
D. Reidel, Martinus Nijhoff, Dr W. Junk and MTP Press.

Sold and distributed in the U.S.A. and Canada
by Kluwer Academic Publishers,
101 Philip Drive, Norwell, MA 02061, U.S.A.

In all other countries, sold and distributed
by Kluwer Academic Publishers Group,
P.O. Box 322, 3300 AH Dordrecht, The Netherlands.

*Printed on acid-free paper*

# THE COSMIC MICROWAVE
# BACKGROUND:
# 25 YEARS LATER

# ASTROPHYSICS AND SPACE SCIENCE LIBRARY

A SERIES OF BOOKS ON THE RECENT DEVELOPMENTS
OF SPACE SCIENCE AND OF GENERAL GEOPHYSICS AND ASTROPHYSICS
PUBLISHED IN CONNECTION WITH THE JOURNAL
SPACE SCIENCE REVIEWS

PROCEEDINGS
VOLUME 164

# TABLE OF CONTENTS

## HISTORICAL REMARKS

## SMALL SCALE CMB ANISOTROPY

CONFERENCE PHOTOGRAPH

# FOREWORD

This book is the result of a Meeting held in L'Aquila (Italy) from the 19th to the 23rd of June 1989. The aim of the Meeting was to gather together the people actively working on the Cosmic Microwave Background radiation, both from an experimental and from a theoretical point of view. In view of the intensive current activity in this field, including ongoing (COBE) and forthcoming (RELIC II, ISO, AELITA, etc.) space missions, a meeting fully dedicated to this important topic was timely. The meeting also celebrated the 25th anniversary of the Microwave Background discovery made in 1964 by the Nobel Prize winners A.Penzias and R.Wilson. We greatly regret that we were not able to have them at the Meeting. There is of course another person whose absence we regret, namely R.H.Dicke, who motivated a generation of experimentalists and theoreticians to open and study this new field of research.

As organizers of the Meeting, we would like to express our gratitude to the people who contributed to its success. We want to thank the members of the Scientific Organizing Committee for their assistance, suggestions and encouragement, the invited speakers for their excellent presentations, and the chairmen for their help in handling the various Sessions. We would like to thank P.Palazzi for her help in secretarial work, dr. L.Marra of the University of L'Aquila for his suggestions, the Consiglio Regionale dell' Abruzzo for generously giving us access to their Conference Hall, and, finally, Athena Traduzioni Congressi for the efficient and professional organization of the logistic of the Meeting and of the social events.

We are also pleased to thank all the various institutions which have provided us with generous financial support, without which the realization of the Meeting would have not been possible. These Institutions are:

- Consiglio Nazionale delle Ricerche

- Universita' dell'Aquila

- Alitalia

- Ministero degli Affari Esteri

- Cassa di Risparmio degli Abruzzi

- Istituto Nazionale di Fisica Nucleare

We believe that these Proceedings, containing the invited talks, summaries of open discussions, and the concluding remarks, represent a good overview of the state of the art in this field at the date of the Meeting. We hope that this book can provide stimulating reading to young students and researchers approaching the subject.

Reno Mandolesi

Nicola Vittorio

## Scientific Organizing Committee:

| | | |
|---|---|---|
| J.R. | BOND | (Canada) |
| P. | CRANE | (ESO) |
| R.D. | DAVIES | (UK) |
| G. | DE ZOTTI | (Italy) |
| H.P. | GUSH | (Canada) |
| S. | HAYAKAWA | (Japan) |
| F. | MELCHIORRI | (Italy) |
| N. | MANDOLESI | (Italy) |
| R.B. | PARTRIDGE | (USA) |
| J. | SILK | (USA) |
| G. | SMOOT | (USA) |
| I.A. | STRUKOV | (USSR) |
| R. | SUNYAEV | (USSR) |
| N. | VITTORIO | (Italy) |
| D. | WILKINSON | (USA) |

## Local Organizing Committee:

| | | |
|---|---|---|
| N. | MANDOLESI | ITESRE/CNR Bologna (Italy) |
| N. | VITTORIO | Università dell'Aquila (Italy) |

**Session Chairmen:**

- Monday, June 19

  - P.J.E. Peebles
  - A.N. Lasenby

- Tuesday, June 20

  - P.L. Richards
  - J.R. Bond

- Wednesday, June 21

  - I.A. Strukov

- Thursday, June 22

  - A.N. Lasenby
  - S. Hayakawa

- Friday, June 23

  - D.W. Sciama
  - D.T. Wilkinson

**Open Discussion Leaders**:

- <u>CMB Anisotropies</u>:

    P. Lubin

- <u>CMB Spectrum</u>:

    C.J. Hogan

    G. Smoot

- <u>CMB and other backgrounds</u>:

    S. Hayakawa

    F. Melchiorri

# LIST OF REGISTERED PARTICIPANTS

**ARGUESO F.**
Institute of Astronomy, Madingley Road, Cambridge, CB3 OHA, UK

**BANDAY A. J.**
Physics Dept., Durham University, South Road, Durham, UK

**BENSADOUN M.**
UC Berkeley, 3125 Lewinston Berkeley, CA 94705, USA

**BERSANELLI M.**
I.F.C.T.R./CNR, via Bassini 15, Milano, Italy

**BIRKINSHAW M.**
Astronomy Dept., Harvard University, 60 Garden Street, Cambridge MA 02138, USA

**BLANCHARD A.**
DAEC, Place Janssen, Meudon France

**BOND J.R.**
CITA, Mc. Lennon Laboratories, University of Toronto, Toronto, Canada

**BOUGHN S.**
Haverford College, Haverford, PA 19041, USA

**BRUNI M.**
S.I.S.S.A., Strada Costiera 11, Trieste, Italy

**CALISSE P.**
Diparirmento di Fisica, Università La Sapienza, P.le A.Moro 2, Roma, Italy

**CARR B.**
Queen Mary College, Mile End Road, London EI 4NS, UK

CAVALIERE A.
Dipartimento di Fisica, Università di Roma, via V.Raimondo, Roma, Italy

CHU Y.
University of Science and Technology, Hefei, Anhui, China

CHURCH S.
Cavendish Laboratory, Madingley Road, Cambridge CB3 OHE, UK

CLEMENTS D.L.
Physics Dept., Imperial College of Science and Technology, Prince Consort Road, London SW7, UK

CRANE P.
European Southern Laboratory, K.Schwarzchild Str. 2, Garching 8046, W.Germany

DALL'OGLIO G.
Dipartimento di Fisica, Università La Sapienza, P.le A.Moro 2, Roma, Italy

DALY R.A.
Physics Dept., Princeton University, Jadwin Hall, Princeton NJ 08544, USA

DANESE L.
Dipartimento di Astronomia, Università di Padova, Vicolo Osservatorio 5, 35122 Padova, Italy

DE AMICI G.
2407 Ward Street, Berkeley, USA

DE BERNARDIS P.
Dipartimento di Fisica, Università La Sapienza, P.le A.Moro 2, Roma, Italy

DE PETRIS M.
Dipartimento di Fisica, Università La Sapienza, P.le A.Moro 2, Roma, Italy

DE SANTIS E.
Dipartimento di Fisica, Università La Sapienza, P.le A.Moro 2, Roma, Italy

DE ZOTTI G.
Osservatorio Astronomico, Vicolo dell'Osservatorio 5, 35122 Padova, Italy

DEL GRANDE P.
Istituto Astronomico, Università La Sapienza, Via Lancisi 29, 00161 Roma, Italy

DEMIANSKI M.
Institute for Theoretical Physics, University of Warsaw, 00-601 Warsaw, Poland

DRAGOVAN M.
Princeton University, Jadwin Hall, Princeton NJ 08544, USA

FOMALONT E.
NRAO, Edgemont Road, Charlottesville, VA 22903, USA

FRANCESCHINI A.
Osservatorio Astronomico, Vicolo dell' Osservatorio 5, 35122 Padova, Italy

GERVASI M.
Dipartimento di Fisica, Università La Sapienza, P.le A.Moro 2, Roma, Italy

GORSKI K.
Los Alamos National Laboratory, T-6 MS B-288 LANL, Los Alamos, USA

GOUDA N.
Physics Dept., Kyoto University, Kyoto 606, Japan

GROMOV V.D.
Space Research Institute, Profsojuznaja 84/32, Moscow, USSR

GUNZIG E.
Service de Chimie-Physique, Universitè Libre de Bruxelles, Bruxelles, Belgium

HAYAKAWA S.
Nagoya University, Furo-Cho Chikusa-Ku, Nagoya, Japan

HOGAN C.J.
Steward Observatory, University of Arizona, Tucson, USA

ISOBE N.
NHR, 4321 Ogawa Machida, Tokyo, Japan

JONES M.
Cavendish Laboratory, Madingley Road, Cambridge, UK

KAISER M.B.
Astronomy Department UCLA, 8105 Math Science Building, Los Angeles CA 94720, USA

KAWASAKY M.
Physics Dept., Tohoku University, Sendai 980, Japan

LASENBY A.N.
Mullard Radio Astronomy Observatory, Cavendish Laboratory, Madingley Road, Cambridge, UK

LAWRENCE C.R.
California Institute of Technology, Mail code 105-24, Pasadena, California, USA

LEVIN S.
Space Science Laboratory, UCB, 1 Cyclotron Road, Berkeley CA 94720, USA

LINDER E.
Max Planck Institute for Astrophysics, Garching 8046, W.Germany

LUBIN P.
Physics Dept., UCSB, Santa Barbara, CA 93106, USA

MANDOLESI N.
TESRE/CNR, via De' Castagnoli 1, 40126 Bologna, Italy

MARTINEZ GONZALEZ E.
Physic Dept., University of Cantabria, Avenida Los Castros S.N., Santander
39005, Spain

MATARRESE S.
Dipartimento di Fisica, Università di Padova, via Marzolo 8, Padova, Italy

MATSUMOTO T.
Astrophysics Dept., Furo-Cho Chikusa-Ku, Nagoya 464-01, Japan

MEINHOLD P.
Physic Dept., UCSB, 623-F de la Viña, St. Santa Barbara, USA

MELCHIORRI F.
Dipartimento di Fisica, Università La Sapienza, P.le A.Moro 2, Roma, Italy

MELCHIORRI OLIVO B.
IFA/CNR, P.zza L.Sturzo, Roma, Italy

MEYER D.
Dept. of Physics & Astronomy, Northwestern University, Evanston IL 60208,
USA

MIRALDA ESCUDÈ J.
Astrophysics Dept., Princeton University, Peyton Hall, Princeton NJ 08544,
USA

MUCIACCIA P.F.
Istituto Astronomico, Universitá La Sapienza, via Lancisi 29, Roma, Italy

xx

MYERS S.T.
California Institute of Technology, Mail code 105-24, Pasadena, California, USA

NATALE V.
CAISMI/CNR, via E.Fermi 5, Firenze, Italy

OCCHIONERO F.
Osservatorio Astronomico, Parco Mellini 84, Roma, Italy

PALAZZI E.
TESRE/CNR, via De' Castagnoli 1, 40126 Bologna, Italy

PARTRIDGE B.
Haverford College, Haverford PA 10941, USA

PEEBLES P.J.E.
Joseph Henry Laboratories, Princeton University, Jadwin Hall, Princeton NJ 08544, USA

PETERSON J.
Princeton University, Box 208, Princeton NJ 08544, USA

PIRO L.
IAS/CNR, via E.Fermi, Frascati, Italy

PROVENZALE A.
Istituto Cosmogeofisica CNR, C.so Fiume 4, Torino, Italy

REPHAELI Y.
Tel Aviv University School of Physics and Astronomy, Tel Aviv 69978, Israel

RICHARDS P.L.
Physics Dept., UCB, Berkeley CA 94720, USA

ROSSETTI E.
TESRE/CNR, via de' Castagnoli 1, 40126 Bologna, Italy

SANCHEZ N.
Observatoire de Paris-Meudon, Principal-Cedex, Meudon, France

SANZ J.L.
Physic Dept., University of Cantabria, Avenida de los Castros, Santander 39005, Spain

SAUNDERS R.
Cavendish Laboratory, Madingley Road, Cambridge, UK

SCARAMELLA R.
Osservatorio Astronomico di Roma, via Frascati 5, Monteporzio Catone 00040, Italy

SCIAMA D.W.
S.I.S.S.A., Strada Costiera 11, Trieste, Italy

SETTI G.
European Southern Laboratory, K.Schwarzschild Str. 2, Garching 8046, W.Germany

SHAFI Q.
Bartol Research Centre University of Delaware, Newark USA

SHAPIRO A.
INFN Gran Sasso, Assergi(AQ), Italy

SHAPOSHNICOV V.A.
Scient. Ind. ASS. Cryogenmach, Lenin St. 67, Balashikka, USSR

SHOLOMITSKII B.D.
Space Research Institute, Profsojuznaja St., Moscow, USSR

SIGNORE M.
Ecole Normale Superieure, 24 Rue Lhomond, Paris, France

SIRONI G.
Dipartimento di Fisica, Università di Milano, via Celoria 16, 20133 Milano, Italy

SKULACHEV D.P.
Space Research Institute, Profsojuznaja St., Moscow, USSR

SMOOT G.
Space Science Laboratory, UCB, 1 Cyclotron Road, Berkeley CA 94720, USA

STEBBINS A.
CITA, 60 St. George Street, Toronto Ontario, Canada

STROUKOV I.A.
Space Research Institute, Profsojuznaja 84/32, Moscow, USSR

SUTO Y.
Department of Physics, Bunkyo Mito 310, Japan

TAKANAYAGI Y.
Japan Broadcasting Corp., 221 Jinnan, Tokyo, Japan

VALENZIANO L.
Dipartimento di Fisica, Università La Sapienza, P.le A.Moro 2, Roma, Italy

VITTORIO N.
Dipartimento di Fisica, Università dell'Aquila, P.zza V.Rivera 1, L'Aquila, Italy

VOLONTÈ S.
ESA, 8-10 Rue Mario-Nikis, Paris, France

WILKINSON D.
Physics Dept., Princeton University, P.O.Box 708, Princeton NJ 08544 USA

Prof. Robert H. Dicke

PRINCETON UNIVERSITY

N.J. - USA

L'Aquila, June 21st 1989

More   than   a hundred of us,   from all over the   world,   are
gathered here in L'Aquila to celebrate 25 years of  research
on the cosmic microwave background.

We   have had the opportunity to review the field you did   so
much to found,   and to reminisce about the early days of the
"fireball".

We join to send you our greetings and very best wishes.

UNIVERSITÀ DEGLI STUDI DELL'AQUILA
CONSIGLIO NAZIONALE DELLE RICERCHE —— L'AQUILA, June 19-23, 1989

Prof. Robert H. Dicke

PRINCETON UNIVERSITY

N.J. - USA


L'Aquila, June 21st 1989


More than a hundred of us, from all over the world, are gathered here in L'Aquila to celebrate 25 years of research on the cosmic microwave background.

We have had the opportunity to review the field you did so much to found, and to reminisce about the early days of the "fireball".

We join to send you our greetings and very best wishes.

Scientific Secretariat:
N. Mandolesi / ITESRE Via de' Castagnoli 1 40126 Bologna, Italy
N. Vittorio

Tel. 39-51-287047 / Telex 511350 CNR BO
Fax 39-51-229702 / E-Mail 38047:: reno

# THE IMPACT OF THE CMB DISCOVERY ON THEORETICAL COSMOLOGY

D.W. Sciama
International School for Advanced Studies, Trieste
International Centre for Theoretical Physics, Trieste

and
Department of Astrophysics, Oxford University

## 1. INTRODUCTION

We are here to celebrate the 25th anniversary of one of the most important scientific discoveries of all time - the cosmic microwave background. Dave Wilkinson will describe in the next lecture the circumstances of its discovery by Penzias and Wilson (1965), its interpretation by Dicke, Peebles, Roll and Wilkinson (1965), and the background against which the discovery was made. My task is to describe the impact it had on our thinking about the universe.

This impact has been so pervasive that it makes one wonder what cosmologists were doing before 1965. I for one started to work in this field around 1950, so I had to struggle for 15 years before we saw the light! One consequence of the primitive nature of our thinking was the low regard in which cosmology was held by the hard-headed physicists of the time. All that is quite different today - thanks entirely to the microwave background and its implications for the early universe and especially, perhaps, the success of pri-

1

*N. Mandolesi and N. Vittorio (eds.), The Cosmic Microwave Background: 25 Years Later, 1–15.*
© 1990 *Kluwer Academic Publishers.*

mordial nucleosynthesis in accounting for the presently observed abundances of certain light isotopes.

In those early far-off days the main of interest were the measured values of the Hubble constant and the deceleration parameter (problems still with us today), the $\alpha\beta\gamma$ theory of the origin of the elements, and the much-publicised battle between the big bang and the steady state theories of the universe. The latter theory had been introduced in 1948 by Bondi, Gold and Hoyle. It relied on the continual creation of matter to maintain the universe in a steady state of constant mean density despite the expansion. This was achieved by invoking the exotic equation of state $p=-\rho$ for the fundamental material substratum of the universe, with which was associated the de Sitter metric for space-time. The reader will recognise immediately that this conception is now enjoying a renaissance under the name "inflation" -ironically tucked away into an extremely brief period in the early high density stage of the big bang!

We may usefully distinguish three ways in which the cosmic microwave background changed our thinking:

a) From its use as a *probe* of conditions in the universe, particularly of its small-scale and large-scale isotropy.

b) *dynamically*, by its direct influence on cosmological phenomena.

c) *fundamentally*, by its significance for our understanding of the origin of the universe.

I will consider these different concequences in turn.

## 2. THE ISOTROPY OF THE UNIVERSE

It was realised early on that a measurement of the dipole anisotropy of the microwave background would give us information about the peculiar velocity of the earth relative to the universe as a whole. Thus already in 1967 Partridge and Wilkinson derived from their observations an upper limit of 300 km/s for the equatorial component of this velocity. In the same year I obtained an estimate for our net velocity of $\sim 400$ km/s towards $l_{II} \sim 33$, $b_{II} \sim 7$ from an analysis of the pattern of red shift data for the galaxies in our vicinity, and the Virgo cluster was considered as a possible Great Attractor (though this name, of course, is of recent origin). A more detailed analysis was given by Stewart and Sciama (1967).

Also in 1967 Sachs and Wolfe published their fundamental paper on the effect of perturbations of Robertson-Walker models on the isotropy of the microwave background. This great paper, the original source of the Sachs-Wolfe effect, was communicated as early as 13 May 1966. It was followed in 1968 by an attempt to calculate the effect of a non-linear localised irregularity by Rees and myself (an attempt which was improved in 1976 by Dyer).

The modern versions of these problems, and of the Sunyaev-Zel'dovich effect which dates from 1970, are the main subject of this conference, so I will consider them no further here.

Our discussions will concentrate on the small-scale isotropy of the background, so I will consider here only its large scale isotropy. This has the fundamental implication that, considered as a whole, the universe is extremely close to a Robertson-Walker model, at least as far as its isotropy is concerned. If, in addition, we assume that we occupy a typical position in the universe, so that it is nearly isotropic about every point, then we can appeal to Schur's theorem (cf. Ehlers, Garen and Sachs, 1968) and conclude that it is also (nearly) homogeneous.

A good account of the relativistic theory of a nearly isotropic universe has been given by Ellis (1971) (for a simplified account see, for example, Sciama (1980)). Essentially one expands the velocity field in the neighbourhood of a point in a Taylor series, as in fluid mechanics. The first-order terms can be classified as expansion, shear and vorticity. In the cosmological case the expansion would correspond to the Hubble effect, while the shear would represent a change of shape without any accompanying rotation. Such a shear would manifest itself both in the form of transverse velocities and in a directional dependence of the expansion rate (Hubble constant).

By contrast, the vorticity would correspond to a rotation of the universe with respect to a local inertial frame. The existence of such a rotation would be denied by Mach's principle, which asserts that the local inertial frame is itself determined by the large scale distribution of matter. However, there do exist exact solutions of Einstein's field equations for a rotating expanding universe, so great interest attaches to a purely empirical determination of the vorticity of the universe, or of upper limits to it. Here the isotropy of the microwave background has provided a spectacular, and indeed crucial, improvement in the previously known upper limit. We shall discuss this point first, and then consider attempts to determine or limit the shear using the microwave background.

If the universe as a whole were rotating relative to our local inertial frame we would expect to see a *transverse* Doppler shift in the spectra of distant galaxies, except along the axis of rotation. Since the traverse shift is of second order in v/c this method is not very sensitive, and we cannot do more than rule out a transverse velocity of order c for a source a Hubble radius away. It is convenient to express this limit in terms of the vorticity $\omega$ to the Hubble constant H, and so we obtain the weak inequality

$$\frac{\omega}{H} \lesssim 1$$

(Kristian and Sachs, 1966).

Nevertheless this is a more stringent limit on $\omega$ than one can deduce from the consideration that the flattening of our Galaxy due to its rotation is compatible with observations of the proper motions of slowly rotation outlying stars and of extragalactic objects. This tells us only that any rotation period of the universe must exceed about $10^9$ years, which is a ten times weaker limit than the previous one. However, this weaker limit is still stronger than one could derive using the best available gyroscopes to determine the local non-rotating reference frame.

This situation has been transformed by the discovery of the microwave background and by the observed limits on its anisotropy on large angular scales. If the universe has large-scale vorticity the last scattering surface of the microwave background would be rotating around us, giving rise to a transverse Doppler effect whose magnitude would depend on the angle between the direction of observation and the rotation axis. This question was analysed by Hawing (1969) and by Collins and Hawking (1973a, b) for homogeneous but anisotropic universes, which are characterised by their so-called Bianchi types. The limits obtained depend on the Bianchi type assumed and on the red shift of the last-scattering surface. The weakest limit is

$$\frac{\omega}{H} \lesssim 10^{-3}$$

while the strongest is

$$\frac{\omega}{H} \lesssim 10^{-12}$$

These limits are very strong, but one must bear in mind that during the matter-dominated phase of the expansion the ratio $\omega/H$ decreases with time. To some extent then, the small value of this ration at the present epoch simply reflects the advanced age the universe has today. If we want to say that the universe is rotating slowly, or not at all, we must consider what limits can be placed on the total number of rotation periods that may have occurred since the big bang. This questions was also consider by Collins and Hawking. Their strongest limit arises for a closed universe (Bianchi type IX), for which they found that the universe could have rotated through at most $2 \cdot 10^4$ seconds of arc since the big bang. Stringent as this limit is, we cannot regard it as necessarily vindicating Mach's principle, since other explanations (such as inflation) exist for the low vorticity of the universe. I cannot pursue this questions further here.

Turning to estimates of large-scale shear $\sigma$, we note first that there is no gross anisotropy observed in the value of the Hubble constant. This gives us the weak limit

$$\frac{\sigma}{H} \lesssim 0.3$$

(Kristian and Sachs, 1966). To obtain a stronger limit we again turn to the microwave background. Its temperature decreases as the universe expands, and if the expansion rate were anisotropic the radiation would continue to

have a black body spectrum in each direction, but the temperature itself
would vary with direction. This question was discussed by Torne (1967) and
in more detail by Collins and Hawking (1973a, b), again in terms of Bianchi
models of the universe and the red shift of the last-scattering surface. Their
weakest limit was

$$\frac{\sigma}{H} \lesssim 10^{-3}$$

while their strongest was

$$\frac{\sigma}{H} \lesssim 10^{-7}$$

In this case a still stronger limit can be obtained from the influence of shear
on the expansion time scale of the universe at the epoch of primordial nucleosynthesis, and so on the abundance of helium produced at that time. Of
corse the relevant nuclear reactions arose from the high temperatures then
associated with the "microwave" background, so that the stronger limit is
still a consequence of the existence of this background. This question was
analysed by Barrow (1976) and by Olson (1978). Again the results are model-
dependent. The strongest limit obtained was

$$\frac{\sigma}{H} \lesssim 10^{-11}$$

We may conclude from this discussion that the existence and properties of the
microwave background show that on a large scale the universe is remarkably
close to a highly symmetrical Robertson-Walker model. The reason for this
high symmetry has still to be established.

## 3. THE DYNAMICAL INFLUENCE
## OF THE MICROWAVE BACKGROUND

We may usefully distinguish between the dynamical influence of the microwave background today and in the early universe. This letter influence is
so well known and so widely discussed, that it is not necessary to go into detail here. I need only recall that the background radiation density increases

into the past faster than the matter density, so that the early universe was radiation dominated, with the temperature increasing without bound as the initial singularity is approached. (For more about this singularity see the next section). We therefore have the concept of the hot early universe as the ultimate high energy laboratory, where in particular neutrinos and possibly more exotic elementary particles were pair produced and thermalised. Many of these particles would have survived to the present day, and would now dominate the universe and even individual galaxies if their rest-masses fall within certain ranges. They would also produce a measurable influence on the abundances of the light elements formed during primordial nucleosynthesis (Peebles 1966), itself a process arising from the thermal effects of the background radiation.

In addition to these rather well-explored phenomena, there is a rich variety of less well-understood processes associated with the hot early universe: the possibility of inflation, baryosynthesis, the quark-hadron transition, the formation of cosmic strings and so on. Later came galaxy formation, itself influenced by the dynamical role of the background and possibly of modern cosmology.

We now turn to the dynamical role of the microwave background today. In this case I shall enter into a little more detail. From a laboratory point of view a temperature of 3 K is very low. Indeed to measure it the observers had to use a reference terminal immersed in liquid helium. Nevertheless from an astrophysical point of view 3 K is a high temperature. A universal black body radiation field at this temperature would contribute an energy density everywhere $\sim 1$ eV/cm$^3$. This is comparable with the energy density in our Galaxy of the various modes of interstellar excitation-starlight, cosmic rays, magnetic fields and turbulent gas clouds. So even in our Galaxy the cosmic microwave background would be for many purposes as important as the well-known energy modes of local origin. In intergalactic space, however, these localised energy densities probably drop off by a factor between 100 and 1000, whereas the black body component would maintain its energy density at $\sim 1$ eV/cm$^3$.

This was recognised as soon as the microwave background was discovered, and in 1965-6 three important consequences were pointed out for high energy astrophysics. These consequences involve the interaction of cosmic ray protons, γ-rays and electrons with the microwave background. The proton interaction was first discussed by Greisen (1966) and by Kuzmin and Zatsepin (1966). They pointed out that a cosmic ray proton of energy $10^{20}$ eV, which has a Lorentz factor $\gamma$ [ $= (1\text{-}v^2/c^2)^{-1/2}$] of $10^{11}$, would in its rest frame, regard a typical microwave photon (energy $\sim 10^{-3}$ eV) as having an energy $\sim 10^8$ eV. Such an energetic photon striking a stationary proton would be close to the threshold for producing a pion (rest mass $\sim$ 137 MeV.). This means that from the terrestrial point of view a cosmic ray proton of $10^{20}$ eV could collide with a black body photon, produce a pion, and so be degraded in energy.

The observed spectrum of cosmic rays shows no cut-off tantalisingly close to the threshold energy for attenuation by the black body background. Once it sets in this attenuation is very severe. According to Stecker (1968) the mean free path of a proton with energy exceeding $10^{20}$ years is less than 50 million light years. Thus if such proton are extragalactic in origin strong attenuation should be observed beyond the threshold. The required experiments are difficult, and the attenuation has not yet been established (for a recent discussion see Hill, Schramm and Walker (1986)).

High energy cosmic γ rays would also be degraded by interaction with microwave photons. We cannot now look at the collision from the rest-frame of high energy object, but we can use a frame in which the γ-ray and the microwave photon have the same energy. If the energy of the γ ray is E relative to the Earth, and we transform to a frame moving away from the γ ray with velocity v, then the energy of the γ-ray becomes $E/\gamma$ and the energy of a typical microwave photon moving towards the γ-ray becomes $10^{-3}\,\gamma$ eV. Setting these equal we have

$$\frac{E}{\gamma} = 10^{-3}\gamma$$

Let us now choose E so that the common energy of each photon is just the rest-energy of an electron. We would then be at the threshold for electron-pair creation. This requires that

$$\frac{E}{\gamma} = 10^{-3}\gamma = m_o c^2$$

Eliminating $\gamma$ we have

$$10^{-3}E = \left(m_o c^2\right)^2$$

or

$$E = 2.5 \cdot 10^{14}\text{eV}$$

The effect is a large one (Jelley (1966), Gould and Schreder (1966)). The cross-section $\sigma$ for the process is roughly the square of the classical electron radius $\sim 10^{-25}$ cm$^2$, and so the mean free path, which would be about $1/n\sigma$ where n is the density of microwave photons, becomes $\sim 10^{22}$ cm. This is smaller than the size of the Galaxy.

It was later pointed out that the electron pairs produced would themselves undergo Compton collisions with the microwave photons, so that a photon-electron cascade would be set up, with the regenerated photons tending to replace the attenuated ones (Bonometto (1971), Bonometto and Lucchin (1971)). This effect in turn would be reduced if an ambient magnetic field deviates the induced electron pairs from the line of sight. In this way it may become possible to detect a magnetic field as low as $10^{-12}$ gauss (Gould and Rephaeli (1978), Prothero (1986), Honda (1989)).

Finally we consider the effect of the microwave background on the relativistic electron component of the cosmic rays. The dominant interaction process is inverse Compton scattering. By working, as before, in the rest frame of the relativistic particle, we see that on average an incident photon of energy E in the terrestrial frame gains an energy E' given by

$$E' \sim \gamma^2 E$$

where $\gamma$ is the Lorentz factor of the electron.

Now consider the electrons that are responsible for the diffuse radio emission of our Galaxy through the synchrotron mechanism. A typical energy for such an electron would be 1 GeV. Its $\gamma$ would then be 2000, and with $E \sim 10^{-3}$ eV we see that the scattered photon would be raised in energy to about 4 keV. This takes us right into the X-ray region, so that the Galaxy would be an X-ray source due to this process (Hoyle (1965), Gould (1965)). Now the inverse Compton effect and the synchrotron mechanism are essentially the same process of a relativistic electron emitting radiation while under the influence of an electromagnetic field (Ginzburg and Syrovatsky (1964), Felton and Morrison (1966)). Consequently they lead to a similar rate of energy transfer from the electrons if the energy density in the photon field and the static magnetic field are comparable. This is just the case for our Galaxy, as we have seen. Thus the known energy flux in the galactic radio background should be of the same order as the energy flux in the emitted X-ray background.

This consideration leads to a stringent upper limit on the microwave background temperature. This limit arises because the rate of energy transfer to the X-rays is proportional to the Compton collision rate, and so to the photon density in the black body radiation field, that is, to the third power of its temperature. Moreover, if the radiation field had a higher temperature, the mean energy of its photons would be greater. Accordingly an electron of lower energy would suffice to produce a given X-ray energy. We must therefore allow for the fact that in our Galaxy there are more electrons at the lower energies. The net result of all this is that if the black body background had a temperature of 6 K, the galactic halo would be a stronger X-ray emitter than is observed.

There is another important aspect to this problem and that is the resulting energy drain on the electrons. For $T \sim 3$ K the drain due to the production of X-rays is of the same order as that due to the radio emission (or to inverse Compton collisions with starlight photons). If we tried to increase T, however, this drain would increase very rapidly, roughly like $T^4$. It would soon become so high that it would be very difficult to understand where the electrons got their energy from in the first place. For $T \sim 10$ K, for instance, a 1 GeV electron would lose half its energy in 10 million years, instead of the billion years characteristic of the synchrotron process alone. A further point is that the observed energy spectrum of the electrons shows no sign of significant energy losses out to at least 300 GeV, at which energy the lifetime of an electron in the 3 K radiation field would be only about $3 \cdot 10^4$ years. This tells us that these electrons must leak out of the local trapping region is a shorter time than this. In a 10 K radiation field the leakage time would have to be less than 300 years. This would have very severe implications also for the propagation of cosmic ray protons and heavier nuclei.

These considerations take on even greater importance when we apply them to a radio source with a substantial red shift z. The temperature of the microwave background at such a source would be increased by the factor 1+z over its local value, and the all-important energy density by $(1+z)^4$. The balance between the synchrotron and Compton effects that holds in our Galaxy would then be disturbed. At a redshift of 4, for example, the inverse Compton lifetime of an electron would be 625 times less than that of a contemporary electron, and the X-ray emission correspondingly larger. These effects are crucial for our understanding of high energy processes in sources at large red shifts (see, for example, Felten and Rees (1969)).

## 4. IMPLICATIONS FOR THE ORIGIN OF THE UNIVERSE

In 1965 when the microwave background was discovered the steady state theory was already in retreat because counts of radio sources indicated that these sources were subject to considerable evolution. Nevertheless the matter was not quite settled, mainly because most of the sources involved had not been optically identified, so that there was still room for argument about the whereabouts of the sources and so about the interpretation of the counts. The coup de grace against the steady state theory came when the thermal spectrum of the background became fairly well established. The key point here is that to thermalise excess radiation in the universe as dilute as it is today would take far longer than the relevant timescale of $10^{10}$ years. One would have to go back to the epoch corresponding to a red shift of at least about $10^5$ in an evolving universe before one reaches a regime dense enough for the thermalisation time to be less than the expansion timescale (Rees 1972). Thus by observing the thermal spectrum today one is directly observing processes which must have occurred at the latest at this red shift of $10^5$.

Attempts to avoid this argument have been made, but have not commanded significant support.

Given that there was a big bang we would like to know more about conditions at or close to the bang, as a prelude to an attempt to understand the origin of the universe. Now we have already seen that the large-scale isotropy of the microwave background implies that the universe is close to Robertson-Walker. If the universe were exactly Robertson-Walker, the big bang would represent a singularity of infinite density and curvature involving the whole universe. The question then arises, is the actual universe sufficiently close to being Robertson-Walker for it to have been singular at the bang?

Curiously enough, the best answer to this question itself involves the microwave background, and that in two distinct ways, namely in its role as a probe and also as a dynamical agent. The argument involves the famous singularity theorems of Penrose and Hawking (Hawking and Ellis (1973)). These theorems delineate the circumstances in which singularities arise in general relativity in generic situations when the presence of irregularities leads to deviations from exact symmetries. The physical basis of these theorems is that self-gravitation can be so strong in Einstein's theory (because of its non-linearity) that singularities arise even in the presence of irregularities. The mathematical statement of these theorems requires the validity of certain assumptions, some of which, though plausible, are difficult or impossible to verify in practice (for instance that the universe should admit a well-defined global Cauchy surface, that is, a spacelike surface on which initial-value data can be defined). However, Hawking and Ellis (1968) showed that one particular theorem is based on an assumption which can be verified using observed properties of the microwave background.

The theorem is the following:

Space-time is not singularity-free if the following conditions hold:

(a) Einstein's field equations

(b) A positive energy condition ( $T_{ab}u^a u^b \geq T/2$ where $T_{ab}$ is the energy-momentum tensor and $u^a$ any timelike unit vector)

(c) Strong causality (every neighbourhood of a point contains a neighbourhood of that point that no nonspacelike curve intersects more than once)

(d) There exists a point P such that all past-directed timelike geodesics through P start converging again within a compact region in the past of P.

The crucial conditions are (b) and (d). The energy condition (b) guarantees that gravity has always been attractive; (d) is a precise statement of the requirement that there was enough gravitation to produce a singularity. It turns out that we can use observations of the microwave background to show that in the actual universe (d) is satisfied, with the point P corresponding to the Earth. The procedure is to use these observations to show that the actual universe is sufficiently like an exactly Robertson-Walker one, where the reconvergence certainly does occur. Now we have already seen from the angular distribution of the microwave background that the universe is isotropic to better than 1 part in 1,000 back to the last-scattering surface. There are two possibilities:

1. The red shift $z_s$ of this surface is small ($\sim 7$). This wold require a relatively large amount of intergalactic scattering material, and direct calculation then shows that the gravitational effect of this material would produce the required reconvergence before the red shift $z_s$ is reached.

2. $z_s$ is appreciably larger than 7. The influence of intergalactic matter may now itself be unable to produce reconvergence. On the other hand, the isotropy of the universe would remain close to ideal out to $z_s$, that is, to a much greater red shift than 7. In this case one can show that the energy density of the microwave background itself is enough to cause reconvergence before $z_s$.

If we wish to avoid a singular origin for the universe we must therefore challenge (a), (b) or (c). A violation of (c) would perhaps be worse than a singularity, being a global rather than a local breakdown of our ordinary physical concepts. Altering Einstein's field equations should perhaps be a last resort, so at the moment people are concentrating on challenging the energy condition. It is generally agreed that nonquantum matter probably always does satisfy this condition. However, in the early universe one would expect the quantum properties of matter (and of gravitation) to be important, and we know that in some circumstances the quantum energy-momentum tensor of matter in a gravitational field can violate the energy condition. No satisfactory resolution of the problem along these lines has yet been achieved, however.

We therefore continue to face a major intellectual crisis: that we do not possess a self-consistent theory of gravitation which can be used to study the earliest stages of the universe. This is a perhaps surprising conclusion to be able to draw from observations made 25 years ago by means of radio telescopes.

## REFERENCES

Barrow, J.D., 1976 *Mon. Not. Roy. Astr. Soc.*, **175**, 359.

Bondi, H., & Gold, T., 1948 *Mon. Not. Roy. Astr. Soc.*, **108**, 252.

Bonometto, S.A., (1971) *Nuovo Cim. Lett.*, **2**, 1299.

Collins, C.B., & Hawking, S.W., 1973a *Mon. Not. Roy. Astr. Soc.*, **162**, 307.

Collins, C.B., & Hawking, S.W., 1973b *Astrophys. J.*, **180**, 317.

Dicke, R.H., Peebles, P.J.E., Roll P.G., & Wilkinson, D.T., 1965 *Astrophys J.*, **142**, 414.

Dyer, C.C. 1976 *Mon. Not. Roy. Astr. Soc.*, **175**, 429.

Ehelers, J., Geren, P. and Sachs, R.K., 1968, *J. Math. Phys.*, **9**, 1344.

Ellis, G.F.R., in *General Relativity and Cosmology*, Ed. R.K. Sachs (Academic Press, New York) 1971.

Felten, J.E. and Morrison, P. 1966 *Astrophys. J.*, **146**, 686.

Felten, J.E. and Rees, M.J. 1966 *Nature*, **221**, 924.

Ginzburg, V.L., and Syrovatsky, S.I., in *The Origin of the Cosmic Rays*, Pergamon, London, 1964.

Gould, R.J. 1965 *Phys. Rev. Lett.* ,**15** ,511.

Gould, R.J. and Schreder, G. 1966 *Phys. Rev. Lett.*, **16**, 252.

Gould, R.J. and Rephaeli, Y. 1978 *Astrophys. J.*, **225**, 318.

Greisen, K. 1966 *Phys. Rev. Lett.*, **16**, 748.

Hawking, S.W. and Ellis, G.F.R., 1968 *Astrophys. J.*, **152**, 25.

Hawking, S.W. 1969 *Mon. Not. Roy. Astr. Soc.*, **142**, 129.

Hawking, S.W. and Ellis, G.F.R., in *The Large Scale Structure of Space- Time*, Cambridge, 1973.

Hill, C.T., Schramm, D.N., and Walker, T.P. 1986 *Phys. Rev. D*, **34**, 1622.

Honda, M. 1989 *Astrophys. J.*, **339**, 629.

Hoyle, F. 1948 *Mon. Not. Roy. Astr. Soc.*, **108**, 372.

Hoyle, F. 1965 *Phys. Rev. Lett.*, **15**, 131.

Jelly, J.V. 1966 *Phys. Rev. Lett.*, **16**, 479.

Kristian, J. and Sachs, R.K. 1966 *Astrophys. J.*, **143**, 379.

Kuzmin, V.A. and Zatsepin, G.T. 1966 *JETP Lett.*, **4**, 78.

Olson. D.W. 1978 *Astrophys. J.*, **219**, 777.

Partridge, R.B., and Wilkinson, D.T. 1967 *Phys. Rev. Lett.*, **18**, 557.

Peebles, P.J.E. 1966 *Phys. Rev. Lett.*, **16**, 410.

Penzias, A.A., and Wilson, R.W., 1965 *Astrophys. J.*, **142**, 419.

Protheroe, R.J. 1986 *Mon. Not. Roy. Astr. Soc.*, **221**, 769.

Rees, M.J. 1972, *Phys. Rev. Lett.*, **28**, 169.

Rees, M.J. and Sciama, D.W. 1968 *Nature*, **217**, 511.

Sachs, R.K. , and Wolfe, A.M., 1967 *Astrophys. J.*, **147**, 73.

Sciama, D.W. 1967 *Phys. Rev. Lett.*, **18**, 1065.

Sciama, D.W. , in *Some Strangeness in the Proportion*, Ed. H.Wolfe, Adison- Wesley Reading, 1980.

Stecker, F.W. 1968 *Phys. Rev. Lett.*, **21**, 1016.

Stewart, J.M., and Sciama, D.W., 1967 *Nature*, **216**, 748.

Sunyaev, R.A., and Zel'dovich, Y.B. 1970, *Astron. and Space Sci.*, **7**, 3.

Thorne, K.S., 1967 *Astrophys. J.*, **148**, 51.

# DISCOVERY OF THE 3 °K RADIATION[1]

D. T. Wilkinson and P. J. E. Peebles
Joseph Henry Laboratories
Princeton University

Perhaps all discoveries in science have some elements of a good mystery story. This one does. In 1964 the major cosmic puzzle was whether the Universe is Evolving (Big Bang) or Steady State; both ideas had merit, neither had proof. However, the Big Bang had left a deliberate clue - a shadowy remnant of its firey youth. Three groups of physicists are on the case, each starting from a different premise, and unaware of the others. One group (in Russia) is putting together published theoretical and experimental evidence; they are very close, but misinterpret a clue. Another group, ignorant of the past, starts from the beginning and plods systematically toward the solution. The third group is looking, very carefully, right at the clue, and wonder what it is. Thermal Radiation from the Big Bang is about to be discovered. Proving again that nothing is new, the Radiation had been predicted 15 years earlier, but not searched for. And at least two published observations - one 25 years

---

[1]Based on an unpublished manuscript written in 1968. It was dusted off in 1983 by DTW for a talk at "Serendipitous Discoveries in Radio Astronomy"- a Green Bank Workshop celebrating the 50th anniversary of Karl Jansky's first detection of cosmic radio waves (Proceedings edited by K. Kellermann and B. Sheets are available from NRAO, unpublished). Remarks about current results have been updated to mid-1989 in the version presented here.

*N. Mandolesi and N. Vittorio (eds.), The Cosmic Microwave Background: 25 Years Later*, 17–31.

old - gave strong evidence for the existence of the Radiation prior to its discovery.

What follows is a worms-eye[2] view of the events leading to this discovery - a textbook example of serendipitous discovery. We are not historians, and most of what we know about this comes from the scientific literature, personal experience, and randomly collected anedoctal stories. In no sense is this intended to be definitive history.

The first glimmer of the fireball came to us via the musings of R. H. Dicke - the head of our research group at Princeton. He reasoned roughly as follows. If the Universe is closed and oscillating[3] (radius a periodic function of time), what happened to all the heavy elements which were cooked up in stars during the previous cycles of the Universe? Most of the matter we see now is hydrogen, so somehow each cycle must destroy the heavy elements before the expansion phase of the next cycle. An attractive way to do this is to say that the matter temperature exceeds $10^{10}$ °K in the highly contracted stage, causing the heavy nuclei to evaporate. A consequence of this assumption is that the Universe tends to relax to thermal equilibrium, producing a sea of blackbody radiation.

As the Universe expands out of this state, photons are red shifted from gamma rays to microwaves, but the spectrum remain thermal, so that we are left now with a sea of residual blackbody radiation. On the basis of Dicke's tenuous (at best) argument, Peter G. Roll and Wilkinson began working on a radiometer to search for the Primeval Fireball - as we dubbed it, and Peebles assumed the task of thinking about theoretical schemes of estimating the present Fireball temperature. This all took place in the late summer of 1964. The instrument generally used to measure the intensity of microwave radiation is known (to radio astronomers) as the 'Dicke radiometer'. This device was invented by Dicke in 1946 [1], and was used to measure the microwave radiation intensity from atmospheric water vapor, the moon, and the sun. Figure 1 shows Dicke and colleagues with their radiometer atop an M.I.T. building. This instrument, on a mountain top (to get above water vapor), could have detected the Fireball radiation, and, in fact, it was used [2] in 1946 to set an upper limit of 20 °K on the temperature of "radiation from cosmic matter". (By 1964 Dicke had forgotten about this result.)

---

[2] In time honored detective story style, we use the first person narrative - with apologies.

[3] One philosophically attractive option, among many.

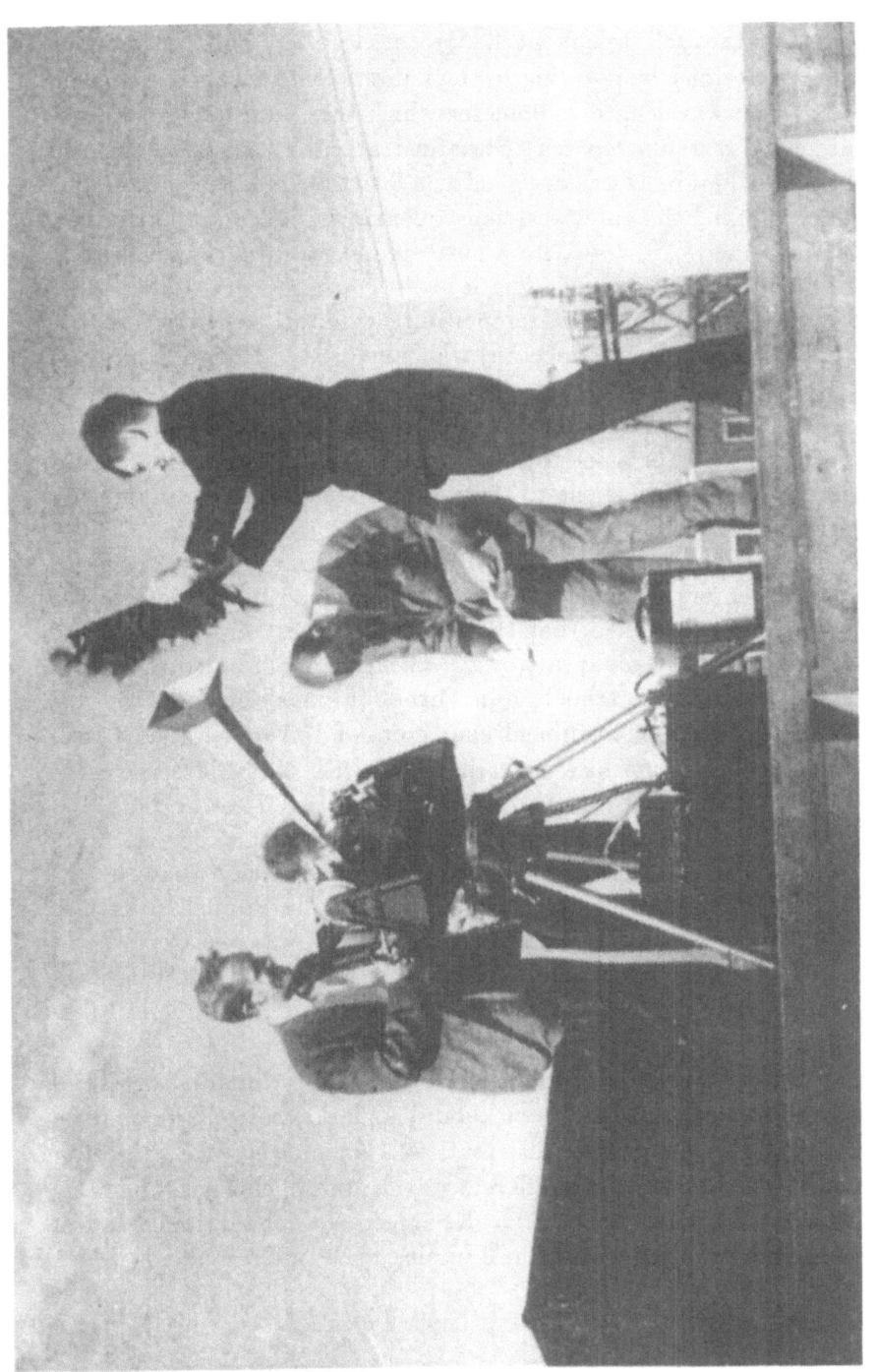

*Figure 1.* Early 1.5 cm microwave radiometer surrounded by (left to right) E. Beringer, R. Kyhl, A. Vane, and R.H. Dicke. Dicke is shaking the "shaggy dog" - a piece of absorbing material used as a calibration source.

While the Princeton radiometer was being built, Arno Penzias and Robert Wilson at the Bell Telephone Laboratories in Holmdel (only about 30 miles from Princeton), were trying to track down some excess noise in the front end of a 7 cm wavelength radiometer which they were using for absolute measurements of radio sources. Their instrument (whose antenna you see behind them in Figure 2) was designed as a low noise receiving station for signals bounced from Echo satellites; consequently, its noise properties had been studied in detail [3], always with a measured excess which was assumed to be back-lobe pick-up of ground radiation.

Penzias built a precision low-temperature calibration source (almost identical to one being built at Princeton) which he used to isolate this excess noise. He and Wilson decided that this noise had to be either anomalous radiation from the antenna walls, or leakage into the antenna of ground radiation (a roaring 300 °K), or an isotropic cosmic background radiation. Their excess noise power was equivalent to thermal radiation with a temperature of a few degrees Kelvin. The plot thickens.

Peebles, in a lecture at the Applied Physics Laboratory, Johns Hopkins University (which enters the story again later) mentioned Dicke's idea about the fireball and also mentioned that the Princeton group was preparing to look for the predicted isotropic microwave background radiation. An old friend and peer in graduate school, Ken Turner (a physicist and radio astronomer at the Carnegie Institution, Department of Terrestrial Magnetism) was in the audience, and Peebles' remark stayed with him. Later he mentioned it to Bernie Burke (then at D.T.M., now at M.I.T.) who passed it along to Penzias via a telephone conversation. Through this devious route the Holmdel and Princeton groups came together and decide that the excess noise in the Holmdel radiometer was very likely the Fireball radiation. Hence, the "excess antenna temperature" (3.5±1.0 °K) was reported [4] and interpreted [5] as "Cosmic Blackbody Radiation". Bob Wilson tells this story from his perspective in "Serendipitous Discoveries in Radio Astronomy" (see earlier footnote).

Of course, the Fireball interpretation needed to be tested by measurements of the spectrum (should be blackbody) and the isotropy (should be at least as isotropic as the matter distribution). Fortunately, the Princeton apparatus had been designed for a different wavelength (3 cm) and when this work was completed the result (3.0±0.5 °K) agreed with the Holmdel result, this supporting a blackbody spectrum. Jumping ahead for a moment we see

*Figure 2.* Robert Wilson (left) and Arno Penzias soon after the discovery of
the microwave background. Their radiometer is in the shack at the small end
of the horn-reflector antenna - a giant-sized version of the ones commonly
seen on microwave relay towers.

in Figure 3 an up-to-date graph of the Fireball temperature measurements [6] over 3 decades of wavelength. On this graph a blackbody radiator is represented as a horizontal line; the more familiar brightness vs. wavelength curve peaks at $\lambda \sim 0.2$ cm for T=2.75°K. Only the exceedingly difficult measurements [7] on the Wien side of the spectral peak show interesting deviations from a blackbody. The isotropy of the radiation has now been checked, and the Fireball is featureless on angular scales from 10 arcseconds to 90 degrees. Figure 4 shows the current results [6]. The upper limits around a few arcminutes are embarrassing to many theories of cosmic structure formation. When the Universe was about $10^6$ years old, the matter started to form clumps, now seen as galaxies, galaxy clusters, etc. That process should have left bumps of magnitude $\Delta T/T \sim 10^{-4}$ to $10^{-5}$, which haven't yet been seen. Incidentally, the Dipole in Figure 4 is mostly due to the Sun's motion through the radiation, that is, with respect to the natural reference frame of the universe. This by itself is an interesting result. But we must resist tempting digressions and get on with the story.

Flashback now to a part of the story, unknown to any of the three groups of physicist. In 1938, Mount Wilson astronomer S.W. Adams discovered in a stellar spectrum lines due to interstellar cyanogen (CN) molecules. Later, Adams [8] observed absorption from the first rotational excited state as well as the ground state of this molecules. Andrew McKellar [9] used the relative intensities of the two absorption lines to obtain the relative populations of the ground and excited states, and hence the excitation temperature of the CN molecules. This temperature was 2.3 °K. Although the excitation could not be associated with a specific production mechanism, it was thought that particle collisions were most likely responsible. By a fantastic stroke of luck, the first state of CN is excited by radiation of 2.6 mm wavelength, very close to the peak of a 3 °K blackbody spectrum. If space is filled with 3 °K radiation, all CN molecules must show excitations; the CN molecules are made-to-order interstellar thermometers and provide a strong existence test for the Fireball. One case of observed absorption from the ground state, unaccompanied by a line from the first excited state, would overturn the Fireball hypothesis.

The connection between the Fireball proposition and McKellar's rotation temperature was made independently by N.J. Woolf and George B. Field, soon after they learned of the proposed microwave radiation. This led to new measurements of the interstellar CN excitation which gave results consistent

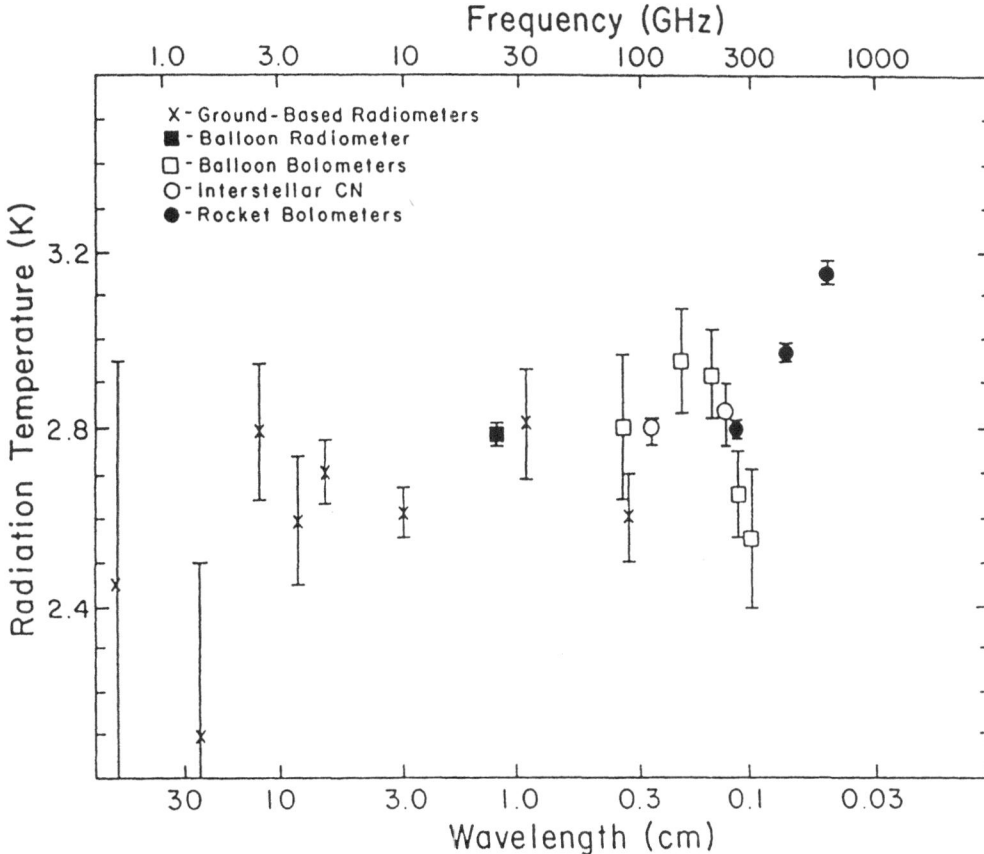

*Figure 3.* Measurements of the temperature of the cosmic microwave background radiation. Measurements, by a variety of techniques, are in reasonable agreement with a 2.75 °K blackbody spectrum, except for the points at $\lambda$=0.5 mm and 0.7 mm.

24

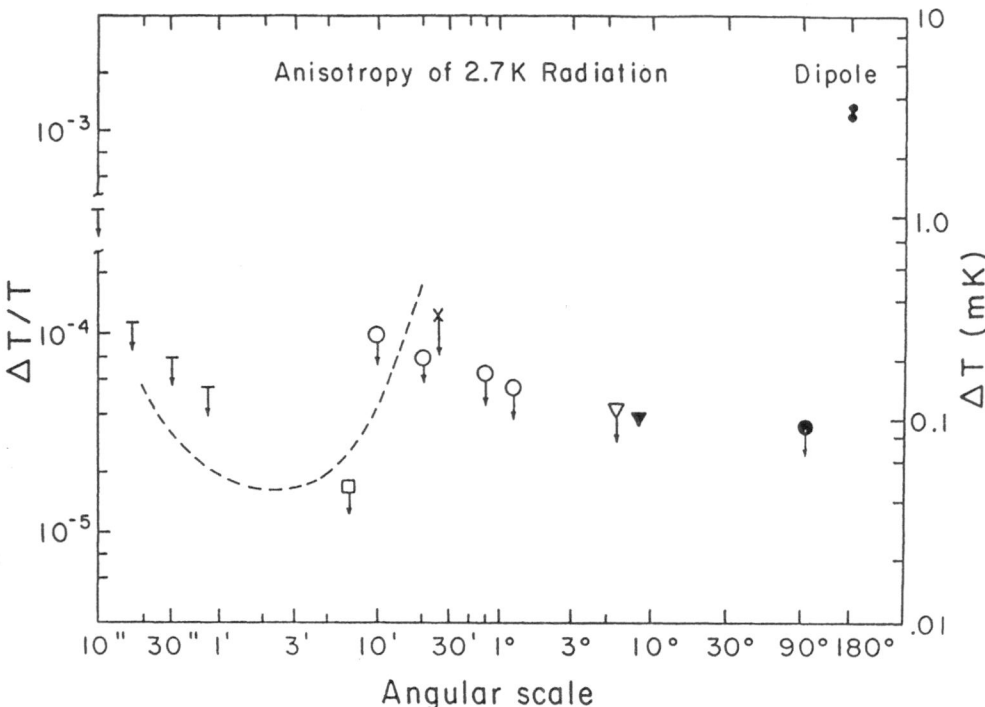

*Figure 4.* Modern upper limits on spatial structure in the Fireball. The latest results[18] at a few arcminutes are straining the limits of current radio astronomy equipment.

with the radiometer measurements (see Fig. 3 for recent temperature measurements using the CN technique). The work of Patrick Thaddeus and John Clauser also indicated that the CN excitation is a universal phenomenon, as it was found in spectra of 8 different stars. This is expected for excitation by Fireball radiation, but not so easy to understand for the collisional excitation model.

The stage is now set for telling the rediscovery episode in this story. It's, of course, an old story in science; physicists, in particular, are known for reinventing the wheel. You will recall that Peebles was supposed to find some sort of theory for the present Fireball temperature. The scheme he hit upon was to relate two "observable" cosmological parameters, the present mean mass density and present radiation temperature, to the amount of helium produced in the early cooling Fireball. (This is evidently a relation of considerable interest if one can deduce the helium abundance before the stars formed; that's an observable quantity.). We later found that this line of reasoning had been used 16 years earlier, by George Gamow and his doctoral student Ralph Alpher (who was then at the Applied Phisics Laboratory, Johns Hopkins University). Gamow and Alpher were trying to cook up the heavy elements in a Big Bang stew, and they found that a necessary ingredient was the sea of blackbody radiation.

Dicke was unaware of this earlier work when he reinvented the Fireball. The story takes a curious twist - the Fireball radiation was independently predicted starting from exactly opposite premises. Gamow and Alpher were trying to make heavy elements in the Big Bang; Dicke was trying to get rid of heavy elements (from the supposed previous cycle). However, the cornerstone of both arguments is a hot Big Bang Universe. And, indeed, the discovery and verification of the 3 °K radiation has led most cosmologists to accept this picture, rejecting the Steady State model. Again, working independently, and from opposite directions, Gamow, Alpher, and Herman, and later Peebles, managed to estimate a temperature for the fireball, before it was discovered. Let's go back and see how they did it [10].

Gamow [11] had pointed out, in 1946, the difficulties with the then popular thermal equilibrium (superstar) element cooker: among other things, general cosmologies don't permit the assumed static highly compressed state of the Universe. In fact, Gamow estimated that the Universe would pass through the pressure cooker phase in something like one second, so that the equilibrium assumption is highly questionable, to say the least. He suggested instead that elements are built up during the expansion by coagulation of cold neutrons.

Gamow and Alpher then found that they could get a better fit to the elements abundance data if they assumed that the elements were built up by the capture of hot neutrons in a radiation-dominated expanding Universe. They were guided to this idea by newly published data on capture cross sections for hot (∼1 MeV) neutrons, which they found showed an inverse correlation with element abundance. Their picture was that the elements were built up by successive neutron capture during the very early rapid expansion of the Universe. The thermal radiation accompanying the hot neutrons dominated the expansion, and made the time scale consistent with the nuclear reaction rates.

As it happened, this later proved not to be the whole story because, as Alpher already foresaw in his thesis, the building up process gets hung up at helium by the gap at atomic mass 5. However, Gamow and Alpher's connection between neutron-capture cross section and element abundance is by now experimentally documented in considerable detail, and their neutron capture process figures prominently in the modern theory of element formation of stars. Some preliminary results were reported in the famous $\alpha$-$\beta$-$\gamma$ [12] paper in 1948. But, for this story, the most important aspect of their work was the prediction that the early Universe should contain thermal radiation.

In his characteristic way, Gamow [13] reduced the problem to its essential physical parts. He concentrated on the first step in element synthesis, deuterium formation. First, Gamow knew that element formation would commence when the temperature fell to $10^9$ °K, for at higher temperatures the radiation photo-dissociates deuterium as fast as it forms. Second, because the mass density contributed by the radiation dominates that of nucleons, he could use Stefan's law to get the mass density from the temperature, and then general relativity to get the expansion rate of the Universe from the mass density. Knowing the rate, and the neutron- capture cross section of protons, Gamow could then observe that if the nucleon density were too low the nuclear reaction wouldn't happen with appreciable probability, and on the other hand if the density were too high the reaction would go too fast, and everything would end up as heavy elements, an equally sorry results. In this way he concluded that when the radiation temperature was $10^9$ °K, the nucleon density should have been about $10^{18}$cm$^{-3}$. Since radiation temperature is proportional to the cube of nucleon density in an expanding Universe, we can extrapolate to a present nucleon density of about $10^{-6}$cm$^{-3}$ and get a fireball temperature of 10 °K. Figure 5 shows a picture of this remarkable physicist taken at about the time of the prediction of Thermal Radiation

Alpher and Herman [14] repeated Gamow's calculation with greater accuracy, using a computer. For the first time, they extrapolated the radiation temperature to the present, and got 5 °K. Remember, all this happened in 1948 - 17 years before the Radiation was discovered.

This story is getting complicated, so we have to summarize it on the flow diagram in Figure 6. The most surprising feature of this diagram is that the "Gamow" box makes no connection with the "Discovery" box. This is hard to explain. The Big Bang element production papers were widely read, and created quite a stir at the time of their publication. If there is any one reason for so many missed connections it is probably this. In the early 50's evidence mounted that heavy elements are built up in stars, thus relieving the Big bang of this burden. As Big Bang element production grew unfashionable, the thermal radiation era faded with it. Even so, Gamow saw fit to write a review paper in 1956 which sets out very clearly the "Importance of Thermal Radiation in Cosmology"[15].

The one connection to the Gamow box was made [16] by Russian astrophysicists, who apparently read the U.S. literature better than we do. They were the only ones to put together published theoretical and experimental results and suggest that microwave measurements "are extremely important for experimental checking of the Gamow theory". They refer to the early Bell Labs work [17] where careful accounting was made of all contributions to the total system noise (typically 20 °K) with an accuracy of about $\pm 1$ °K. In retrospect the 3 °K radiation was probably included in the $2\pm 1$ °K usually attributed to ground radiation into the antenna back-lobes. "This estimate is based on the temperature 'not otherwise accounted for' in a previous experiment; it is somewhat larger than the calculated temperature expected from back lobes measured on a similar antenna." [17]. A radio astronomer, looking for evidence for a few degrees of Fireball radiation, would have leaped to her feet upon reading this because it is very difficult to distinguish between back-lobe radiation and isotropic background radiation in the main beam. Both are approximately independent of antenna position. Doroshkevich and Novikov apparently missed this technical point, and only commented that the measured atmospheric emission agrees with theory.

We close with two more remarks, which may contain a message about

*Figure 5.* George Gamow at about the time he and Alpher introduced thermal radiation into Big BAng cosmology. He is probably saying "...when the universe was so big...".

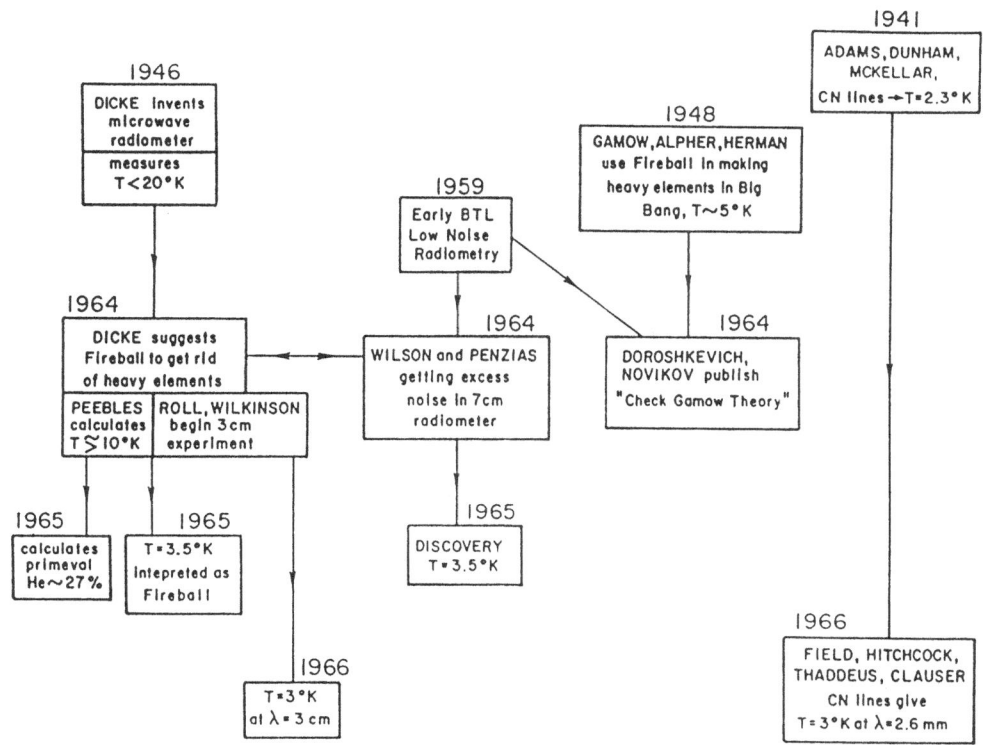

*Figure 6.* Summary of the discovery of the Primeval Fireball, showing little cross fertilization. A vertical classification scheme might be (left to right) physicist, radio astronomers, theoretical physicist, and astronomers.

discovery processes. The observed excitation of interstellar CN was a quite well- known puzzle in astrophysics, yet the physicists and radio astronomers, wondering about Fireball radiation, missed it. Generally, astronomers know, and follow, the literature better than physicists do. We are impressed that at least two astronomers, learning of the Fireball idea, remembered a funny business in the absorption spectrum of the star $\zeta$-Ophiuchi. There seems to be more coupling between physicists and radio astronomers, but the "Dicke" and "Gamow" boxes in Figure 6 were not connected prior to the discovery. Both groups published in the Physical Review, at a time when it was a relatively thin single volume. Still, the connection between a well- known instrument - the Dicke radiometer - and a prediction of the widely discussed "Gamow theory" waited for 15 years, and didn't contribute directly to the discovery.

This research was supported in part by the National Science Foundation.

## 4. REFERENCES

1. Dicke, R.H., *Rev. Sci. Instruments*, **17**, 268 (1946).
2. Dicke, R.H., obert Beringer, R., Kyhl R.L., and Vane, A. B. *Phys. Rev*, **70**, 340 (1946).
3. DeGrasse, R.W., Hogg, D.C., Ohm, E.A., and Scovil, H.E.D., *Proceedings of the National Electronics Conference*, **15**, 370 (1959).

4. Penzias, A.A., and Wilson, R.W., *Astrophys. J.*, **142**, 419 (1965).
5. Dicke, R.H., Peebles, P.J.E., Roll, P.G., and Wilkinson, D.T., *Astrophys. J.*, **142**, 414 (1965).
6. Wilkinson, D.T., *Measurements of the Cosmic Microwave Radiation*, Proceedings of the Berkeley Workshop on Particle Astrophysics, ed. E. Norman (1989).
7. Matsumoto, T., Hayakawa, S., Matsu, H., Murakami, H., Sato, S., Lange, A.E., and Richards, P.L., *Astrophys. J.*, **329**, 567 (1988).
8. Adams, S.W., *Astrophys. J.*, **93**, 11 (1941).
9. McKellar A., *Publ. Dominion Astrophys. Obs.*, Victoria, B.C., **7**, 251 (1941).
10. Our account is gleaned mostly from reading the early papers,

and from conversation and correspondence with Gamow. For a first hand account see: Alpher, A., and Herman, R., *Cosmology, Fusion and other Matters* George Gamow Memorial Volume, ed. F. Reines (Assoc. Univ. Press, Boulder, CO., 1972) p. 1.

11. Gamow, G., *Phys. Rev* ,**70** ,572 (1946).

12. Alpher, R.A., Bethe, H.A., and Gamow, G., *Phys. Rev* ,**73**, 803 (1948).

13. Gamow, G., *Phys. Rev.* ,**74**, 505 (1948).

14. Alpher, R.A., and Herman, R.C., *Nature*, **162**, 774 (1948).

15. Gamow, G., *Vistas in Astronomy* ed. by A. Beers (Pergamon Press, New York, 1956) **Vol. 2**, 1726.

16. Doroshkevich, A.G., and Novikov, I.D., *Sov. Phys-Doklady* **9**, 11 (1964). This paper was brought to our attention by C.H. Townes after the 1968 Washington APS meeting.

17. Ohm, E.A., *Bell Syst. Techn. J.*, **40**, 1065 (1961).

18. Readhead, A.C.S., Lawrence, C.R., Myers, S.T., Sargent, W.L.W., Hardebeck, H.E., and Moffet, A.T., *Astrophys. J.*, October (1989).

# SMALL SCALE ANISOTROPY MEASUREMENTS

C. R. Lawrence
Owens Valley Radio Observatory
California Institute of Technology
Pasadena, CA 91125, USA

## 1. INTRODUCTION

The horizon size $\theta_*$ at the redshift of last scattering $z_*$ is given by

$$\theta_* \approx \left(\frac{\Omega_0}{z_*}\right)^{1/2}$$
$$\geq \left(\frac{0.1}{1000}\right)^{1/2} \approx \frac{1}{2}^\circ .$$

Thus small scale anisotropy measurements are sensitive both to primordial fluctuations and to fluctuations produced by the development of structure at intermediate redshifts. In fact, given the statistical cancellation of primordial fluctuations below about 10', we expect fluctuations from intermediate redshifts to dominate on small scales. However, in this 25th year of the Cosmic Background era, we don't have to worry about the redshift of fluctuations, because none has been detected.

The task of reviewing small scale measurements is also made easy by the fact that observers have learned the lessons of previous experiments very well. For an excellent recent review that includes a complete history of anisotropy measurements, see Partridge (1988). I will discuss only the most sensitive measurements; these turn out to be the most recent as well.

## 2. ANGULAR SCALES LESS THAN 1'

### 2.1 VLA results at 5 GHz

The most sensitive measurements on this scale have been made by Martin and Partridge (1988) and Fomalont et al. (1988), both with the VLA at a frequency of 5 GHz. The observations are summarized in the table.

Both groups followed roughly the same procedure. An image was made of a region of sky much larger than the main beam of the individual 25 m antennas of

*N. Mandolesi and N. Vittorio (eds.), The Cosmic Microwave Background: 25 Years Later, 33–44.*
© 1990 *Kluwer Academic Publishers.*

## VLA Observations at 6 cm

| | Fomalont *et al.* | | Martin & Partridge 1984 |
|---|---|---|---|
| | 1984 | 1987 | |
| Field R. A. . . . . . . . . . . . . . . . . . | $00^h15^m24\overset{s}{.}0$ | $14^h16^m15\overset{s}{.}5$ | $03^h10^m00\overset{s}{.}0$ |
| Field Decl. . . . . . . . . . . . . . . . . | 15°35′00″ | 52°40′56″ | 80°08′00″ |
| Integration time . . . . . . . . . . . . | 54 hours | 46 hours | 25 hours |
| Frequency. . . . . . . . . . . . . . . . . . | 4.86 GHz | 4.86 GHz | 4.86 GHz |
| Bandwidth . . . . . . . . . . . . . . . . . | 2 × 50 MHz | 2 × 50 MHz | 2 × 50 MHz |
| Theoretical rms noise . . . . . . . . | 4.6 $\mu$K | 5.0 $\mu$K | 6.8 $\mu$K |

the array. After obvious sources were removed, the distribution of pixel values was determined.

In the region of the image where the antennas are sensitive to radiation from the sky, positive pixel values are produced by instrumental noise, discrete radio sources too weak or confused to be detected individually and removed, or by fluctuations in background levels. Negative pixel values are produced by the same things, except that only the (negative) sidelobes of discrete sources show up. Well outside the primary beam area, where the antennas are insensitive to any radiation from the sky, the pixel values are determined entirely by instrumental noise. In both the Martin and Partridge and the Fomalont *et al.* data, the distribution of pixel values in the central region of the image was broader than the distribution from the outer region, and with a noticeable positive tail.

Both groups concluded that at least most of the positive tail was due to unsubtracted, weak sources, but they differed in the techniques they used to account for such sources. Martin and Partridge extrapolated 20 cm and 6 cm source counts to low flux density levels, while Fomalont *et al.* "fit" a 2-parameter source count function by simulating data sets, then processing them in the same way as the observations.

The numbers from the two groups on various angular scales are summarized below:

Martin & Partridge:  18″–160″  $\frac{\Delta T}{T}$  $= 2 \times 10^{-4}$
$< 4 \times 10^{-4}$

Fomalont *et al.*:  12″  $< 8.5 \times 10^{-4}$  "95% confidence"
18″  $< 1.2 \times 10^{-4}$
30″  $< 0.8 \times 10^{-4}$
60″  $< 0.6 \times 10^{-4}$

Martin and Partridge concluded that all fluctuations in excess of the instrumental noise level could not be accounted for by sources. Hence their first value above is not an upper limit. The second value is a conservative upper limit based on the same data. In contrast, Fomalont *et al.* concluded that excess negative fluctuations in both their own and the Martin and Partridge data sets are accounted for by negative sidelobes of weak (positive) sources. Hence all of their values above are upper limits.

Interesting questions in image processing and extragalactic source counts are raised by these papers (see Partridge 1989 for a discussion). All of these questions have not been resolved; however, Martin and Partridge now agree that negative sidelobes of weak sources have a significant effect on the distribution of pixel values. At this meeting Fomalont showed the results of additional tests that reinforce this point.

If the Fomalont *et al.* position that no background fluctuations have been detected in the VLA data holds up, then the VLA has done all it can do at 5 GHz. To break through the confusion barrier there are only two options: either higher resolution but with the same surface brightness sensitivity is required, or observations must be made at a higher frequency where discrete sources are weaker compared to the background. The second option is being actively pursued.

## 2.2 VLA results at 15 GHz

To avoid the 5 GHz confusion limit, Hogan and Partridge (1989) observed a field at R.A. = $08^h41^m42^s0$, Decl. = $+44°42'45''$ known to be free of strong sources at 1.4 GHz with the VLA for 24 hours at 14.94 GHz. At this frequency receiver noise is much higher than at 5 GHz, and the thermal noise limit in the image is about $25\,\mu$K/beam. However, *no discrete sources* are seen in the image, therefore no corrections must be made. Hogan and Partridge calculate upper limits on various angular scales of

$$
\begin{array}{ll}
5\overset{''}{.}4\text{--}48'' & \frac{\Delta T}{T} < 6.3 \times 10^{-4} \\
10''\text{--}48'' & < 3.2 \times 10^{-4} \\
18''\text{--}50'' & < 1.6 \times 10^{-4}
\end{array}
$$

The first limit is the best achieved so far on such a small angular scale. The last limit supports the conclusion that at 5 GHz the VLA is limited by confusion.

These 15 GHz measurements are limited by receiver noise; however, given the recent rapid progress of HEMT technology, it is not unreasonable to hope for sensitivity improvements in the high frequency VLA receivers by a factor of two or even three over the next decade. Until then it is likely that VLA background work will be concentrated at 8.4 GHz, where a superb new receiving system (a legacy of Voyager's encounter with Neptune) is now available. The combination of higher resolution and weaker sources at this frequency will give a confusion limit ~5 times lower than at 5 GHz, and the low receiver noise will make it relatively easy to reach that limit.

## 2.3 IRAM results at 230 GHz

The lowest limits on anisotropies made at high frequencies started out as observations of a set of radio quiet IRAS quasars. Kreysa and Chini (1989) observed 25 known quasars at 230 GHz with the 30 m IRAM telescope and a bolometer receiver with instantaneous bandwidth of 50 GHz. The sensitivity of the system under ideal atmospheric conditions was 50 mJy s$^{1/2}$; however, the average sensitivity during the actual observations was 70 mJy s$^{1/2}$. Only four of the quasars were detected. Kreysa and Chini analyzed the data from the empty fields using the "standard" (for single-dish background work) likelihood-ratio technique (e.g., Boynton and Partridge 1973; Uson and Wilkinson 1984), and found a 95% confidence limit of

$$\Delta T/T < 2.6 \times 10^{-4},$$

assuming that the background consists of uniform, uncorrelated patches of emission 30″ across (see § 3.2 for a more precise statement of the assumptions).

This limit is not competitive with the latest VLA results on the same angular scale as far as the cosmic background is concerned. However, at 230 GHz the putative submillimeter excess would contribute a much larger fraction of the total background level than at 5 or 15 GHz. Thus this high frequency limit translates into an important constraint on anisotropy of the submillimeter excess. Kreysa and Chini estimate that it would take $10^4$ times as many unresolved sources of submillimeter radiation as there are galaxies to produce a submillimeter background as smooth as they measure. The implications of these measurements are discussed in more detail elsewhere in this volume.

The Kreysa and Chini limit was obtained from 35 hours of data, a modest investment of telescope time by the normal standard of single-dish background measurements. Moreover, the quasar fields were spread over the sky, rather than near the celestial pole where certain systematic errors are minimized, and the number of fields is larger than optimum for placing limits on Gaussian fluctuations. Thus there are no technical obstacles to a considerable improvement in millimeter-wavelength limits over the next few years.

## 3. ANGULAR SCALES GREATER THAN 1′

### 3.1 Ratan 600 m Results

Limits ranging from $\Delta T/T < 1.3 \times 10^{-5}$ to $8.0 \times 10^{-5}$ on angular scales from a few arc minutes up to $2°5$ have been claimed from observations with the Ratan 600 m telescope (Parijskij 1973a, b; Parijskij, Petrov, and Cherkov 1977). Unfortunately, few details of the observations and analysis are available in the West. Partridge (1980a, b; 1983) and Lasenby (1981) have converted these limits to 95% confidence limits and corrected for certain statistical problems so that they can be compared more easily with other measurements, obtaining limits of $5.4 \times 10^{-5}$ on scales of 75′ to $1 \times 10^{-4}$ on scales of 10′. Berlin et al. (1983, 1984) give a $1\sigma$ limit of $1 \times 10^{-5}$ on scales from 4′.5 to 9′.5. At 3.9 GHz, however, Ratan observations require large corrections for discrete radio sources. Amirkyanyan (1987) claims that the corrections made by Parijskij, Berlin, and collaborators are too low by a factor of 10. Given the problems caused by discrete sources in the higher frequency VLA data described above, I think one is justified in maintaining a sceptical attitude to the Ratan limits, at least until full details of the analysis and confusion correction are available.

### 3.2 OVRO 40 m Results

Using the 40 m altazimuth telescope of the Owens Valley Radio Observatory, Readhead et al. (1989) observed eight fields at $\delta = 89°$, $\alpha = 1^h, 3^h, \ldots, 15^h$ at 20 GHz. Two beams of $1′.8$ FWHM were separated by $7′.15$ on the sky. During cold dry winter conditions the equivalent noise temperature of this system including the atmosphere was $\sim 40$ K, giving a sensitivity on the sky of $\sim 9$ mK s$^{1/2}$.

In 24 hours we observed each of the eight fields for 2 hours at upper culmination, and the middle four fields for another 2 hours at lower culmination. Fields close to the pole can be observed near transit with only small motions of the telescope, especially in elevation. This minimizes the effects of differential ground and atmospheric emission in the two feeds. In addition, we used a standard (for single-dish work) double switching scheme (see e.g., Lake and Partridge 1980; Uson and Wilkinson 1984) to eliminate offsets between the two feeds and atmospheric gradients to first order. Thus we measured

$$\Delta T = T_{\mathrm{M}} - \frac{1}{2}(T_{\mathrm{R}1} + T_{\mathrm{R}2}), \tag{1}$$

where M, R1, and R2 refer to patches of sky with $1\!\!.\!8$ FWHM separated by $7\!\!.\!15$, with M in the middle. In effect, we measure the second derivative of the sky temperature at each of the eight different positions.

The measurements are given in the table. Note that the $1\sigma$ errors on individual fields are about $30\,\mu\mathrm{K}$. There is a known extragalactic radio source in field 7. (This was clear early in the observations, but we continued to observe the field because detection of the source provided a useful check on proper operation of the receiver.) Therefore field 7 was excluded from further analysis.

OVRO OBSERVATIONS AT 20 GHz

| Field | $\overline{\Delta T} \pm \sigma\ (\mu\mathrm{K})$ |
|---|---|
| NCP1 ........ | $-64 \pm 35$ |
| NCP3 ........ | $20 \pm 34$ |
| NCP5 ........ | $-29 \pm 27$ |
| NCP7 ........ | $217 \pm 28$ |
| NCP9 ........ | $34 \pm 26$ |
| NCP11 ....... | $-23 \pm 26$ |
| NCP13 ....... | $-20 \pm 32$ |
| NCP15 ....... | $-36 \pm 39$ |

From a Bayesian statistical analysis using a uniform prior distribution we derived an upper limit on $\sigma_{\mathrm{sky}}$, the dispersion of Gaussian sky fluctuations for the triple beam, of $58\,\mu\mathrm{K}$. From a uniformly most powerful likelihood ratio test (the "standard" technique) we obtain a 95% confidence limit of $\sigma_{\mathrm{sky}} < 52\,\mu\mathrm{K}$. We use $58\,\mu\mathrm{K}$ as the official OVRO limit. The assumption of Gaussian fluctuations is convenient, in that the majority of models produce them, and also reasonable, given that a fluctuation spectrum with a flat power-law tail (specifically, with power greater than $-2$) is better constrained by observations of a large number of fields.

The statistics of a Gaussian random field are completely specified by the 2-point correlation function $C(\phi)$. We can write equation 1 for $\Delta T$ in terms of this autocorrelation function as follows:

$$\overline{\Delta T^2} = \frac{3}{2}C(\phi_0, 0) - 2C(\phi_0, \phi_s) + \frac{1}{2}C(\phi_0, 2\phi_s), \tag{2}$$

where $C(\phi_0, \phi)$ is the true autocorrelation function $C(\phi)$ suitably smeared by the non-zero dispersion of the observing beam ($\phi_0 = 0.4247 \times \phi_{\mathrm{FWHM}} = 0\!\!.\!78$ for OVRO), and $\phi_s$ is $7\!\!.\!15$, the beam separation. Define the *coherence angle* as

$$\phi_c \equiv \left(-\frac{C(0)}{C''(0)}\right)^{1/2}.$$

If the coherence angle of the fluctuations is much smaller than the beam separation (i.e., $\phi_c \ll \phi_s$), equation 2 reduces to

$$\overline{\Delta T^2} \approx \frac{3}{2} C(\phi_0, 0).$$

If, in addition, $\phi_0 \gg \phi_c$,

$$\overline{\Delta T^2} \approx \frac{3}{2} C(\phi_0),$$

where now the "beam-smeared" autocorrelation function is no longer needed. Real telescopes are unlikely to achieve the conditions $\phi_0 \ll \phi_c \ll \phi_s$, because control of systematic errors generally limits the ratio $\phi_s/\phi_0$ to values near three. However, for $\phi_c \approx \sqrt{\phi_0 \phi_s} \approx 2\rlap{.}'3$ (for the OVRO beams) the approximation is generally quite good. In this case the OVRO result is

$$"\delta T/T" = \frac{C^{1/2}(0)}{T} = \sqrt{2/3}\,\sigma_{\text{sky}}$$
$$< 1.7 \times 10^{-5}.$$

Although anisotropy observations have often been reduced to a single number in this way, the (sometimes unstated) assumption that $\phi_0 \ll \phi_c \ll \phi_s$ is quite restrictive. Ideally, one should use the complete model autocorrelation function in equation 2 to compare theory and experiment. However, the much less restrictive approximation that the autocorrelation function is Gaussian,

$$C(\phi) = C_0 \exp\left(-\frac{\phi^2}{2\phi_c^2}\right),$$

still allows easy comparison with a wide range of models. For model autocorrelation functions that go negative on large scales, a Gaussian approximation is poor; however, on small and intermediate scales, over not too wide a range of coherence angles, a Gaussian approximation is quite reasonable. Figure 1 shows the limits placed on $C_0^{1/2}$ as a function of $\phi_c$ by the VLA observations of Fomalont et al. described above and by the OVRO data, and well as by the data of Davies et al. on a larger scale. In both cases sensitivity is reduced at small coherence angles because a beam or resolution element averages over many independent fluctuations, and at large angles because only small segments of single fluctuations are sampled.

Figure 1 provides some model-dependent support for the position that the VLA results of Martin and Partridge should not be interpreted as detections of background fluctuations; however, not too much should be made of this.

## 4. A FEW STATISTICAL COMMENTS

While we work toward a future of 10 and $20\sigma$ measurements of the cosmic background, we should remember some of the dangers of statistical living.

1. Neither measurements, subject to errors both random and systematic, nor models, based on sometimes uncertain data (i.e., the galaxy-galaxy correlation function used for normalization), should be taken too literally. Also, the usual 95%

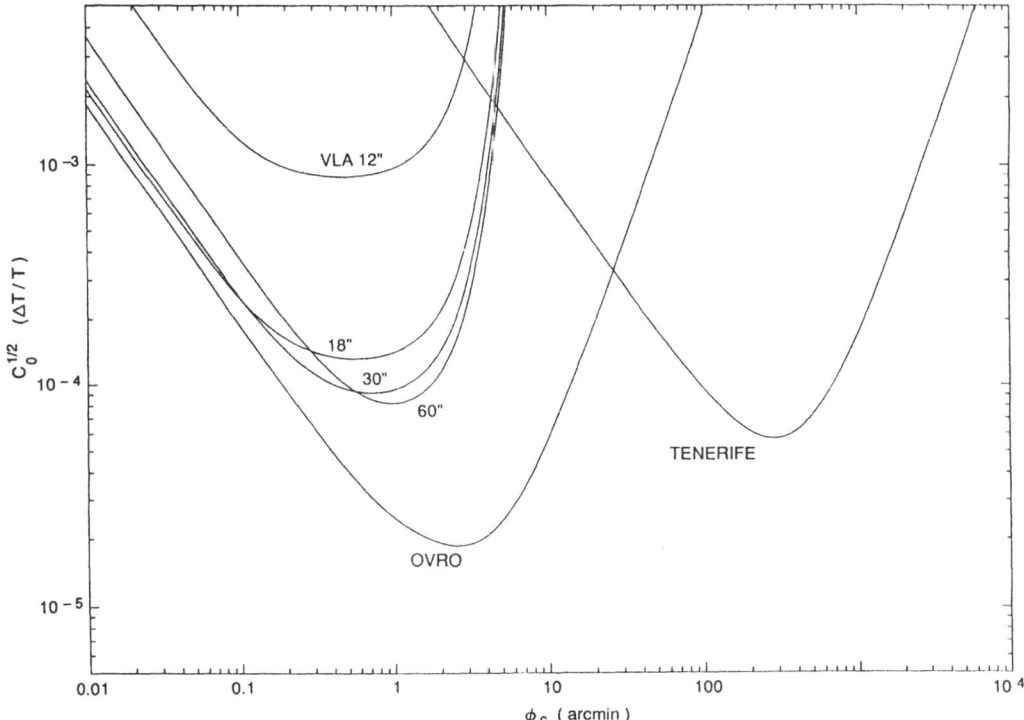

FIG 1.—Limits placed on a Gaussian autocorrelation function by the measurements of Fomalont *et al.* (1988), Readhead *et al.* (1989), and Davies *et al.* (1987) as a function of the coherence angle $\phi_c$. From Readhead *et al.* (1989).

confidence limits quoted for observations are less stringent than the "3$\sigma$" rule of thumb we often use to distinguish believable detections from noise in other kinds of experiments.

2. Both under- and overestimation of measurement errors are bad. Underestimation leads to spurious detections (if it isn't noise, it must be signal!). Overestimation of errors can lead to missed detections. Somewhat paradoxically, when upper limits are calculated using the likelihood ratio method overestimation of errors can lead to spuriously low upper limits (see Readhead *et al.* 1989 for details).

3. Suppose we observe $N$ fields (i.e., bits of sky) to a given noise level. It doesn't matter whether the fields are continguous, as in VLA images or continuously scanned strips of sky, or discrete, as in many single-dish experiments. In general, statistical conclusions become more stringent as $N$ increases. But there are problems in using the large-$N$ statistical hammer to infer limits or detections well below the noise level of individual fields. Most importantly, it is possible that

$$\text{level of inference} \ll \text{systematic errors} \ll \text{noise per field},$$

in which case undetectable systematic errors can completely dominate the conclusions. This is a serious worry in some space experiments where the whole sky is scanned, giving thousands of independent fields.

4. Statistical results are valid only to the extent that the assumptions used to obtain them are valid.

## 5. WHEN LIMITS BECOME DETECTIONS, WHAT WILL WE DETECT?

We can be confident that over the next 25 years anisotropies will be detected in background observations. Then we will have to answer the more interesting but more difficult question, what have we detected? Figure 2 divides the universe into astronomically interesting and useless regions. Experiments so far have been dominated by the useless part, although as we have seen the VLA results at 5 GHz seem to have run into the discrete source confusion limit already. Davies *et al.* (1987) may have detected anisotropy in galactic emission on degree scales as well, although this remains uncertain. The Sunyaev-Zel'dovich effect has been detected in several clusters (see Birkinshaw, this volume). The integrated SZ effect along radom lines of sight depends sensitively on the redshift of formation of hot cluster gas. If this redshift is large enough, the integrated SZ effect could be large (Sunyaev 1977, 1978; Rephaeli 1981, this volume).

Figure 3, from Wilkinson (1989), gives estimates of some foreground sources of anisotropy as a function of frequency and angular scale. Emission from H$_{II}$ regions has a $\nu^{-0.1}$ frequency dependence, so it shows up here with a slope of $-2.1$. Upper limits, based on previous anisotropy observations, are given for high galactic latitudes and two beam sizes. Clearly, microwave background work will be impossible at low galactic latitudes.

Radio source confusion has been estimated in detail by Fransceschini *et al.* (1989). The estimates in Figure 3 come from a preliminary version of that work. Notice that the slope is once again about $-2$. The low brightness temperature sensitivity of large beams gives low resolution instruments a clear edge for the purposes of background anisotropy detection. At low frequencies source counts are well-known, and the lines are determined directly. At higher frequencies things are less well-determined. According to Fransceschini *et al.*, high frequency observations will be fundamentally limited by dust emission from foreground galaxies, estimated from IRAS 60 $\mu$m measurements. They conclude that only at frequencies between roughly 30 and 200 GHz, and on angular scales either less that about 20″ or greater than about 1°, will it be possible to reach a level of $\Delta T/T \approx 10^{-6}$.

The galactic dust emission (IRAS cirrus) line in Figure 3 is an extrapolation from measurements by Matsumoto *et al.*, and represents a large scale average away from the galactic plane. Although this line is based on average emission, rather than anistotropy directly, cirrus fluctuations are typically 100%, and the line is a reasonable estimate of the likely effects of this dust.

Figure 3 will change as our knowledge of these confusing sources of radiation improves. Nevertheless, it is clear that detection of anisotropies in the *cosmic* background radiation will not be easy.

## 6. $10^{-6}$ IS NO PROBLEM

Over the first quarter century of anisotropy measurements upper limits have fallen by at least two orders of magnitude, due to a combination of improved instruments, improved techniques, and a lot of hard work. The smoothest universes that have been theoretically imagined are still an order of magnitude below the current limits. Can we reach this level in the next 25 years?

Instrumentally, I see no problem. Within a few years, low noise telescopes or arrays with low sidelobes in good sites (e.g., space, Antarctica), equipped with low noise receivers or detectors with large bandwidths, should be able to reach noise

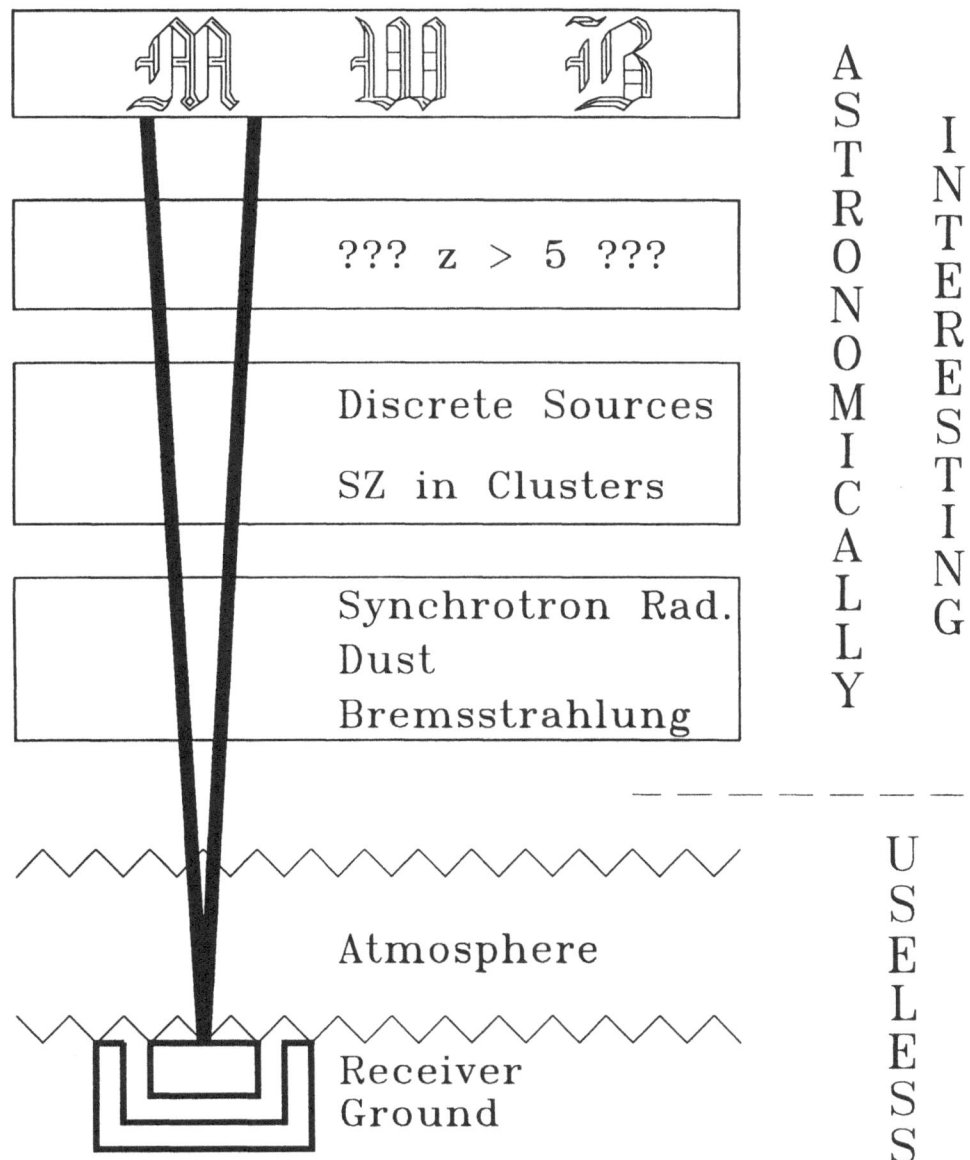

Fɪɢ 2.—Block diagram of the universe. Everything comes between us and the microwave background radiation.

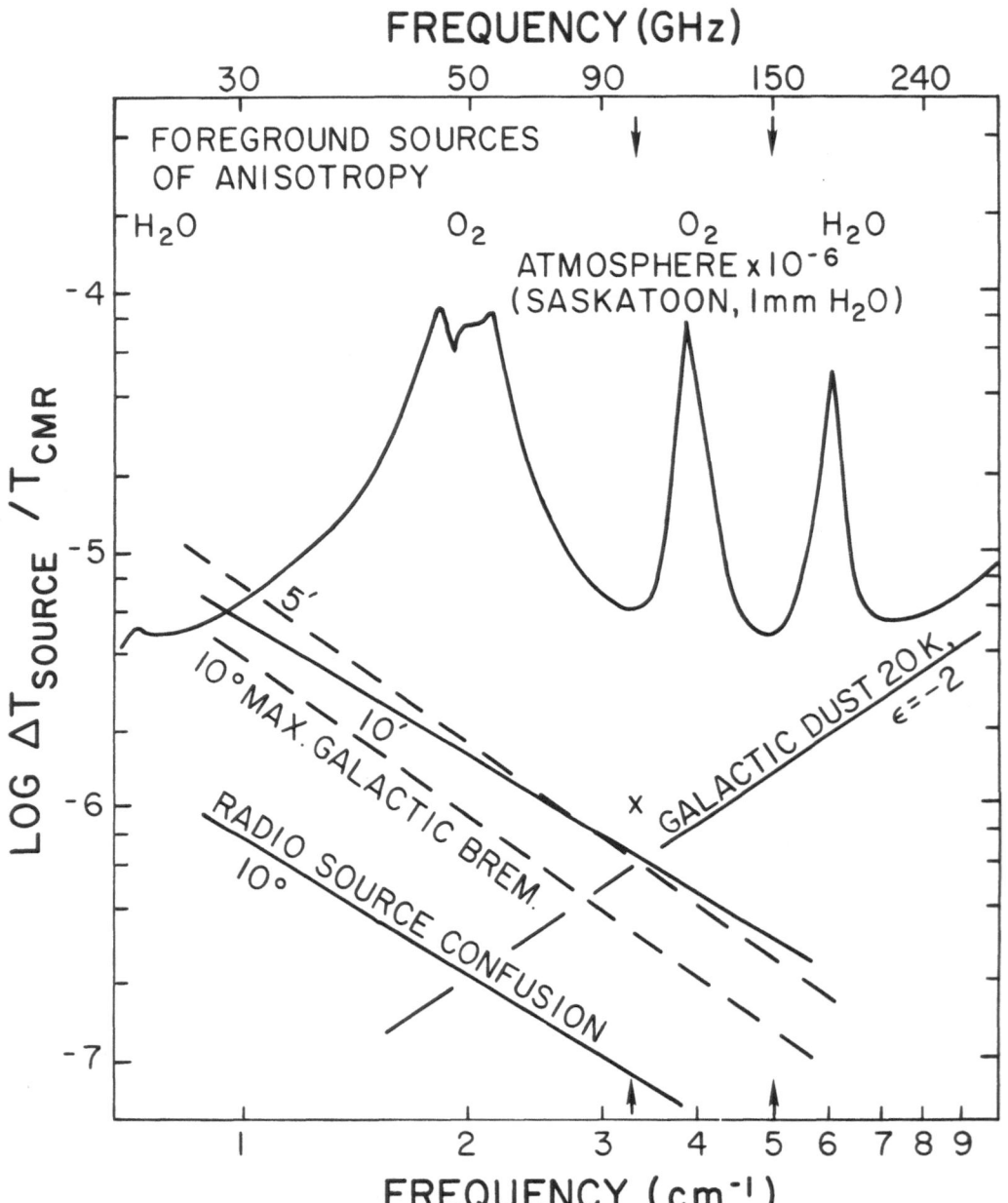

FIG 3.—Estimates of foreground sources of anisotropy as a function of frequency and angular scale, from Wilkinson (1989).

levels of a microKelvin or two. "No problem" does not mean "no effort", of course! Neither does it mean cheap.

Most importantly, achieving a given noise level is not the same thing as knowing the cosmic background temperature to that level, because everything else in the universe will be detected along with the cosmic background. The frequency and angular scale of experiments will have to be chosen for insensitivity to confusing backgrounds, perhaps with simultaneous higher angular resolution experiments for discrete source removal. Spectral information undoubtedly will be required to distinguish one source of radiation from another, and multifrequency observations will become the norm.

## 7. REFERENCES

Amirkhanyan, V. R. 1987, *Soobshch. Spec. Astrofiz. Obs.*, **53**, 96.

Berlin, A. B., Bulaenko, E. V., Vitkovsky, V. V., Kononov, V. K., Parijskij, Yu. N., and Petrov, Z. E. 1983, in *IAU Symposium 104, The Early Evolution of the Universe and its Present Structure*, ed. G. Abell and G. Chincarini (Dordrecht: Reidel), p. 121.

Berlin, A. B., Gassanov, L. G., Gol'nev, V. Ya., Korol'kov, D. V., and Parijskij, Yu. N. 1984, *Soob. Spec. Astrofiz. Obs.*, **41**, 5.

Boynton, P. E. and Partridge, R. B. 1973, *Ap. J.*, **181**, 243.

Davies, R. D., Lasenby, A. N., Watson, R. A., Daintree, E. J., Hopkins, J., Beckman, J., Sanchez-Almeida, J., and Rebolo, R. 1987, *Nature*, **326**, 462.

Fomalont, E. B., Kellermann, K. I., Anderson, M. C., Weistrop, D., Wall, J. V., Windhorst, R. A. and Kristian, J. A. 1988, *A. J.*, **96**, 1187.

Franceschini, A., Toffolatti, L., Danese, L., and De Zotti, G. 1989. *Ap. J.*, **344**, 35.

Hogan, C. J., and Partridge, R. B. 1989, *Ap. J. (Letters)*, **341**, 29.

Kreysa, E. and Chini, R. 1989, in *Proc. of the Third ESO-Cern Symposium, Astronomy, Cosmology and Fundamental Physics*, ed. M. Caffo, R. Fanti, G. Giacomelli, and A. Renzini (Dordrecht: Kluwer), p. 433.

Lake, G., and Partridge, R. B. 1980, *Ap. J.*, **237**, 378.

Lasenby, A. N. 1981; Ph. D. thesis, University of Manchester.

Martin, H. M., and Partridge, R. B. 1988, *Ap. J.*, **324**, 794.

Matsumoto, T., Hayakawa, S., Matsuo, H. Murakami, H., Sato. S., Lange, A. E. and Richards, P. L. 1988, *Ap. J.*, **329**, 567.

Parijskij, Yu. N. 1973a, *Ap. J. (Letters)*, **180**, L47.

Parijskij, Yu. N. 1973b, *Soviet. Astr.*, **17**, 291.

Parijskij, Yu. N., Petrov, Z. E. and Cherkov, L. N. 1977, *Soviet. Astr. Letters*, **3**, 263.

Parijskij, Yu. N., Berlin, A. B. and Vitkovskij, V. V. 1987. *Soob. Spec. Astrofiz. Obs.*, **53**, 99.

Partridge, R. B. 1980a, *Physica Scripta*, **21**, 624.

Partridge, R. B. 1980b, *Ap. J.*, **235**, 681.

Partridge, R. B. 1983, in *NATO Advanced Study Institute, The Origin and Evolution of Galaxies*, ed. B. L. T. Jones and J. E. Jones (Dordrecht: Reidel), p. 121.

Partridge, R. B. 1988, *Rep. Prog. Phys.*, **51**, 647.

Partridge, R. B. 1989, in *Proc. of the Third ESO-Cern Symposium, Astronomy, Cosmology and Fundamental Physics*, ed. M. Caffo, R. Fanti, G. Giacomelli,

44

and A. Renzini (Dordrecht: Kluwer), p. 105.

Readhead, A. C. S., Lawrence, C. R., Myers, S. T., Sargent, W. L. W., Hardebeck, H. E., and Moffet, A. T. 1989, *Ap. J.*, **346**, ;.

Rephaeli, Y. 1981, *Ap. J.*, **245**, 351.

Sunyaev, R. A., 1977, *Pis'ma Astr. Zh.*, **3**, 491.

Sunyaev, R. A., 1978, *Soviet Astr. Letters*, **3**, 268.

Uson, J. M. and Wilkinson, D. T. 1984, *Nature*, **312**, 427.

Wilkinson, D. T. 1989, in *Proc. of the Workshop on Particle Astrophysics: Forefront Experimental Issues*, ed. Eric B. Norman (New Jersey: World Scientific), p. 111.

# THEORETICAL MAPS OF CMB ANISOTROPIES

J. Richard Bond
CIAR Cosmology Program
Canadian Institute for Theoretical Astrophysics
University of Toronto
Toronto, Ontario
M5S 1A1, Canada

**ABSTRACT**  Patterns of the fluctuations in the temperature of the CMB that arise in various theoretical models of structure formation are used to illustrate the potential richness of the CMB sky on all scales: from arcseconds, for primeval dust emission from dwarf galaxies; to arcminutes, for Compton cooling anisotropies from hot gas in groups and clusters; through tens of arcminutes to a few degrees, where primary anisotropies from linear gas flows and photon bunching arise; to tens of degrees and radians where the scale invariant nature of fluctuations generated during inflation may be discovered. Across this broad range the experiments are getting dangerously close to the levels predicted for favourite theories. The casualty rate is increasing and it may soon become an epidemic. A discovery must be near to change the prognosis.

## 1. INTRODUCTION

In the past decade, the theory of cosmic structure formation has been greatly developed, stimulated by the importation of such particle physics concepts into cosmology as inflation, quantum–generated density fluctuations, cosmic strings and walls, baryosynthesis, and massive elementary particle dark matter candidates. Equally powerful as a stimulant is the enormous expansion of our cosmic data–base that any viable theory must contend with. This is especially true for CMB anisotropies. Most often the models I shall treat derive their structure from linear

45

*N. Mandolesi and N. Vittorio (eds.), The Cosmic Microwave Background: 25 Years Later, 45–65.*
© 1990 *Kluwer Academic Publishers.*

perturbations that formed a homogeneous and isotropic Gaussian random field. In this case, a single function fully specifies the nature of the perturbation field, the density fluctuation spectrum, $d\sigma_\rho^2/d\ln k \equiv \frac{k^3}{2\pi^2} \langle |(\delta\rho/\rho)(k)|^2 \rangle$, giving the contribution of (statistically independent) modes of comoving wavenumber $\vec{k}$ to the $rms$ linear density fluctuations $\sigma_\rho$. However, although Gaussian perturbations are the most likely outcome of inflation, initially non-Gaussian perturbations may arise in models with more than one scalar field. A major goal of cosmic structure research is therefore to ascertain whether the perturbations were initially Gaussian-distributed, and, if so, to determine the two-point function $d\sigma_\rho^2/d\ln k$. Although the form of the statistics is maintained in the linear regime, as the Universe evolves into the nonlinear regime the coupling of modes causes high order correlations to develop, obscuring the simplicity of the initial state of the perturbations. Linearity *is* an appropriate approximation for anisotropies in the cosmic microwave background radiation and for some aspects of large scale galaxy flows and clustering. Herein lies their importance.

Primordial Gaussian perturbations lead to a Gaussian primary CMB radiation pattern. A non-Gaussian primary radiation pattern is a signature of cosmic string theories (Bouchet, Bennett and Stebbins 1988), as well as of exotic inflation models. Indeed, the cleanest way to decide on the statistics of the primordial fluctuations would rely on primary CMB anisotropy observations (*e.g.,* Bond and Efstathiou 1986), since one does not have to untangle the initial underlying correlations from the complex dynamical correlations that develop with nonlinearity. If we are lucky, the CMB anisotropies will reveal to us the full mass distribution when the universe was simpler, and galaxy formation and its epoch will be derivable, albeit with some difficulty, from a known initial state.

It is worthwhile to emphasize that *all* current theories of structure formation need to invoke inflation, whether the theory relies on scalar field fluctuations, normal or superconducting cosmic strings, explosions, or isocurvature baryon perturbations. No other viable mechanism to ensure flatness and smoothness has been proposed. The strong implication is that our local patch of the universe (the region we can see within about $10^4$ Mpc) is likely to have $\Omega \approx 1$, if possible vacuum energy (non-zero $\Lambda$) contributions are included in the tally. A weaker implication relies on the specific model of fluctuation generation. If the perturbations that grew into the observed cosmic structure were from quantum oscillation modes of scalar fields, then it is most likely that the perturbation field was initially Gaussian with a fluctuation spectrum that was scale invariant. (The power law index $n$ defined by $d\sigma_\rho^2/d\ln k \propto k^{3+n}$ is 1 for the adiabatic case). Although this has focussed our attention on this specific case it should be recognized that disproof of initial scale invariance or initial Gaussian-ness by CMB observations will not disprove inflation: scale invariance might be broken and/or the statistics might not be Gaussian (Kofman and Linde 1987, Grinstein and Wise 1987, Kofman, L.A. and Pogosyan, D.Y. 1988, Salopek, Bond and Bardeen 1989). And although adiabatic models are the most natural, the isocurvature mode might also be generated in inflation models, with fluctuations in some matter density component (baryons or cold dark matter) compensated by fluctuations of opposite sign in the radiation (photons, quark-antiquark pairs, gluons *etc.* ) at the epoch of generation. For isocurvature baryon perturbations (once referred to as isothermal or entropy perturbations), $d\sigma_{n_B}^2/d\ln k$ would be proportional to the oscillations in some field to which the baryon number density $n_B$ couples. The result is likely scale invariant, $n = -3$, with equal power per decade.

**Figure 1.1** *Fluctuation spectrum $[d\sigma_\rho^2/d\ln k]^{1/2}$ for the standard $\Omega = 1$ CDM model. A critical ingredient of the CDM model is its hierarchical character: peaks of the unevolved density field on a sequence of scales of decreasing resolution can be associated with a hierarchy of collapsed structures. That is, objects such as the first stars, galaxies, groups and clusters are (roughly) associated with specific wavenumber bands of the spectrum. Also plotted are the wavenumber ranges probed by various observational tests. Note how the CMB tests (RELICT — and COBE — Tenerife, OVRO and, in between, the Lubin and Meinhold experiment) cover more of k-space than the conventional approach of galaxy clustering observations ($\xi_{gg}, w_{gg}, \xi_{gc}, \xi_{cc}$, large scale texture (LST) and streaming (LSSV)). Further, Sunyaev-Zeldovich observations can probe the intermediate group to cluster scale range and anisotropies from dust emission associated with pregalactic stars or primeval galaxies can potentially probe even shorter distances. For comparison with CDM, an isocurvature baryon model with spectral index $n = -1$, $\Omega = \Omega_B = 0.4$, and $h = 0.5$ is shown as well. For this model, the scales which the CMB experiments probe should all shift by a factor $\Omega^{-1/2}$ to longer wavelengths.*

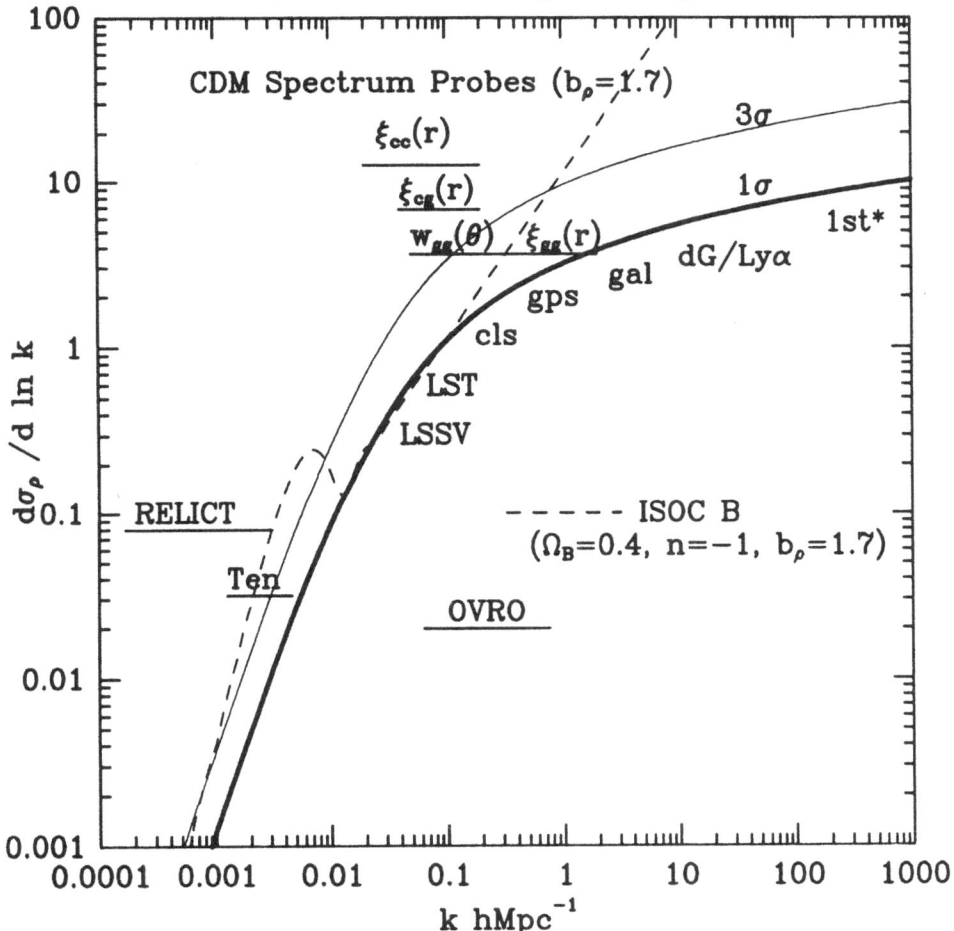

The combination of the post-inflation spectrum and the transfer function gives the pre-nonlinear shape of $d\sigma_\rho^2/d\ln k$. There is still an overall normalization amplitude to determine. In principle, this would be set by knowing the details of the inflaton potential. In practice, we must introduce a parameter to characterize the amplitude of the linear spectrum. It has now become conventional to define a biasing factor $b_\rho$ to do this (Bardeen et al. 1986), which is one if mass traces light and greater than one if galaxies are more clustered than the mass distribution. The technical definition Efstathiou and I have used in our calculations is '$J_3$-normalization' (see Bond and Efstathiou 1984 for details). The other widely used normalization, to mass fluctuations in $8\,h^{-1}$Mpc spheres, agrees to within 10% with this $J_3$-normalization for non-pathological spectra. (Here $h \equiv H_0/100\,\mathrm{km\ s^{-1}\ Mpc^{-1}}$.)

The minimal CDM model of Fig. 1.1 provides our standard illustration of structure in the CMB because it is so definitive in its predictive power. Assuming an initially scale–invariant spectrum of fluctuations from inflation, there is really only one free parameter in the theory. It is not $\Omega$ which must be within $\varepsilon$ of unity, with $\varepsilon \lesssim 10^{-5}$ in order that enough e-foldings occur to ensure homogeneity and isotropy. It is not the Hubble parameter $H_0$, which must be in the 40 to 50 range, or else there would be a time crisis for globular clusters and nuclear cosmochronology. It is not $\Omega_B$, the baryon abundance, for it must be $\lesssim 0.2$ in order to maintain the successes of primordial nucleosynthesis. (If $\Omega_B$ is near the upper range, the transfer function and the amplitude of CMB anisotropies are modifed (Bardeen et al. 1987).) The one free parameter is $b_\rho$. Currently it is thought to lie in the range 1.4-2.6.

To put the scales in Fig. 1.1 into context, I now list the values of some relevant physical scales for h = 0.5 universes with a fraction $\Omega_{nr}$ of the critical density in nonrelativistic particles (CDM, baryons): the horizon at the epoch when the energy densities in relativistic and nonrelativistic matter are equal, $k_{Heq}^{-1} = 10\Omega_{nr}^{-1}\,h^{-1}$Mpc, which determines the shape of the CDM spectrum; the horizon scale at recombination, $k_{Hrec}^{-1} = 41\Omega_{nr}^{-1/2}\,h^{-1}$Mpc; the viscous (Silk) damping scale, $k_{Silk}^{-1} \sim 1.3\ \Omega_B^{-1/2}(\Omega_{nr}^{-1/4})^{10/9}\,h^{-1}$Mpc; the baryon Jeans length at recombination, $k_{JBrec}^{-1} = 0.0016\ \Omega_{nr}^{-1/2}\ h^{-1}$Mpc; the collisionless damping scale for hot dark matter, $k_{\nu damp}^{-1} = 6\ (\Omega_{nr}\Omega_{m\nu})^{-1/2}\,h^{-1}$Mpc. As well, some scales of relevance to CMB fluctuations are: the curvature scale for open universes (in which case $k$ is not exactly wavenumber), $k_{curv}^{-1} = 3000\,(1-\Omega_{nr})^{-1/2}\ h^{-1}$Mpc; the fuzziness of the last scattering surface (for normal recombination) below which destructive interference damps CMB anisotropies, $k_{LS}^{-1} \approx 5\Omega_{nr}^{-1/2}\,h^{-1}$Mpc; and of course $k_{Hrec}^{-1}$. The other scales have a smaller impact on CMB anisotropies. Associated with these CMB scales are the angular scales $\theta_{LS} \approx 3'\Omega_{nr}^{1/2}$ and $\theta_{Hrec} \approx 2°\Omega_{nr}^{1/2}$ evaluated using the angle-distance relation $\theta(d) = 0.95°\Omega_{nr}d/100\,h^{-1}$Mpc appropriate for an $\Omega = \Omega_{nr} = 1$ universe and also for an $\Omega = \Omega_{nr} \ll 1$ universe.

The microwave background pattern can be expanded in multipole moments

$$\frac{\Delta T}{T}(\hat{q}) = \sum_{\ell m} a_{\ell m} Y_{\ell m}(\hat{q}). \tag{1.1}$$

Here $\hat{q}$ is the angular direction of the incoming photons. For Gaussian theories such as the CDM models, the multipole coefficients $a_{\ell m}$ are Gaussian-distributed with zero mean and variance

$$C_\ell = \langle |a_{\ell m}|^2 \rangle, \quad \text{with} \ \langle a_{\ell'm'}^* a_{\ell m} \rangle = 0 \ \text{for} \ \ell \neq \ell', \ m \neq m', \tag{1.2}$$

**Figure 1.2** *Although in general the comparison of models with theories involves complex calculations in multipole space, a simple way to illustrate how effective an experiment is in probing $C_\ell$ is to plot its filter function $F_\ell$ defined so that the rms amplitudes expected theoretically are given by*

$$\langle (\Delta T/T)^2_{expt} \rangle = \sum_\ell \frac{2\ell + 1}{4\pi} F_\ell C_\ell. \tag{1.3}$$

*The COBE filter is the same as the RELICT filter (§4). Tenerife is for their 8° beam (with 8° throw), not their more sensitive 5° beam. OVRO is Owens Valley with a 1.8′ beam and 7.15′ throw. IRAM is the Kreysa and Chini (1989) experiment, good for probing dust emission anisotropies. The Lubin and Meinhold (1989) South Pole experiment has a multipole range ideal for optimizing the signal from models like CDM. The dotted curve is $10^{10}b_p^2 \, \ell^2 C_\ell/(2\pi)$ for a CDM model with $\Omega = 1$, $\Omega_B = 0.03$ and h=0.75. The h = 0.5 model looks similar but has a higher amplitude. To translate to an angular region probed in high $\ell$ experiments, $\ell$ can roughly be considered to be in inverse radians, so the $\ell$–pole probes angles around $3438/\ell$ arcminutes.*

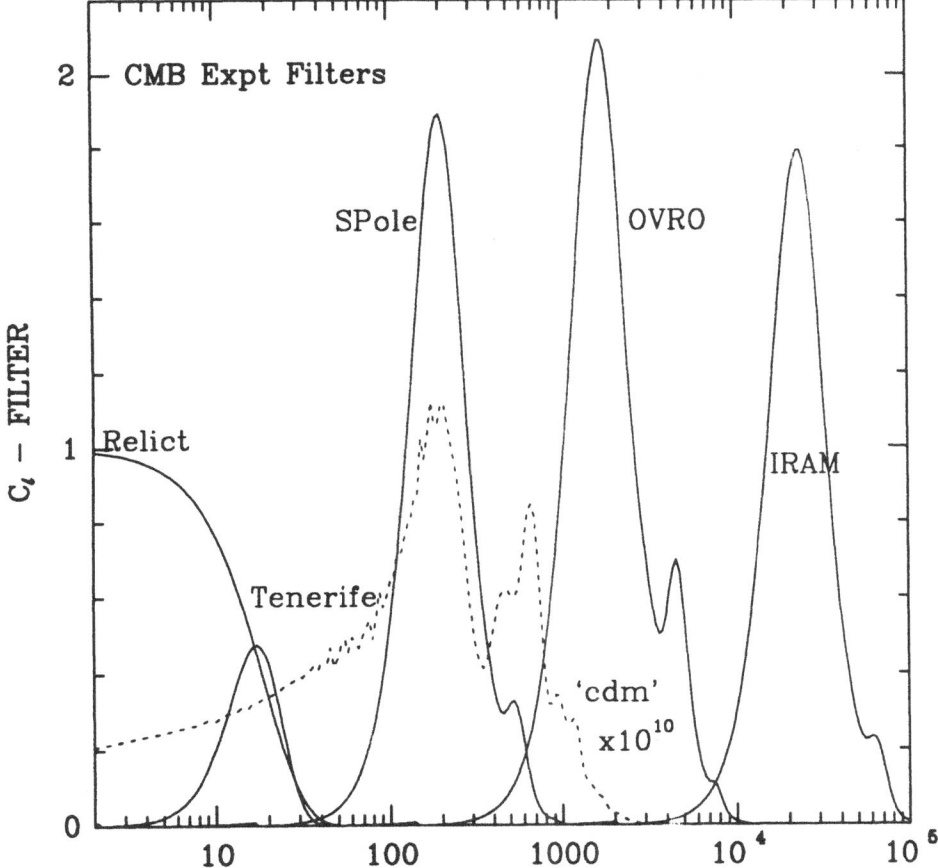

where the angular power spectrum $C_\ell$, is the goal of our calculating. (Note that $C(\theta) = \frac{1}{4\pi} \sum_\ell (2\ell + 1) C_\ell P_\ell(\cos\theta)$.) It is very instructive to compare the region of $\ell$-space that various anisotropy experiments probe with the power concentration for the theory under consideration. This is done for CDM in Figure 1.2.

## 2. MAPS OF PRIMARY ANISOTROPY PATTERNS

The maps of this section are all $10^o \times 10^o$ and are generated for these small patches using a $256 \times 256$ Fast Fourier transform. That this is sufficient to accurately represent the multipole expansion is demonstrated in §4. All of these Gaussian maps have the same random number seed so the patterns can be directly compared. A filter with $\theta_{fwhm} = 1.8'$ has been imposed. This corresponds to the OVRO beam (Readhead et al. 1989). In most cases the natural coherence scale associated with the thickness of the photon decoupling surface is much larger. The calculations were all done in collaboration with George Efstathiou and the perturbed transport and gravitational field equations solved are described in Bond and Efstathiou (1984, 1987) and Efstathiou and Bond (1986,1987).

Source terms in the linearized photon transport equations which drive the development of primary anisotropies are the Sachs–Wolfe source, most important on large angles, the photon density source, important on intermediate angles and, for isocurvature perturbations, on large angles as well, and the electron velocity source due to the Doppler effect in the Thomson scattering terms, an effect which is significantly diminished by the fuzziness of the last scattering surface. Destructive interference arises from opposing contributions to $\Delta T/T$ from the troughs and crests that straddle the photon decoupling surface. This damping is thus important for waves with $k > k_{LS}$, and defines a 'coherence' scale $\theta_c \approx \theta_{LS}$ for primary anisotropies, the characteristic size of the hills and valleys of the $\Delta T/T$ pattern. However, as Vishniac (1987) has pointed out, non-negligible small scale anisotropy can develop from quadratic nonlinearities in the scattering ($\propto \int \sigma_T \delta n_e \, v_e \, dt$) which do not suffer dramatic destructive interference, since different wave-modes are coupled. This effect leads to significant anisotropies for the isocurvature baryon models as is demonstrated in Figure 2.2 below.

Maps for the CDM model are shown in Figure 2.1. The angular CMB power spectrum for neutrino-dominated universes is quite similar in shape to that, but the overall amplitude must be higher to ensure that enough structure forms. Indeed current constraints from the OVRO and Lubin and Meinhold experiment strongly rule against these models (for adiabatic scale invariant spectra).

Anisotropies for the classic isocurvature baryon models with initial spectral index $n = 0$ (Poisson seed model) or $n = -1$ (phenomenological) that were very popular in the 1970s, and recently advocated by Peebles (1987), have now been calculated by a number of groups who seem to be in basic agreement. An example of a density spectrum for one is given in Fig. 1.1. It obviously has very impressive large scale structure properties as well as an earlier redshift of galaxy formation than CDM, both of which might better describe the observations. Some sample maps (with h=0.5) are shown in Figure 2.2. It is conventional to take the biasing factor $b_\rho$ to be unity in these models, since dynamical clustering is likely to completely dominate any statistical clustering of galaxies. Since there is a great deal of power at short distances in these models, star formation is expected to occur very early, hence it seems likely that, even if the Universe underwent standard recombination at

$z \sim 1000$, it would have been photoionized shortly after. Thus the no recombination models are more appropriate. With $b_\rho = 1$, a vast region in $\Omega_B$—$n$ space is ruled out by the observations, especially when the South Pole results are included. For isocurvature baryon perturbations to be compatible with inflation, we would require $\Omega = 1$, and, unless scale invariance is broken, $n = -3$ (a flat spectrum in $\delta n_B$), which is even more strongly ruled out than the $n = -1$ models.

In Figure 2.3, we compare the non-Gaussian structure formation theory whose particle physics is almost as well motivated as the inflaton fluctuation theory: the cosmic string model, which is also not very far from observability.

## 3. SECONDARY ANISOTROPIES FROM NONLINEAR STRUCTURE

Secondary backgrounds arise once fluctuations become nonlinear, turnaround and collapse. On fairly broad theoretical grounds we expect backgrounds such as the sub-millimetre excess found by the Berkeley-Nagoya team (Matsumoto $et\ al.$ 1988) to arise. Although current observations allow large energy releases, the surprising part of the BN data is its apparent magnitude. The radiation energy density $\Omega_{RT}$, in units of the critical energy density, is expected from various sources such as nuclear energy from stars and accretion energy on black holes to be up to $10^{-6}h^{-2}$; for comparison the total CMB energy is $\Omega_{cmb} \approx 25 \times 10^{-6}h^{-2}$. The most popular models for the distortion are emission from primeval dust, in which case the total energy density is $\Omega_{RT}h^2 \approx 6 \times 10^{-6}$, and Compton cooling of hot gas, in which case $\Omega_{RT}h^2 \approx 2 \times 10^{-6}$ ($e.g.$, Bond, Carr and Hogan 1990, BCH2). For both generation mechanisms, the energy must be injected at high redshift to explain the data and also to avoid anisotropy limits. However, a history of Wein distortion false alarms in the seventies and early eighties invites caution. COBE, launched in November 1989, will be much more sensitive, capable of probing to $\Omega_R h^2 \sim 3 \times 10^{-8}$ over a spectral range $500 - 10^4 \mu$.

Once a non-Gaussian component to the random density field develops as structures turn-around and collapse, non-Gaussian $secondary$ anisotropies arise from Thomson scattering by nonlinear bulk-flow currents, from Compton cooling of inhomogeneous hot gas, and by redshifted emission from primeval dust. These will be the only significant fluctuations in structure formation theories based on explosions, whether supernova-driven, radiation-driven or driven by the electromagnetic energy from superconducting strings. Hot gas in collapsed groups and clusters will also give rise to a non-Gaussian component to the microwave sky.

In this section, I first focus on anisotropies in dust emission models. In Bond, Carr and Hogan (1986 [BCH1]), we showed that if galaxies (even dwarfs) exist at $z \sim 10$ then they cover the sky, and if they are dust-laden then $all\ energy$ from the near IR to the X gets absorbed and re-emitted in the far IR, with a peak wavelength $\lambda \sim 700\mu$ which is relatively model insensitive. McDowell (1986) and Negroponte (1986) came to similar conclusions. BCH1 showed that energy sources would necessarily deliver a distortion energy $\Omega_R/\Omega_{cmb} \sim 10^{-3}$, and that $\Omega_R/\Omega_{cmb} \sim 10^{-1}$ was plausible. In BCH1 and Bond (1988) we showed that the intensity fluctuations in the sub-mm range would typically be $\Delta I_\nu/I_\nu \sim 0.01 - 0.1$ from galaxies for a wide range of models. In BCH2 we considerably extended the treatment of the emission and the expected anisotropy levels. I refer the reader to those papers for the details. Here I concentrate on visualizations of possible models, with the emphasis on pattern. The model I adopt is a shot noise model, where the shots are dwarf galaxies with some simple dust profile assumed. The shots may be clustered as well. The relative fluctuation level $(\Delta I_\nu / I_\nu)_{map}$ of the distortion is not very sensitive to the energetics, depending mostly upon the shot density and distribution. However, the observability very much depends upon the energetics since it is $\Delta I_\nu$ which is measured, not $\Delta I_\nu / I_\nu$. The translation to $\Delta T/T$ is very wavelength dependent, resulting in small effects at cm wavelengths (depending upon the dust law), but plausibly large ones in the sub-millimetre band. However, if we assume a distortion energy at the BN level, then the current anisotropy data can be used to place strong limits on the spacing of dust emitting objects and on their clustering length (BCH2). The source of the anisotropies are fluctuations in the dust density and in the luminosity of the sources irradiating the grains. For the pictures shown here the fluctuations in the optical depth of the dust along the line of sight were assumed to dominate.

Figure 3.1 shows the radiation pattern in a $1^o \times 1^o$ patch for a typical dust emission model of the BN experiment. It is designed to just satisfy the 1300 micron IRAM constraint obtained by Kreysa and Chini, a chop and wobble experiment with beamsize $\theta_{fwhm} = 11''$, and a throw of $30''$. The anisotropy pattern derives from fluctuations in the optical depth of the dust along the line of sight. The corresponding $1^o \times 1^o$ patch for the standard CDM model (with $b_\rho \sigma_{map} = 1.7 \times 10^{-5}$), the short distance structure of Fig. 2.1(a), is very smooth by contrast: experiments designed to look at very small angles are sensitive only to secondary anisotropies, a result of the strong destructive interference at the last scattering surface.

An example of possible energetics for a model such as that in Fig.3.1 is provided by Model P1 of BCH2, Table 2b, which reproduces well the BN data points: normal dust with the rather large mean density $\Omega_{dust} = 10^{-4}$ is assumed; the energy is the required $\Omega_{RT} h^2 = 6.2 \times 10^{-6}$, emitted between redshift 24 and now, but with an assumed luminosity per mass strongly favouring the higher redshifts. The result is an emissivity function which peaks at $z_e = 11$, but with a large spread $(\Delta z/z \sim 0.8)$ about it, corresponding to a $630 \, h^{-1}$Mpc comoving thickness of the emission shell. The average redshifted dust temperature is $aT_d = 4K$ ($a$=expansion factor), with a spread of 0.46K. The optical depth across the shell at redshifted 709 microns is 0.07.

The Pic de Valeta data constrains this model to have a comoving density of galaxies $n_G > 100 (h^{-1}$Mpc$)^{-3}$, and, if clustering is described by a $\gamma = 1.8$ power law correlation function, a clustering length $r_0 > 38 \, h^{-1}$kpc. Within these bounds we are free to vary the clustering properties of the radiation. For example if we place galaxies of radius 1 kpc with comoving density $n_G = 100 (h^{-1}$Mpc$)^{-3}$ down at random, but with these P1 emission properties, we have a large fluctuation level, $(\Delta I_\nu / I_\nu)_{map} = 0.13$. However, when convolved with the Kreysa and Chini experimental configuration, we find that (as designed) their experimental bound is obtained. The model actually shown in Fig.3.1 is considerably more complex, looking like a white noise pattern on that scale, but with interesting substructure on $1' \times 1'$, as shown in Figure 3.2. Thus even with the specificity of BCH2 model P1 there is an extremely wide variety of emission textures that satisfy the available constraints. Although the intensity fluctuations are at the percent level in this example, one can easily invent models which have much smaller ones or larger ones. In part this is because, over the short distances involved, gas dynamical processes are certain to be effective, and could equally well wash out as enhance anisotropies. Another unresolved issue is whether gravitational lensing will effectively smear over a region of this size, meaning that the fluctuations in the far infrared and submillimetre will not be an effective way to probe the pregalactic universe.

In Figure 3.3, another type of model is shown which is much more conservative than that demanded by the BN data, but with a large fluctuation level which allows it to be observed through its fluctuations even if not through the CMB spectral distortion it generates. There are 230 dwarf galaxies in this $1' \times 1'$ patch. The average properties correspond to that of BCH2 Model F6, Table 2(b), one that had a conservative dust abundance $\Omega_{dust} = 10^{-5}$, with a normal dust emission law and size, modest energetics, $\Omega_{RT}h^2 = 10^{-6}$ (4% of the CMB), with an extended emission shell, $\delta z/z \sim 60\%$, around a mean emission redshift of 4.7. The number density of galaxies, $n_G = 0.23\,h^{-1}\mathrm{Mpc}^{-3}$, an order of magnitude above that of bright galaxies, was chosen to ensure that the Kreysa and Chini bound is just saturated.

The final secondary anisotropy figure, 3.4, shows what the world might look like if one could use the CMB to image hot gas through its Compton cooling. In the map shown there are $10^4$ small groups, 170 large groups, 9 poor clusters and 1 rich cluster. The model is based upon the Gaussian peak models for hierarchical virialized structures given in Table 3, §7.2.7, of Bond (1988), which uses top hat model parameters to describe the temperature and density of the gas. Unfortunately the parameters of the peak model are sufficiently uncertain at the present time that the number of objects can vary by factors of a few (and more for the rarer events such as rich clusters) from map to map. Clustering of the objects has not been included, although this is a reasonably significant effect on intermediate angles. The observability of the $rms$ SZ anisotropies is sensitive to $b_\rho$. For example, a map constructed for a hot dark matter (massive neutrino) model with nonlinear redshift 3, corresponding to $b_\rho = 0.25$, also described in Table 3, §7.2.7, of Bond (1988), has 470 rich cluster scale peaks and 8 extra-rich clusters leading to $\bar{y} \sim \times 10^{-4}$ and $\sigma_{map} = 4 \times 10^{-4}$. Such a strongly dynamical model is convincingly ruled out by primary anisotropy constraints derived from the OVRO and Lubin and Meinhold data.

## 4. THE FUTURE

In the last two figures, we look forward to the large angle results from COBE. It is not too optimistic to think that COBE will be able to see the scale invariant Sachs Wolfe anisotropies predicted for inflation models. These two maps are realizations of the fluctuations for a scale invariant spectrum, for which the power spectrum is given by $C_\ell = C_2\,6[\ell(\ell+1)]^{-1}$ for small $\ell \lesssim 30$. In this language, the Strukov et al. (1987) 95% confidence limit on the quadrupole amplitude for such a spectrum was $C_2^{1/2} < 2.5 \times 10^{-5}$. For the biased CDM model it is $C_2^{1/2} = 10^{-5}/b_\rho$, a level which COBE should achieve provided $b_\rho \lesssim 2-3$.

Both maps are equal area projections but, as in all cases of mapping the sphere onto the plane, there is distortion somewhere or other. The first map (a) is in the usual all sky coordinates $x = \varphi \sin\theta$, $y = b$, where the polar coordinates $(\theta, \varphi)$ are related to Galactic longitude and latitude by $\ell = \varphi$ and $b = \pi/2 - \theta$. The $rms$ map amplitude is $b_\rho \sigma_{map} = 1.3 \times 10^{-5}$. The highest hot spot is $3.39\sigma_{map}$ and the lowest cold spot is $-2.88\sigma_{map}$.

The second map (b) uses $x = \varpi \sin\varphi$, $y = \varpi \cos\varphi$, where the radial coordinate s $\varpi = 2\sin(\theta/2)$. Although the entire range up to $\varpi = 2$ is shown, the extreme distortion for $\varpi > 1$, corresponding to $\theta > 60°$, makes these maps unappealing for full sky representations. However patches centered about the polar axis (the origin) are very well represented, since $x$ and $y$ reduce to Cartesion angle coordinates. If we keep enough multipoles to represent the radiation pattern in the small patch of Fig 2.1, we would just get a realization equivalent to the Fourier transform generated maps of §2. For Fig.(4.1), a Gaussian beam profile with the resolution of the COBE beam, $\theta_{fwhm} = 7°$, was adopted. Spherical harmonics up to $L = 35$ were included; the smearing ensures higher ones were not needed. The monopole and dipole components are also not included. (a) has 20897 pixels and (b) has 51433 pixels.

No attempt at complete references was made in this sightseeing tour. I hope these maps of theoretical models for CMB anisotropies illustrate how rich at all angular scales the CMB may be, and how much cosmology we can learn when its fluctuations are finally found. Was it really 1982 when George Efstathiou and I started on our primary trek and 1984 when Bernard Carr and Craig Hogan and I followed a secondary path down dusty lanes into the far IR? And the next decade can only lead us into more exciting territory. Support of a Canadian Institute for Advanced Research Fellowship, a Sloan Foundation Fellowship and the NSERC of Canada is gratefully acknowledged.

## 4. REFERENCES

Bardeen, J.M., Bond, J.R., Kaiser, N. and Szalay, A.S. 1986, *Ap. J.*, **304**, 15.

Bardeen, J.M., Bond, J.R. and Efstathiou, G. 1987, *Ap. J.*, **321**, 28.

Bond, J.R., Carr, B.J. and Hogan, C.J. 1986, *Ap. J.* **306**, 428 [BCH1]; 1990, *Ap. J.*, April 20. [BCH2].

Bond, J.R. and Efstathiou, G. 1984, *Ap. J. Lett.* **285**, L45; 1987, *M.N.R.A.S.* **226**, 655.

Bond, J.R. 1988, in *The Early Universe*, Proc. NATO Summer School, Vancouver Is., Aug. 1986, ed., Unruh, W.G. (Dordrecht:Reidel).

Bouchet, F., Bennett, D.P. and Stebbins, A. 1988, *Nature* **335**, 410.

Efstathiou, G. and Bond, J.R. 1986, *M.N.R.A.S.* **218**, 103; 1987, *M.N.R.A.S.* **227**, 33P.

Kofman, L. A. and Linde A. D. 1987, *Nuc. Phys.* **B282**, 555.

Kofman, L.A. and Pogosyan, D.Y. 1988, *Phys. Lett.* **214B**, 508.

Kreysa, E. and Chini, A. 1989, *Proc. Particle Astrophysics Workshop, Berkeley*, (Singapore: World Scientific).

Lubin, P.M. and Meinhold, P.R. 1989, in *The Cosmic Microwave Background 25 Years Later*, L'Aquila, Italy, ed. N. Mandolesi and N. Vittorio

Matsumoto, T., Hayakawa, S., Matsuo, H., Murakami, H., Sato, S., Lange, A.E. and Richards, P.L. 1988, *Ap. J.*, **329**, 567. [BN]

McDowell, J. C. 1986, *M.N.R.A.S.* **223**, 763.

Negroponte, J. 1986, *M.N.R.A.S.* **222**, 19.

Peebles, P.J.E., 1987, *Ap. J. Lett.* **277**, L1.

Readhead, A.C.S., Lawrence, C.R., Myers, S.T., Sargent, W.L.W., Hardebeck, H.E. and Moffet, A.T. 1989, *Ap.J.*, **346**, 556.

Salopek, D.S., Bond, J.R. and Bardeen, J.M. 1989, *Phys. Rev.* **D40**, 175.

Strukov, I.A., Skulachev, D.P. and Klypin, A.A. 1987, in Proceedings I.A.U. Symposiu 130, ed. Audouze, J. and Szalay, A.S. (Dordrecht: Reidel).

**Figure 2.1(a)** $10° \times 10°$ *CMB grey scale map for the standard adiabatic* **CDM model** *with* $\Omega_B = 0.1$. *Pure white is hot* $(> 3.5\sigma_{map})$ *and pure black is cold* $(< -3.5\sigma_{map})$. **(a)** *has* **standard recombination**, *with photon decoupling at* $z \sim$ 1000. *The rms amplitude is* $\sigma_{map} = 3.2 \times 10^{-5} \, b_\rho^{-1}$. *(See also Fig. 2.3b for a contour map.) Varying the Hubble parameter* h *does not change the texture very much, but does change the amplitude,* (h = 0.75, $\Omega_B = 0.03$ *has* $\sigma_{map} = 1.5 \times 10^{-5} \, b_\rho^{-1}$). *Similarly there are relatively modest changes in the spectrum* $C_\ell$ *as* $\Omega_B$ *varies, but there are big changes in the amplitude:*

| $\Omega_B$ | 0.01 | 0.03 | 0.1 | 0.2 | 0.5 | 0.1 (no rec) |
|---|---|---|---|---|---|---|
| $b_\rho \sigma_{map}/10^{-5}$ | 2.1 | 2.4 | 3.2 | 4.4 | 11 | 1.5 |

**Figure 2.1(b)** *is the standard CDM model, but with* **no recombination** *(very early reionization would give a similar result) and* $\sigma_{map} = 1.5 \times 10^{-5} \, b_\rho^{-1}$, *showing the very large effect this has upon the small angle anisotropies. On the other hand, for standard recombination models, lowering* $\Omega$ *concentrates the power to smaller scales than in the standard CDM model. Open CDM models are by now largely ruled out. Sample map fluctuation levels for* h = 0.5 *CDM models with* $\Omega = 0.2$ *are* $b_\rho \sigma_{map} = 3.8 \times 10^{-4}$ *for* $\Omega_B = 0.1$ *and* $b_\rho \sigma_{map} = 1.8 \times 10^{-4}$ *for* $\Omega_B = 0.03$. *One way to make such models compatible with inflation is through the (unpalatable) assumption of a nonzero cosmological constant, which can be thought of as contributing a 'vacuum energy' density* $\Omega_{vac} = 1 - \Omega$. *Although the structure of the* k-*space fluctuations is identical in both models, the* $\ell$-*space spectra differ, with more power at small angles in the open models. Vacuum–dominated models are now strongly constrained by the data: an* $\Omega_{vac} = 0.8$, $\Omega_{nr} = 0.2$ *map has* $b_\rho \sigma_{map} = 1.9 \times 10^{-4}$ *for* $\Omega_B = 0.1$ *and* $b_\rho \sigma_{map} = 8.7 \times 10^{-5}$ *for* $\Omega_B = 0.03$.

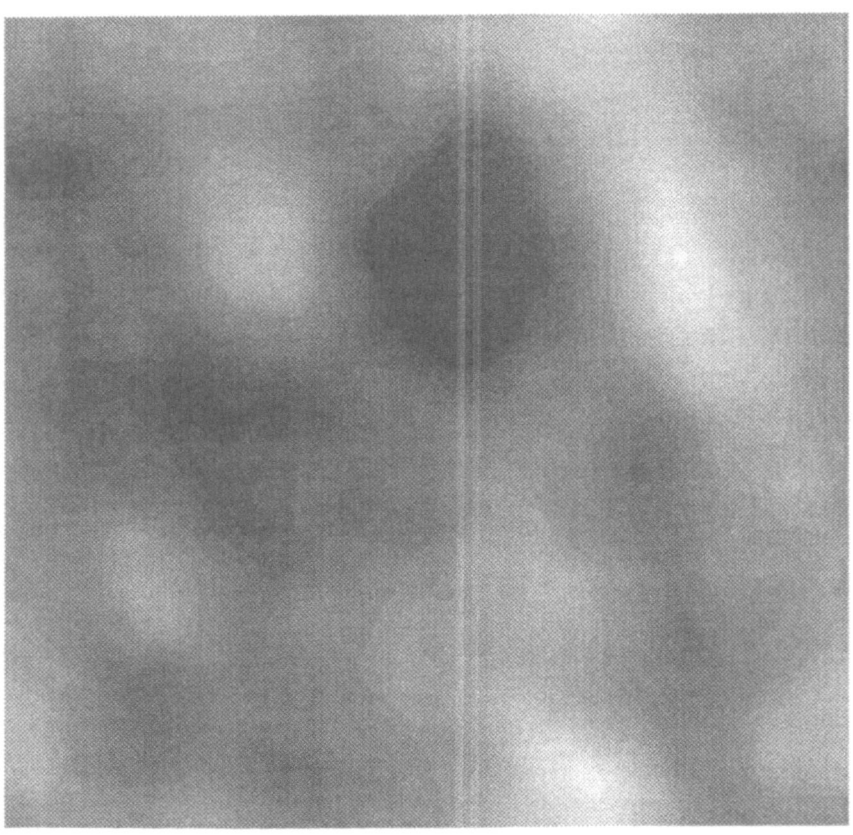

**Figure 2.2(a)** *A $10^o \times 10^o$ map for an $n = 0$, $\Omega = \Omega_B = 0.2$* **Isocurvature Baryon** *model with* **(a) no recombination**, *calculated using only linear perturbation theory. The* **standard recombination** *model ($b_\rho \sigma_{map} = 8.7 \times 10^{-5}$) is not shown since it rather looks like Figure 3.1. (All roughly white noise models look the same in their texture.)*

58

**Figure 2.2(b)** *More relevant than (a) is the no recombination case which includes some nonlinearities. Only quadratic nonlinearities were included (**Vishniac effect**), justified by the smallness of the dynamical amplitude when these very small scale anisotropies are generated, around the period when the photon depth is unity. The (b) map is made assuming the induced anisotropies arising from nonlinearities in the electron velocity source term are Gaussian. This is likely to be approximately the case because the superposition of many contributions along each line of sight tends to give Gaussians even if the underlying source isn't. The map amplitude is $b_\rho \sigma_{map} = 2.5 \times 10^{-5}$. For comparison, the same model, but with $n = -1$, has $b_\rho \sigma_{map} = 9.1 \times 10^{-5}$; standard recombination has $b_\rho \sigma_{map} = 1.6 \times 10^{-4}$. The $\Omega_B = 0.4$ $n = -1$ no recombination model of Fig.1.1 has $b_\rho \sigma_{map} = 7.8 \times 10^{-5}$ (without quadratic nonlinearities), while its $n = 0$ counterpart has $b_\rho \sigma_{map} = 2.1 \times 10^{-5}$.*

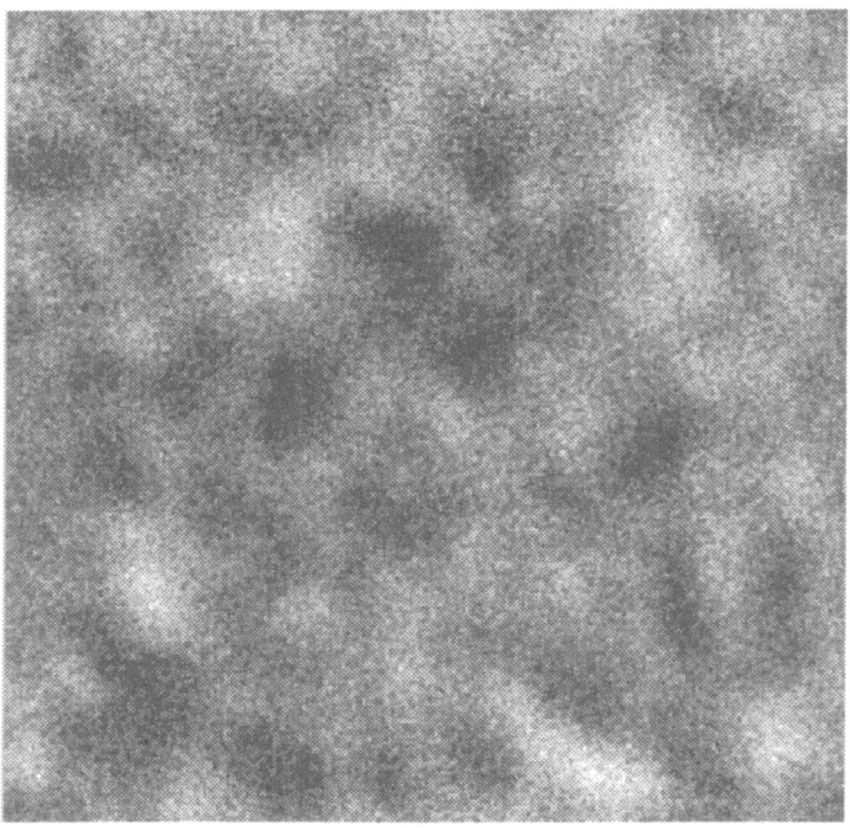

**Figure 2.3(a)** *A 7.2° × 7.2° CMB grey scale map of anisotropies in a cosmic string model, as computed by Bouchet, Bennett and Stebbins (1988). $\sigma_{map} \approx 15.4G\mu$ is $\sim 7 \times 10^{-6}$ if the mass per length $\mu$ of the strings is a typical grand unified energy scale; the current trend is to smaller values however. Notice the sharper features in the string model and the open places where there is little anisotropy. Experiments which make maps will be ideal for determining whether the anisotropy pattern is non-Gaussian, as in the string picture, or Gaussian, as in the CDM picture.*

**Figure 2.3(b)** $7.2^o \times 7.2^o$ *CMB contour maps for the adiabatic CDM case of Fig. 2.1(a), running from* $-4\sigma_{map}$ *to* $4\sigma_{map}$ *in steps of* $1\sigma_{map}$, *where* $\sigma_{map} = 3.1 \times 10^{-5}b_\rho^{-1}$ *which is a large fraction of the total* $3.6 \times 10^{-5}b_\rho^{-1}$. *The analogous no recombination model on this scale has* $\sigma_{map} = 1.3 \times 10^{-5}b_\rho^{-1}$, *c.f. the total* $2.5 \times 10^{-5} b_\rho^{-1}$. *These models will be clearly distinguishable from string models even with relatively large beam smearing.*

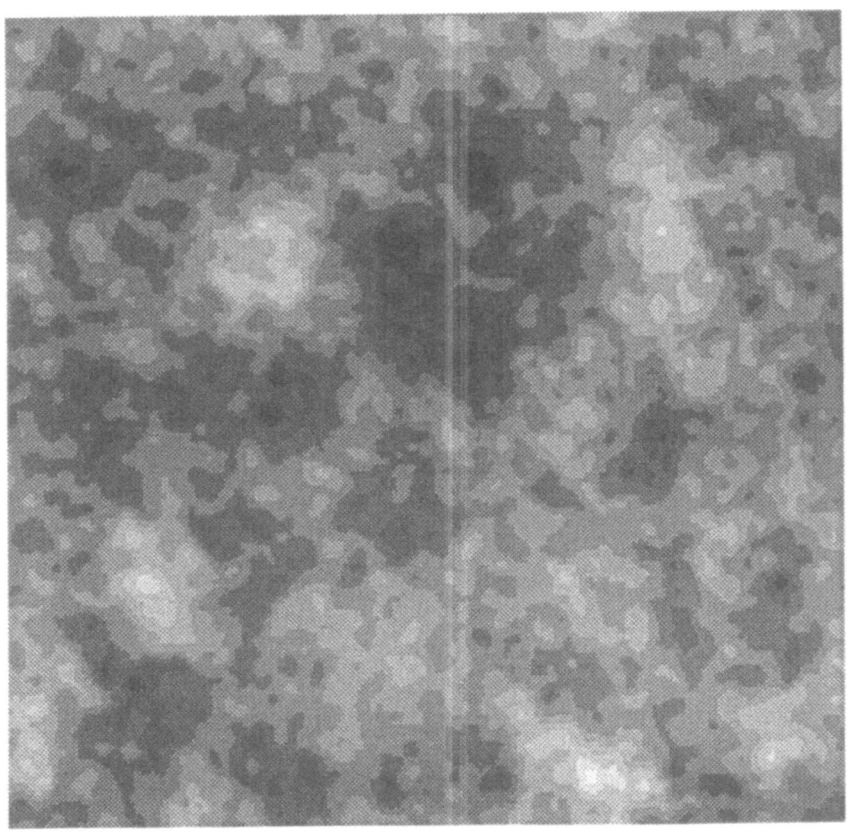

**Figure 3.1:** *A Fractal/Poisson Hybrid Model of* **dust emission** *in a $1^o \times 1^o$ patch. The radiation power spectrum $C_\ell$ has a Poissonian component and one from clustering. The basic shot noise entities are 'dwarflet' galaxies with a very high $n_G = 10^4 (\, h^{-1} Mpc)^{-3}$ density, corresponding roughly to the spacing of $10^6 \, M_\odot$ scale peaks of a Gaussian density field. A power spectrum with index $n = -2.5$ describes the 'dwarflet (and dust) clustering out to a coherence scale $r_{coh} = 350 \, h^{-1} kpc$, beyond which power is truncated to $n = 0$. The index choice was motivated by the CDM power spectrum at subgalactic scales. The amplitude of the power is characterized by a (comoving) clustering scale $r_0 = 10 \, h^{-1} kpc$ at the redshift of maximum emission, $z_e = 10.8$. The Poisson fluctuations of the 'dwarflets' give very small anisotropies, but the anisotropies from the fractal clustering are sufficiently large as to saturate the Kreysa and Chini bound. Above the coherence scale corresponding to an angular scale $\sim 17''$, the structure looks just like white noise, but with a Poissonian number density that of the correlated clumps, not that of the dwarflets. Even so, the rms intensity fluctuations are only $(\Delta I_\nu / I_\nu)_{map} = 0.01$ for this map which has an effective short distance cutoff at the Nyquist frequency $\ell_N^{-1} = 4.5''$.*

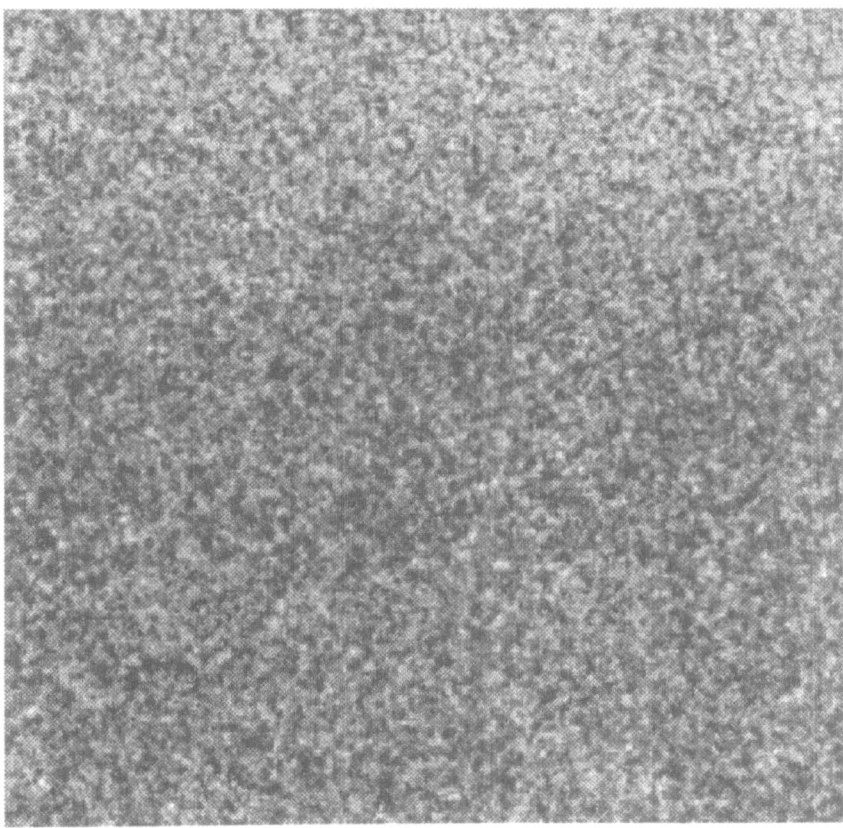

62

**Figure 3.2:** $1' \times 1'$ *map showing the even shorter distance structure of the dust emission model of Fig (3.1) gives a surprise: what is white noise at poor resolution is complex at shorter resolution exhibiting the details of how dust may be correlated in galactic scale patches even though it originates in entities of the very high $10^4 (\mathrm{h}^{-1}\mathrm{Mpc})^{-3}$ space density objects of $\sim 10^6\,M_\odot$. The fluctuation level in the map is $(\Delta I_\nu / I_\nu)_{map} = 0.016$. Another sample model with a narrow emission shell having active energy injection between redshifts 40 and 20 from dwarflets with the same space density and radii 0.3 kpc has a fluctution level $(\Delta I_\nu / I_\nu)_{map} = 0.06$, with, on average, 149 of these 'dwarflets' along any one line of sight.*

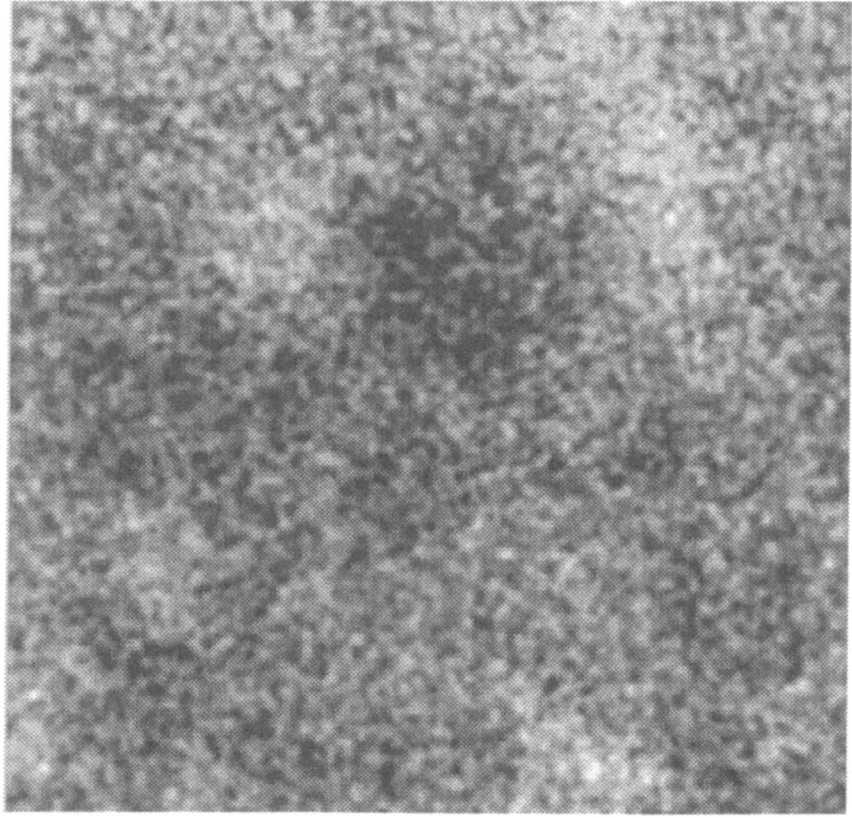

**Figure 3.3** *A $1' \times 1'$ patch consisting of 230 dwarf galaxies, derived from a population with $n_G = 0.23(\mathrm{h}^{-1}\mathrm{Mpc})^{-3}$ and galactic radii 5 kpc radiating with a constant mass-to-light ratio between $z = 14$ and $z = 4$. The galaxies do not cover the sky: there are only 0.39 on an average line of sight. This leads to the very large fluctuation level $(\Delta I_\nu/I_\nu)_{map} = 1.1$ on these small scales. On the other hand a SIRTF style beam would encompass all 230 of these objects, so confusion would reign, but nonetheless the fluctuation level would still tell us interesting things. In this model, if dust were uniformly covering the sky, 89% of the radiation would be absorbed by dust and reemitted in the far infrared, with the rest getting through unimpeded, to be seen as a near infrared background whose magnitude will depend upon the self absorption in the parent dwarf galaxies. If the galaxies are like primeval starbursts, which is what motivated this model, the self absorption could be nearly complete. If we are even more conservative, and include objects with a bright galaxy density $n_G = 0.02\,\mathrm{h}^{-1}\mathrm{Mpc}^{-3}$, and radii 10 kpc, radiating in the IR between $z = 7$ and 3, then the covering factor is only 5% and the intensity fluctuations are $(\Delta I_\nu/I_\nu)_{map} = 3.2$.*

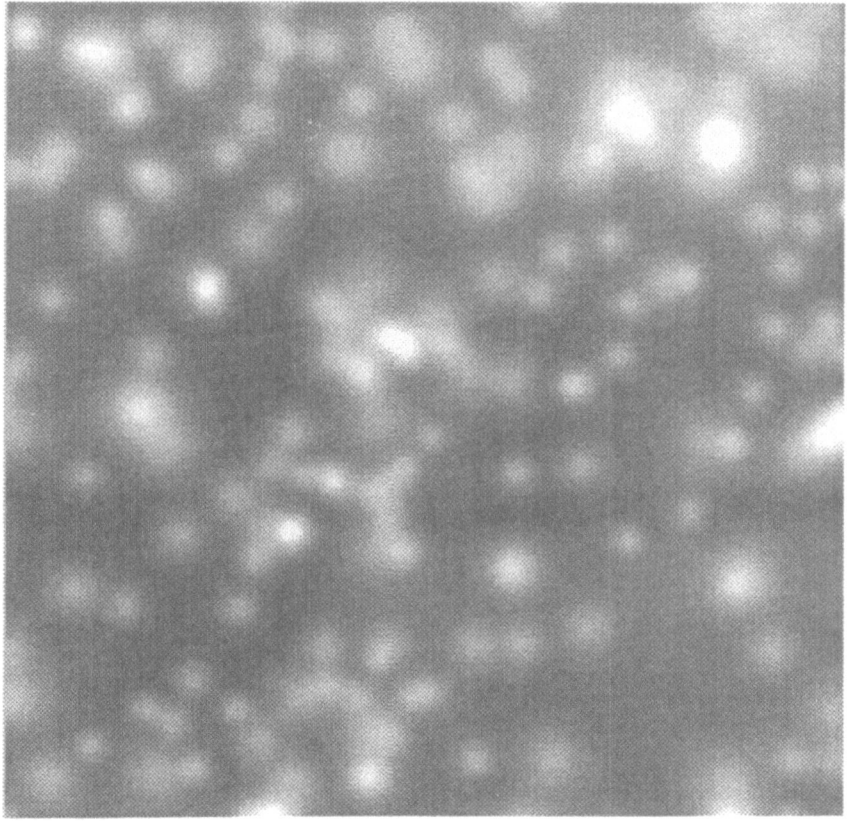

**Figure 3.4** *A $1° \times 1°$ realization of Sunyaev-Zeldovich anisotropies from Compton cooling of inhomogeneously distributed hot gas in groups and clusters in a standard CDM model, except that the biasing factor is taken to be the low value $b_\rho = 1.4$ to give larger effects. The largest object is a rich cluster residing in the lower left corner, but the dominant texture (albeit with very small amplitude) comes from the groups which cover the sky in only moderately hot gas and from poor clusters. The average value of the y-parameter for this model is $\bar{y} = 2 \times 10^{-6}$ and the rms fluctuations in the map are $\sigma_{map} = 3 \times 10^{-6}$ at cm wavelengths, at rather too low a level to observe except in the most non-Gaussian concentrations of the radiation pattern, in the rich cluster.*

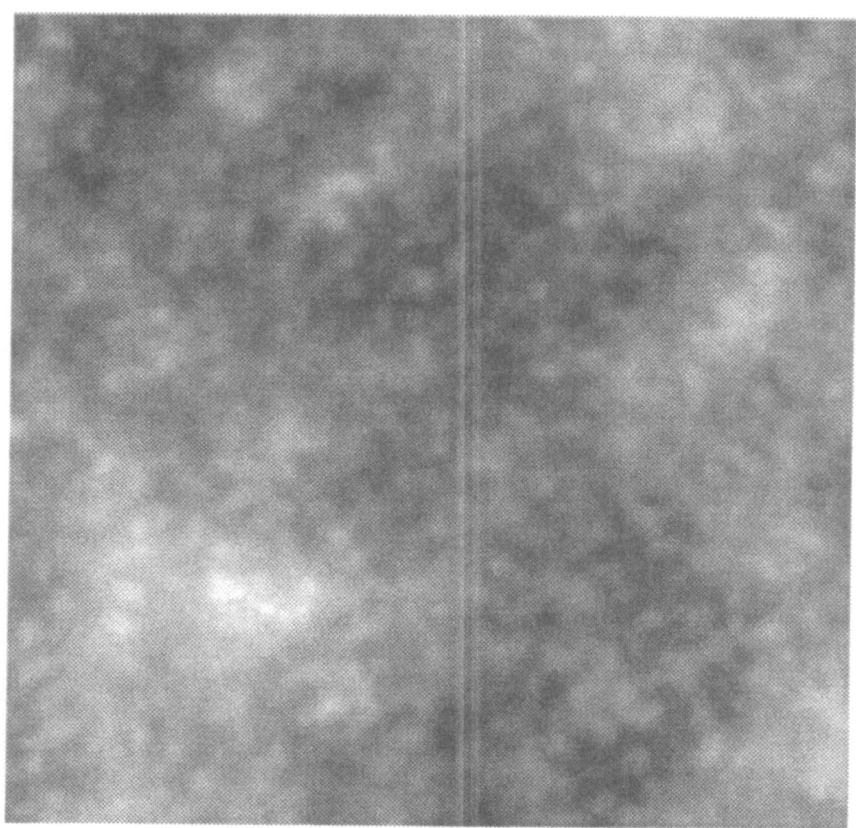

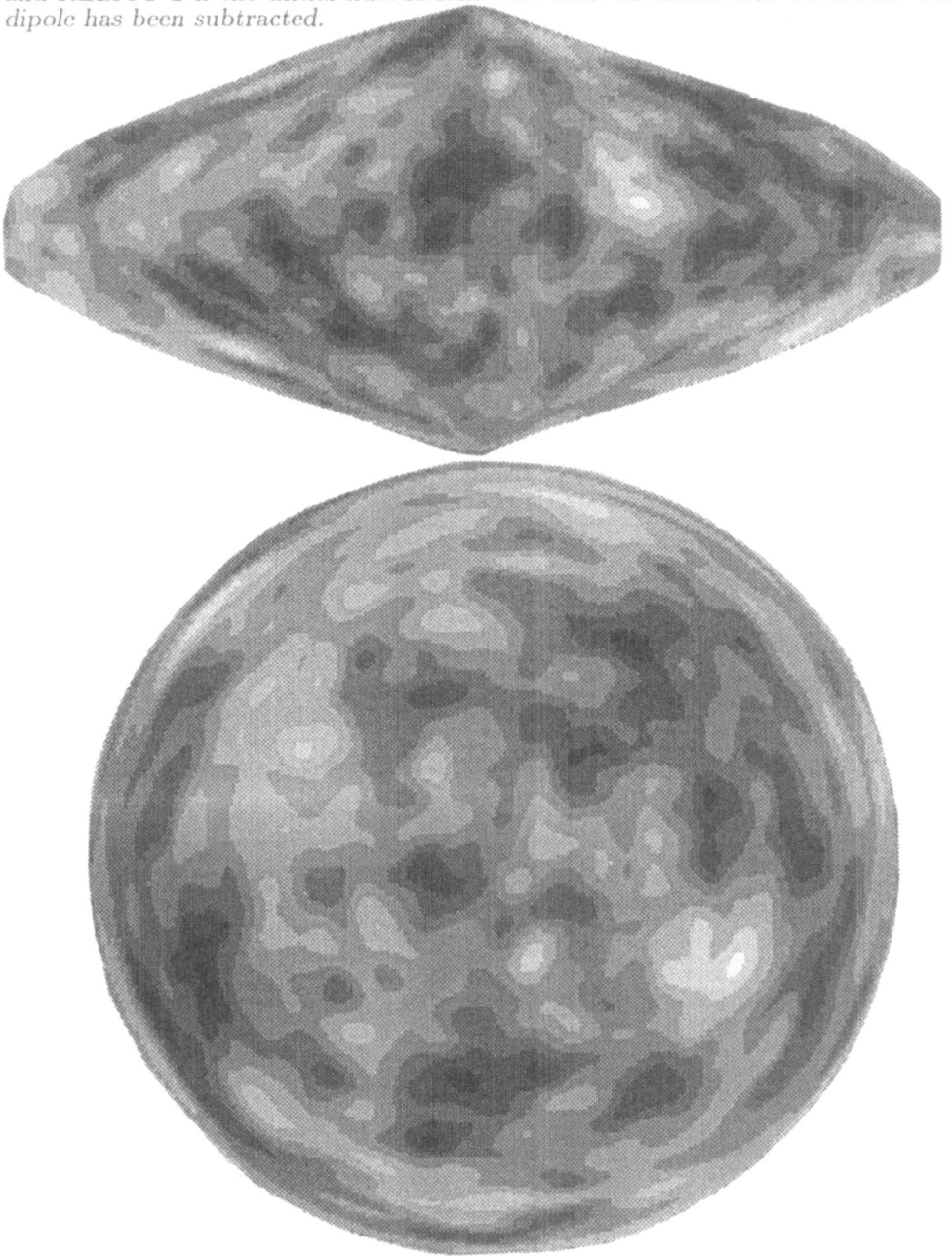

**Figure 4.1a,b** *Two maps of the large angle sky as it should be seen by COBE and RELICT 2 if the initial fluctuations were scale invariant and Gaussian. The dipole has been subtracted.*

# THEORY AND SIGNIFICANCE OF THE SUNYAEV-ZELDOVICH EFFECT

Yoel Rephaeli
Raymond and Beverly Sackler Faculty of Exact Sciences
School of Physics and Astronomy, Tel Aviv University
Tel Aviv, Israel

The theoretical framework of Comptonization of the spectrum of the Cosmic Microwave Background radiation by intracluster (the Sunyaev-Zeldovich effect) and intergalactic gas is briefly reviewed. The Planckian spectrum of this radiation field may had been Comptonized by the recombining matter in the early Universe, as seems to be supported by recent measurements of the submillimeter excess. This affects the traditional treatment of the Sunyaev-Zeldovich effect in ways which we point out. The effect is a unique probe which will eventually be applied to obtain extremely valuable astrophysical and cosmological information; we discuss some of the likely uses of the effect.

## 1. INTRODUCTION

Compton scattering of the Cosmic Microwave Background (CMB) radiation by hot electrons results in a systematic shift of photons from the Rayleigh-Jeans (R-J) to the Wien side of the spectrum. This may occur in various environments; of particular interest is scattering by hot intergalactic (IG) or intracluster (IC) electrons. The change of the CMB spectrum along a line of sight to a cluster of galaxies is often referred to as the Sunyaev-Zeldvich (SZ) effect. This paper reviews the theory and significance of the effect. A review of the current status of measurements of the SZ effect is given in these proceedings by Birkinshaw (1989). Sunyaev and Zeldovich (1980a, 1981) have written excellent, comprehensive reviews of the subject matter.

Interest in the effect of Compton scattering on the CMB radiation arose soon after the discovery of the radiation. Weymann (1966), and Zeldovich and Sunyaev (1969) calculated the Comptonized spectrum due to IG gas which had been heated at various cosmological epochs. The SZ effect has originally been proposed as a test of the emission mechanism of IC X-rays, and as a diagnostic tool for the determination of the properties of the IC gas (Sunyaev and Zeldovich 1972). There

67

*N. Mandolesi and N. Vittorio (eds.), The Cosmic Microwave Background: 25 Years Later, 67–75.*
© 1990 *Kluwer Academic Publishers.*

have been many attempts to measure the effect along lines of sights to a number of clusters of galaxies, with a possibility of it being detected in A401, A665, A2218, and 0016+16 (see a discussion and references in Birkinshaw 1989).

As originally suggested by Zeldovich and Sunyaev (1969), and eversince by others, the Compton signature imprinted on the CMB spectrum by hot IG and IC gas is of considerable cosmological and astrophysical interest. In the following, we will briefly review the theory and some of the possibly significant applications of measurements of this signature on the CMB spectrum. For a more extensive discussion, see the reviews by Sunyaev and Zeldovich (1980a, 1981).

## 2. COMPTONIZATION OF THE CMB SPECTRUM

An accurate description of the interaction of an isotropic radiation field with a hot and isotropic electron gas involves the calculation of the exact frequency re-distribution function in the context of a relativistic formulation. Since the possibility that the IG medium may have been heated up to very high temperatures ($T > 10^9$ K), a non-relativistic (NR) approximation may be inadequate. However, in clusters, where IC gas temperatures are $O(10^8$ K), the NR approximation is generally acceptable, though relativistic corrections to the Comptonized spectrum may not be negligible at very high frequencies. We begin by a discussion of NR Comptonization.

a. Comptonization by a Nonrelativistic Electron Gas

Compton scattering of a **Planckian** radiation field by a NR Maxwellian electron gas can be well approximated by means of a solution of the Kompaneets (1957) equation. In clusters the electrons are much hotter than the CMB radiation, so the change in the (spectral) intensity due to Compton scattering is (Zeldovich and Sunyaev 1969; Gould and Rephaeli 1978)

$$\Delta I_T = \frac{2(kT)^3}{(hc)^2} y g(x),$$ (1)

where $T_0$ is the CMB temperature, $x = h\nu/kT_0$, and

$$g(x) = \frac{x^4 e^x}{(e^x - 1)^2} [x coth(x/2) - 4].$$ (2)

The spectral form of this thermal effect is contained in the function $g(x)$, which is zero at the critical value $x_0 = 3.83$, or $\nu_0 = 218$ GHz for $T_0 = 2.74$ K. The Comptonization parameter, $y$, is

$$y = \int (kT_e/mc^2) n \sigma_T dl,$$ (3)

where $n$ and $T_e$ are the electron density and temperature, $\sigma_T$ is the Thomson cross section, and the integral is over a line of sight through the cluster. The net effect of the scattering by hot electrons is the systematic shift of photons to higher energies, which amounts to heating the radiation field. The relative change in the CMB energy density is $e^{4y}$.

The expression in equation (1) for the change in the spectral intensity is valid only if before traversing the cluster the photons have an exact Planckian distribution. We have considered elsewhere (Rephaeli 1980) the more general case of an additional distortion due to hot, but NR, IG gas. The recent measurements of submillimetric excess emission (Matsumoto et al. 1988) give some evidence for the existence of such a gas. While the origin of this excess is not yet known, a possible interpretation is that it is due to gas not processed in galaxies. As mentioned in the previous section, this theoretical possibility was indeed the original motivation for the investigations of Zeldovich and Sunyaev (1969). If the measured excess is due to Compton scattering by IG gas, the corresponding Comptonization parameter is $y < 0.03$, so the relative change in intensity may not be small on the Wien side (where the effect is of particular importance - Gould and Rephaeli 1978). Strictly speaking, therefore, the total intensity change along a line of sight to a cluster cannot, in general, be represented as a linear superposition of the effects of IG and IC electrons. Nonetheless, this is what has been done so far. The general expression for the IC intensity change of an already Comptonized Planckian (CP) was derived by Rephaeli (1980; see also Fabbri 1981). The intensity change of the CMB due to Comptonization by both IG and IC electrons is shown in Figure 1 of Rephaeli (1980).

The more general treatment of the SZ effect of a CP shows that the magnitude of the effect deviates significantly from that of a pure Planckian, even if $y = O(10^2)$ (Rephaeli 1980). In the R-J region the relative temperature change is reduced by a factor $e^{2y}$. The value of the critical frequency where $g(x)$ vanishes, which has to be accurately known in some applications of the effect, is also affected. If current measurements are fit by a CP, the spectrum can be characterized by Comptonization parameter $y = .02$ and a temperature $T_0 = 2.81$ K (Smoot et al. 1988). Using the formulae in Rephaeli (1980), we calculate that $x_0 = 3.96$, or $\nu_0 = 232$ GHz, for a non-relativistically Comptonized spectrum with $y = .02$. Excluding the rocket measurements of Matsumoto et al. (1988), the spectral (Smoot et al. 1988), for which $\nu_0 = 218$ GHz.

The above equations describe the SZ effect due to thermal motions of the electrons, an effect which has been searched for eversince it was proposed. Much less attention has been given to the intensity change due to the motion of the cluster with respect to the CMB frame (Sunyaev and Zeldovich 1972). This kinematic effect is proportional to the line of sight component of the peculiar velocity, $v_r$. As has been discussed by Sunyaev and Zeldovich (1980b), under typical conditions in clusters the kinematic intensity change, $\Delta I_K$, is small in comparison with $\Delta I_T$, unless $v_r$ is $O(10^3)$ km/s. Interest in the possibility that some clusters might be moving at such velocities has increased recently, as a result of observational work stemming from the growing realization of the significance of the velocity field in studies of the large scale structure.

While the Doppler change in the temperature of the radiation along a line of sight through the cluster is frequency independent, $\Delta I_K$ is not,

$$\Delta I_K = -\frac{2(kT)^3}{h^2 c^2}\frac{v_r}{c}\tau h(x),$$ (4)

where $\tau = \sigma_T \int n dl$, and

$$h(x) = \frac{x^4 e^x}{(e^x - 1)^2}.$$ (5)

The function $h(x)$ is very different from $g(x)$; a most interesting feature is that $h(x)$ attains its maximum value exactly at the point where $g(x)$ vanishes (see Sunyaev and Zeldovich 1980b; Rephaeli and Lahav 1989). The expression for $\Delta I_K$ is modified when the incident radiation does not have an exact Planckian form. Treating a Compton distortion as a small change, we obtained a more general, $y$-dependent form for $\Delta I_K$. Using this form, we find that the frequency at which $\Delta I_K$ of a CP attains its maximum is essentially the same as that where $\Delta I_T$ vanishes.

The above kinetic change in the CMB intensity depends linearly on $v_r$ and $\tau$. The motion of a cluster trnsverse to the line of sight, with a tangential velocity $v_t$, induces linear polarization of the CMB along lines of sights through the cluster (Sunyaev and Zeldovich 1980b). However, the degree of polarization is extremely small as it is determined by terms proportional to $v_t^2 \tau$ or $v_t \tau^2$. Thus, the determination of the tangential motion of a cluster through CMB polarization measurements is unrealistic.

All dependence on the gas properties is contained in y, which is a line integral of (essentially) the electron pressure. The gas density, temperature and their radial profiles are deduced from X-ray spectral and spatial measurements. The gas is usually modeled as a polytrope $T_e \propto n^{\gamma 1}$ , with the density profile commonly taken to be of the form $n = n_0 (1 + r^2/a^2)^\alpha$ , where $\gamma$, $\alpha$ and the core radius, $a$, are best fits to the X-ray data (Gull and Northover 1975; Cavaliere and Fusco-Femiano 1976; Henriksen and Mushotzky 1986; Sarazin 1986). In rich clusters the values of these three parameters usually fall in the (rough) ranges $a = 0.2 - 0.5$ Mpc, $\gamma = 1 - 5/3$, and $\alpha = 1 - 1.5$ (see, e.g., Sarazin 1986). Since $n(r)$ and $T_e(r)$ are monotonically decreasing functions of (the radial coordinate) $r$, $y$ is maximal for a line of sight through the center of the cluster. As an example for the magnitude of the SZ effect, consider the Coma cluster which may be typical of the rich Abell clusters with moderately high X-ray luminosity. Using the values of the above parameters deduced from the X-ray data (Henriksen and Mushotzky 1986), we calculate $y = 1.25 \cdot 10^4$ , so that the magnitude of the thermal effect in the R-J region $\Delta T/T = -2y = 2.5 \cdot 10^4$ (Rephaeli 1987). Note, however, that the relative intensity change on the Wien side is significantly higher; e.g., $\Delta I_K/I = 6.7 \cdot 10^4$ at $x = 5$.

b. Comptonization by a Trans-Relativistic Electron Gas

As mentioned already, it is possible (though somewhat doubtful from energetic considerations) that the IG gas was heated to temperatures $T > 10^9$ K at recent cosmological epochs. Electron velocities in such a gas would be high enough to require a more proper relativistic formulation of the kinematics of Compton scattering than that of Kompaneets (1957). Wright (1979) generalized the Comptonization calculations to account for relativistic electron velocities, and also for the likely case of a very small IG optical depth to Compton scattering, $\tau$. Wright calculated the exact photon re-distribution function in frequency space. Unlike the description of the Comptonized spectrum based on the Kompaneets equation, no simple expression can be given in the relativistic limit. The spectrum in this limit is specified by $T_0$, $\tau$, and $kT_e/mc^2$, as compared to only $T_0$ and $y$ in the NR limit. The most prominent feature distinguishing a relativistically Comptonized spectrum from a NR one, is the steeper increase of the intensity on the Wien side.

The possibility of a relatively dense and very hot IG gas arises when the residual (i.e., the part which is thought not to be due to discrete sources) cosmic X-ray background (CXB) is hypothesized to be thermal bremsstrahlung emission. The obvious link between a CP spectrum and a thermal origin for the CXB has been pointed out all along (e.g., Field and Perrenod 1977), with a more realistic treatment given by Guilbert and Fabian (1986). The announcement of a submillemtric CMB excess by Matsumoto et al. (1988) has renewed interest in a relativistic description of a CP spectrum. It now seems that the Matsumoto et al. upper limit on the CMB intensity at $262\mu$ rules out Cmptonization by sufficiently hot IG gas that can also account for a substantial fraction of the CXB spectrum (Hayakawa et al. 1987; Taylor and Wright 1989). The baryonic mass fraction in an IG gas hot enough to account for about half of the CXB intensity, has to be around a quarter (Guilbert and Fabian 1986).

## 3. SIGNIFICANCE OF THE SUNYAEV-ZELDOVICH EFFECT

As we have already pointed out (Rephaeli 1980), an unequivocal measurement of the SZ effect will constitute convincing, direct evidence for the truly cosmological origin of the CMB. Moreover, the effect can potentially be an important, general cosmological probe. In the following we briefly mention some of the knowledge which may be gained through spectral and spatial measurements of the effect. The list below is not meant to be complete, but rather reflects some judgement as to what proposed uses of the effect are currently feasible, or may become so in the not too distant future.

a. Physical Processes in the Earlier Universe

Various processes which may have occured in the early Universe may have left their effect on the CMB spectrum and anisotropy. Due to incomplete thermalization, substantial heating of the matter at epochs $z < 10^6$ is bound to distort the CMB spectrum. The shape and magnitude of the spectral distortion determine the epoch and the degree of heating (ZS 1969; Danese and De Zotti 1977; Smoot et al. 1988). A CP spectrum is but the simplest possibility; the spectrum may also have a Bose-Einstein (B-E) form, or a combination of the latter with excess at very long wavelengths due to thermal bremsstrahlung emission from the gas. The recent Matsumoto et al. (1988) measurememts set the limit of $1.4 \cdot 10^2$ on the value of the chemical potential characterizing a B-E distortion.

The most extensive analysis of all recent (i.e., eversince 1980) measurements of the CMB spectrum was carried out by Smoot et al. (1988). These authors find that a NR Comptonized spectrum is an acceptable fit to the data. The Comptonization parameter is $0.02 \pm 0.002$, and heating of the matter took place at $z = 10^3 - 10^5$ (and with a best-fit temperature of 2.81 K). This analysis does not yield a tight constraint on the value of the baryonic mass fraction of the Universe.

Various schemes have been proposed recently for the source of heating at $z = O(10^3)$. Fukugita, Kawasaki and Yanagida (1989) conjecture a radiative decay of a species of massive, $O(10keV)$ neutrinos. Daly (1989) suggests the damping of acoustic waves associated with the transition from an unionized to neutral phase at the recombination epoch.

b. Determination of $H_0$ and $q_0$

Measurements of the SZ signal combined with X-ray measurements of thermal bremsstrahlung emission from the cluster can, in principle, yield $H_0$ and $q_0$ through the use of the definitions of the angular diameter and luminosity distances (Cavaliere, Danese and De Zotti 1977, 1979; Silk and White 1978). For this, the spatial profiles of the SZ effect and the X-ray brightness have to be accurately measured. The required measurements are difficult and constitute a challenge (much greater than the more modest and immediate goal of conclusive evidence for the detection of the effect, at last). The method was nonetheless applied to data from a few clusters (White and Silk 1980; Birkinshaw 1989), but in view of the observational uncertainties, the resulting values of $H_0$ do not improve on those from optical measurements. (The determination of $q_0$ requires measurements of very distant clusters and is at present unrealistic.)

### c. Anisotropy of the CMB on Arcminute Scales

The SZ distortion imprinted on the CMB spectrum along a line of sight to a cluster translates to an anisotropy due to the population of clusters when the intensity is measured in various parts of the sky (Rephaeli 1981). Since these fluctuations are dominated by nearby clusters, the typical angular scale is a few arcminutes. The level of this source of anisotropy is determined by the number density of clusters, and it is interesting and important that $\Delta T/T$ turns out to be of order $10^5$ (Rephaeli 1981). At the time when this source of fluctuations was proposed, it was the prevailing expectation that primordial fluctuations (which gave rise to the formation of the large scale structure) necessarily have related anisotropies which are much larger on arcminute scales. With the ever decreasing observational bounds on $\Delta T/T$, which is now $1.7 \cdot 10^5$ at 2 arcminutes (Readhead et al. 1989), it should be finally clear that **IC gas in clusters may be the dominant source of fluctuations on these scales.** Indeed, this possibility attracted more interest and has by now been explored by Cavaliere et al. (1986), Schaeffer and Silk (1988), and Cole and Kaiser (1988). In these papers, a wider range of evolutionary histories for IC gas has been considered.

The need to distinguish fluctuations due to IC gas from those associated with the primordial spectrum of fluctuations is obvious. Fortunately, this is possible due to the unique Compton signature of the former type of fluctuations. Measurements at two different frequencies on the Wien side can, in principle, isolate the contribution by IC gas to CMB arcminute anisotropy (Rephaeli 1981).

### d. Determination of Peculiar Velocities of Clusters

Peculiar velocities of galaxies and clusters are a basic feature of the process of growth of the large scale structure. In recent years, there has been a growing effort to determine peculiar velocities of galaxies from optical catalogs (Lahav, Rowan-Robinson and Lynden-Bell 1988), and from the IRAS database (Strauss and Davis 1988; Yahil 1988). It is obviously very important to observationally determine the velocities of clusters with respect to the Hubble flow. The only way this has been estimated so far is through the use of Tully-Fisher relations (Tully and Fisher 1977). Peculiar cluster velocities determined by using these relations are below 2000 km/s.

Another method is the measurement of the kinetic CMB intensity change along a line of sight to the cluster (Sunyaev and Zeldovich 1980b), as is clear from equation (4). For this, we have to know $\tau$, which can be estimated from X-ray measurements. However, $v_r$ has to be sufficiently high for a measureable intensity change. Moreover, $\Delta I_K$ cannot be negligible as compared with $\Delta I_T$, or else the total measured intensity change will be dominated by the thermal effect. From equations (1) and (4), it follows that $\Delta I_K = \Delta I_T$ for $vh(x)/c = kT_e g(x)/mc^2$. Since IC gas temperatures are typically 5 keV and higher, the frequency has to be optimally chosen so that the shape of the spectral functions can be exploited. Fortunately, as was pointed out in the previous section, $h(x)$ attains its maximum, at $x_0$, exactly when $g(x)$ is zero. This suggests $x_0$ to be the optimal frequency for the determination of $v_r$. Rephaeli and Lahav (1989) have compiled a list of high velocity clusters which can serve as good candidates for the measurement of $\Delta I_K$ and the deduction of $v_r$.

Note that the motion of galaxies in the gravitational well of the cluster can also constitute a source of noise in the measurement of the thermal intensity change. Typical velocities of galaxies in rich clusters can be as high as 2000 km/s. It is clear, therefore, that a gas-rich spiral galaxy may contribute appreciably to the total intensity change, especially so if the galaxy covers a nonneglibile fraction of the telescope beam. This source of signal contamination has also to be taken into account along with other considerations (e.g., see Gould and Rephaeli 1978; Lake and Partridge 1980; Rephaeli 1987) which will not be discussed further here.

e. Properties of Intracluster Gas

Comptonization of the CMB spectrum by IC gas depends essentially on the gas pressure and spatial extent. Measurements of the SZ signal across the cluster then determine the product $nTa$, where $a$ is the core radius of the gas spatial distribution. X-ray mesurements yield a different combination of these quantities, $n^2 T^{1/2} a^3$, so the CMB measurements give us an independent determination of the basic quantities which characterize IC gas. At present, no meaningfull comparison can be made between the values of these quantities as determined from these two sets of measurements. X-ray data are used, of course, in the selection of good cluster candidates and in the planning of a viable strategy for the detection of the SZ effect (choice of beam size, chopping angle, etc. - Rephaeli 1987).

Compton scattering of the CMB by IC gas heats the radiation (but, of course, conserves the number of photons). Because the effct amounts to a diminution of the CMB spectrum in the R-J region and a brightening on the Wien side, clusters may practically be viewed as sinks for radiation at frequencies such that $x < 3.83$, and as sources of emission for $x > 3.83$ (Sunyaev and Zeldovich 1981).

f. Comptonization of Other Background Radiations

So far, we have considered the effects of the interaction of CMB photons with IC ang IG electrons. Compton scattering also affects the radiation at other parts of the electromagnetic background (Gould and Rephaeli 1978). Thus, measurements of the Compton signature can be valuable in the study of some of other background radiations. For example, through the effect we may be able to discern a truly cosmological from a local component. However, at a given frequency there will be a detectable effect along a line of sight through a cluster, only if the background dominates the emission from the cluster itself. Gould and Rephaeli (1978) calculated the effect on the nonthermal radio background at low frequencies. Sunyaev and Zeldovich (1981) also considered possible Comptonization of the CXB at hard

X-ray energies (higher than, roughly, 50 keV). But because of the smallness of the effect, no useful deductions can be made before the intensity in these regions is known more accurately than at present.

Other interesting uses of the effect are the determination of the CMB temperature through observational determination of $x_0$ (Fabbri, Melchiorri and Natale 1978), or by measurements of the effect at two frequencies on the Wien side (Rephaeli 1980), and the detection of intra-supercluster gas (Persic,Rephaeli and Boldt 1988). Obviously, the effect may also be used as a mean for the identification of clusters (e.g., as lensing objects - Ostriker and Vishniac 1986). In principle, when relative temperature changes of order $10^6$ can be detected, it will be possible to also use the effect in the study of hot, gaseous galactic halos, and for measurements of peculiar velocities of galaxies.

## 4. CONCLUSIONS

The SZ effect is of fundamental importance in astrophysics and cosmology. It can serve as a unique tool for the determination of basic cosmological quantities, and as a probe of extended gaseous environments in the early and present Universe. Its unequivocal detection is of prime importance. To this end, it is essential to carry out more extensive studies of high frequency radio and microwave emission from sources in clusters of galaxies. The sensitive experiments aboard NASA's COsmic Background Explorer (COBE) will give us soon detailed information on the CMB spectrum and (large scale) anisotropy. COBE measurements can settle the issue of the reality of the submillemtric excess. In particular, if this excess is due to Comptonization, we may have to revise our estimates of the baryonic mass fraction of the Universe.

## 5. REFERENCES

Birkinshaw, M., and Gull, S.F. 1984, Mon. Not. Roy. Astron. Soc., **206**, 359.
Birkinshaw, M. 1989, these proceedings.
Cavaliere, A., Danese, L., and De Zotti, G. 1977, Astron. Astrophys., **217**, 6.
Cavaliere, A., Danese, L., and De Zotti, G. 1979, Astron. Astrophys., **75**, 322.
Cavaliere, A., and Fusco-Femiano, R. 1976, Astron. Astrophys., **49**, 137.
Cavaliere, A., Santangelo, P., Tarquini, G., and Vittorio, N. 1986, Astrophys. J., **305**, 651.
Cole, S., and Kaiser, N. 1988, Mon. Not. Roy. Astron. Soc., **233**, 637.
Daly,R. 1989, talk at this meeting.
Danese, L., and De Zotti, G. 1977, R.iv. Nuovo Cimento, **7**, 277.
Fabbri, R., Melchiorri, F., and Natale, V. 1978, Astrophys. Sp. Sci., **59**, 223.
Fabbri, R. 1981, Astrophys. and Space Sci., **77**, 529.
Field, G., and Perrenod, S.C. 1977, Astrophys. J., **21**, 717.
Fukugita, M., Kawasaki, M., and Yanagida, T. 1989, Kyoto preprint RIFP-787.
Guilbert, P.W., and Fabian, A.C. 1986, Mon. Not. Roy. Astron. Soc., **22**, 439.
Gull, S.F., and Northover, K.J.E. 1975, Mon. Not. Roy. Astron. Soc., **173**, 585.
Gould, R.J., and Rephaeli, Y. 1978, Astrophys. J., **219**, 12.

Hayakawa, S., Matsumoto, S., Matsuo, H., Murakami, H., Sato, H., Lange, A.E., and Richards, P.L. 1987, Pub. Astr. Soc. Japan, **39**, 941.

Henriksen, M.J., and Mushotzky, R.F. 1986, Astrophys. J., **302**, 287.

Kompaneets, A.S. 1957, Soviet Phys.-JETP, **4**, 730.

Lahav, O., Rowan-Robinson, M., and Lynden-Bell, D. 1988, Mon. Not. Roy. Astron. Soc., **234**, 677.

Lake, G., and Partridge, R.B. 1980, Astrophys. J., **237**, 378.

Matsumoto, S., Hayakawa, S., Matsuo, H., Murakami, H., Sato, H., Lange, A.E., and Richards, P.L. 1988, Astrophys. J., **32**, 567.

Ostriker, J.P., and Vishniac, E.T. 1986, Nature, **322**, 804.

Persic, M, Rephaeli, Y., and Boldt, E. 1988, Astrophys. J. (Letters), **327**, L1.

Readhead, A.C.S., Lawrence, C.R., Myers, S.T., Sargent, W.L.W., Hardebeck, H.E., and Moffet, A.T. 1989, Astrophys. J., submitted.

Rephaeli, Y. 1980, Astrophys. J., **241**, 858.

Rephaeli, Y. 1981, Astrophys. J., **245**, 351.

Rephaeli, Y. 1987, Mon. Not. Roy. Astron. Soc., **228**, 29p.

Rephaeli, Y., and Lahav, O. 1989, preprint.

Sarazin, C. 1986, Rev. Mod. Phys., **58**, 1.

Schaeffer, R., and Silk, J. 1988, Astrophys. J., **333**, 509.

Silk, J., and White. S.D.M. 1978, Astrophys. J. (Letters), **226**, L3.

Smoot, G.F., Levin, S.M., Witebsky, C., De Amici, G., and Rephaeli, Y. 1988, Astrophys. J., **331**, 653.

Strauss, M., and Davis, M. 1988, in Large-Scale Motions in the Universe, eds. V.C. Rubin and G.V. Coyne (Princeton: Princeton University Press), p. 256.

Sunyaev, R.A., and Zeldovich, Y.B. 1972, Comm. Astrophys. Sp. Phys., **4**, 173.

Sunyaev, R.A., and Zeldovich, Y.B. 1980a, Ann. Rev. Astron. Astrophys., **18**, 537.

Sunyaev, R.A., and Zeldovich, Y.B. 1980b, Mon. Not. Roy. Astr. Soc., **190**, 413.

Sunyaev, R.A., and Zeldovich, Y.B. 1981, Astrophys. Sp. Phys. Rev., **1**, 1.

Taylor, G.B., and Wright, E.L. 1989, Astrophys. J., **339**, 619.

Tully, R.B., and Fisher, J.R. 1977, Astron. Astrophys., **54**, 661.

Weymann, R. 1966, Astrophys. J., **145**, 560.

White, S.D.M., and Silk, J. 1980, Astrophys. J., **241**, 864.

Wright, E.L. 1979, Astrophys. J., **232**, 348.

Yahil, A. 1988, in Large-Scale Motions in the Universe, eds. V.C. Rubin and G.V. Coyne (Princeton: Princeton University Press), p. 219.

Zeldovich, Y.B., and Sunyaev, R.A. 1969, Astrophys. Sp. Sci., **4**, 301.

# Observations of the Sunyaev-Zel'dovich Effect

M. Birkinshaw
Department of Astronomy, Harvard University
60 Garden Street, Cambridge, MA 02138, USA

Reliable detections of the Sunyaev-Zel'dovich effect exist for only a few clusters of galaxies although many observations exhibit random errors < 0.5 mK. Several observational techniques are in use and particular promise for future improvements in the measurements are shown by sub-millimetre work with single dishes and centimetre-wavelength observations with small, dedicated interferometers. The three best-observed clusters display effects which are clearly associated with the X-ray emitting intracluster gas, and which can be used to investigate the properties of this gas and to estimate the value of the Hubble constant.

## 1. THE OBSERVABLE EFFECT

The Sunyaev-Zel'dovich (SZ) effect (Sunyaev & Zel'dovich 1972) is produced by inverse-Compton interactions between electrons of the hot, X-ray emitting, gas in clusters of galaxies and photons of the microwave background radiation (see Rephaeli, this Workshop) and represents one of the most obvious structures that should appear in the microwave background radiation. Consider, for example, the SZ effect produced by the Coma cluster. X-ray observations (Mushotzky & Smith 1980; Abramopoulos & Ku 1983; Hughes $et$ $al.$ 1988) indicate that the intracluster medium in Coma has an electron temperature $T_e \approx 7.9$ keV, and that the electron concentration of the gas can be represented by a $\beta$-model

$$n_e(r) = n_e(0)(1 + r^2/r_{\mathrm{cx}}^2)^{-\frac{3}{2}\beta} \tag{1}$$

(Cavaliere & Fusco-Femiano 1976, 1978), where the core radius $r_{\mathrm{cx}} \approx 500$ kpc (from the angular core radius, $\theta_{\mathrm{cx}} \approx 12.8$ arcmin and $H_0 = 50$ km s$^{-1}$ Mpc$^{-1}$), $\beta \approx 0.68$, and the central electron concentration $n_e(0) \approx 3 \times 10^3$ m$^{-3}$. The optical depth to inverse-Compton scattering through the cluster centre, $\tau_e = \int n_e \sigma_T \, dl \approx 0.015$. On average an electron-photon scattering changes the frequency of a photon by $\Delta\nu/\nu = kT_e/m_e c^2 \approx 0.016$, so that the change in brightness of the background

N. Mandolesi and N. Vittorio (eds.), The Cosmic Microwave Background: 25 Years Later, 77–94.
© 1990 Kluwer Academic Publishers.

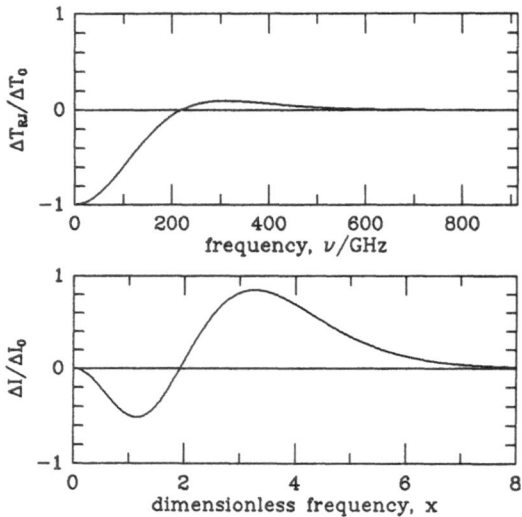

Figure 1. The spectrum of the Sunyaev-Zel'dovich effect expressed as a brightness temperature change, $\Delta T_{\mathrm{RJ}}$, and as an intensity change, $\Delta I$.

radiation in the Rayleigh-Jeans part of the spectrum (where $I_\nu \propto \nu^2$) is $\Delta I/I = -2\tau_e \frac{\Delta \nu}{\nu} \approx -4.8 \times 10^{-4}$, which corresponds to a brightness temperature change $\Delta T_{\mathrm{RJ}} = -1.4$ mK. This is the Sunyaev-Zel'dovich effect. A signal this large should be easily measured in 1 hour using a radio telescope equipped with a (heterodyne) receiver with a noise temperature $\approx 50$ K and bandwidth $\approx 300$ MHz, but unfortunately the observations are subject to significant systematic errors (Sec. 2).

This result for $\Delta T_{\mathrm{RJ}}$ is correct in the low-frequency limit, but in general the magnitude of the effect is related to the shape of the spectrum of the microwave background radiation. If the SZ effect is observed with a conventional heterodyne radiometer the observable quantity is the brightness temperature change in the microwave background radiation, $\Delta T_{\mathrm{RJ}}$. Then if the microwave background radiation has a black-body spectrum characterised by a radiation temperature $T_{\mathrm{r}}$ ($= 2.74$ K; Kogut $et\ al.$, 1988) the spectrum of the SZ effect is

$$\Delta T_{\mathrm{RJ}}(\nu) = \Delta T_0 \left( x^2 \operatorname{cosech}^2 x \left( x \coth x - 2 \right) \right) \qquad (2)$$

where the dimensionless frequency $x = h\nu/2kT_{\mathrm{r}} = 8.3 \times 10^{-3} (\nu/\mathrm{GHz})$ and $-\Delta T_0$ is the size of the effect at $\nu = 0$ (Figure 1). $\Delta T_{\mathrm{RJ}}$ appears as a decrease in the brightness of the background radiation at frequencies below 219 GHz (the largest negative effect occurs at zero frequency), and as an increase at high frequencies, with a peak of $0.10\Delta T_0$ at $\nu = 311$ GHz. Observations of the SZ effect made using a bolometer system measure the corresponding change in the flux of the microwave background radiation, $\Delta I$, with spectrum

$$\Delta I(\nu) = \Delta I_0 \left( x^4 \operatorname{cosech}^2 x \left( x \coth x - 2 \right) \right) \qquad (3)$$

(Fig. 1). Again, the effect is negative at low frequencies, and positive at high frequencies, but the maximum intensity change is $+0.85\Delta I_0$ at $\nu = 372$ GHz. The largest negative effect, $-0.52\Delta I_0$, occurs at $\nu = 129$ GHz.

If the structure of the cluster atmosphere is represented by an isothermal $\beta$-model (equation 1) with $\beta > \frac{1}{3}$, the X-ray surface brightness is given by

$$l_x(\theta) \propto (1 + \theta^2/\theta_{cx}^2)^{\frac{1}{2}-3\beta} \tag{4}$$

and the profile of the Sunyaev-Zel'dovich effect is

$$\Delta T_{RJ}(\theta) \propto (1 + \theta^2/\theta_{cx}^2)^{\frac{1}{2}-\frac{3}{2}\beta} \tag{5}$$

so that the SZ effect is more extended than the X-ray surface brightness. The ratio of the full-width at half maximum in $\Delta T_{RJ}$ to that in $l_x$ decreases from 3.9 at $\beta = \frac{1}{2}$ to 1.6 at $\beta = \frac{3}{2}$. The broadness of the SZ profile has important consequences for searches for the effect (Sec. 2).

Since the SZ effect arises from the propagation of the background radiation through a distant cluster of galaxies, the observation of the effect from a cluster at redshift $z_c$ proves that the background radiation originates at redshifts $> z_c$. Measurements of the spectrum of the SZ effect can be used to measure the spectrum of the background radiation, since the intensity of the effect depends on the shape of this spectrum. This is of particular interest in the frequency range 200 – 400 GHz, now that there is evidence for a strong deviation from a thermal shape (Matsumoto et al. 1988).

The SZ effect can also provide information on the properties of a cluster atmosphere. $\Delta T_{RJ}$ is proportional to the line-of-sight integral of the electron pressure whereas the X-ray surface brightness is approximately proportional to the line-of-sight integral of $n_e^2 T_e^{1/2}$, so that a comparison of the angular structures of the two effects should provide information on the distributions of electron temperature and electron concentration in the cluster gas. This allows a test of the isothermality of the gas that is independent of the X-ray spectrum, which is useful because the X-ray data are biased to high-density regions by the $n_e^2$ factor, whereas the SZ data are more sensitive to lower-density regions, where the path lengths are longer.

Another use of the SZ effect is as a cosmological probe, to measure the value of the Hubble constant (Gunn 1978; Silk & White 1978; Birkinshaw 1979; Cavaliere et al. 1979). In its simplest form, the central SZ effect

$$\Delta T_{RJ}(0) \propto n_e(0)\, T_e(0)\, D_A\, \theta_{cx} \tag{6}$$

and the total X-ray flux density

$$S_X \propto n_e(0)^2 \frac{(D_A \theta_{cx})^3}{D_L^2} \Lambda(T_e(0)) \tag{7}$$

are compared. $D_A$ and $D_L$ are the angular diameter and luminosity distances of the cluster, $\Lambda(T_e)$ is the cooling function of the gas, and the constants of proportionality encode the structure of the gas. If $n_e(0)$ is eliminated then

$$\frac{D_L^2}{D_A} \propto \frac{\Delta T_{RJ}(0)^2}{S_X} \theta_{cx}\, f(T_e(0)) \tag{8}$$

where all the quantities on the right hand side of this equation can be measured. $f(T_e)$ is a known function of the electron temperature, which can be measured from the X-ray spectrum. Thus $D_L^2/D_A$, and the value of the Hubble constant, can be determined when the constant of proportionality in this relation has been found.

Finally, it is to be expected that the integrated SZ effects of clusters will provide a background of confusing signals in the background radiation on arcminute angular scales that may be difficult to distinguish from primordial fluctuations arising from processes near recombination. This integrated effect will also appear as a small inverse-Compton distortion of the spectrum of the background radiation.

## 2. TECHNIQUES FOR OBSERVING THE SUNYAEV-ZEL'DOVICH EFFECT

### 2.1 Single-dish radiometry

The technique used most frequently to search for the SZ effect is single-dish radiometry, where the microwave background radiation is observed using a heterodyne receiver mounted on a large single dish. The receiver usually provides two beams arranged symmetrically about the centre line of the telescope, and the beam-switched signal (proportional to the difference in the antenna temperature in the beams) is recorded. Beam-switching is effective at removing atmospheric noise, but since the beam separation is usually only 4 – 6 beam-widths, an extended source like the Sunyaev-Zel'dovich effect tends to appear in both the 'main' (on-source) and 'reference' (off-source) beams. Thus the beam-size and beam-separation constrain the clusters that can be observed efficiently. Distant clusters, with angular sizes much less than a beam-width, are subject to strong beam dilution in the main beam. Nearby clusters, for which the beam dilution is small, have both the main and reference beams deep within the Sunyaev-Zel'dovich effect, and hence beam-switching subtracts much of the effect. For example, the OVRO 40-m radio telescope has a beam-separation of 7.2 arcmin and a beam-width (FWHM) of 1.8 arcmin at 20.3 GHz. If the SZ effect is modelled by equation (5) with $\beta = 0.8$, then reasonable sensitivity is achieved for clusters with $0.5 \lesssim (\theta_{cx}/\text{arcmin}) \lesssim 6.5$, which eliminates from study the nearby clusters for which good X-ray data are available.

Simple beam-switching removes most of the sky noise, but better results are obtained when a second level of differencing is added. This is commonly provided by position-switching, where the two beams are directed alternately at the point under observation, and the data obtained in different parts of the switching cycle are subtracted. It is popular to position-switch in azimuth, so that the elevation of the telescope changes only slightly during the switching cycle. An alternative to position-switching is drift-scanning: the telescope is stopped ahead of the point to be observed, and observations are taken as the sky rotates through the stationary beam pattern. This method is attractive since the spillover signals from the ground remain constant during each scan, but it tends to be inefficient because the fraction of each scan spent on source is small. Scans where the telescope is driven along a line through the target point have also been used.

Position-switching radiometry is often used at frequencies $\lesssim 30$ GHz, where both the 'on' and 'off' beam positions are subject to significant radio source confusion. Sources near the 'off' positions bias the measurements in the sense of producing a fake SZ effect. Sources near the 'on' position tend to obscure real effects. Radio source confusion is particularly severe for low-frequency observations ($\lesssim 10$ GHz) and where the pattern of beam-switching causes the 'off' positions to be

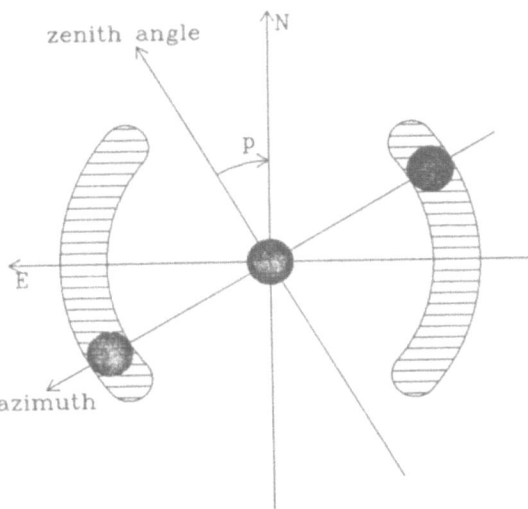

Figure 2. The development of reference arcs by position-switching in azimuth. As the sky rotates, and the telescope is switched so that it spends time $\tau/4$ with the main beam on source (and the reference beam off source to the west), then $\tau/2$ with the reference beam on source (and the main beam off source to the east), then $\tau/4$ with the main beam on source again, so the off-source points rotate about the on-source point. After a long observation the net signal is the difference between the brightness on the source and on the reference arcs. The parallactic angle $p$ is used to specify the locations of beams on the reference arcs.

---

fixed in right ascension and declination. When beam-switching in azimuth is used, the 'off' positions rotate around the central, 'on' position, to create a 'reference arc' (Fig. 2) so that the average measured signal is the difference between the central brightness and the mean over this arc. Strong sources in the reference arcs may be located from the modulation of their signal by the rotation of the sky, but to avoid biasing the SZ measurements independent radio source surveys are needed to locate confusing sources and provide source corrections. Individual galaxies within the cluster may also contribute to the measured signal, through their weak radio emission or scattering effects in their intragalactic atmospheres (Rephaeli, this Workshop), and thermal bremsstrahlung from gas in spiral galaxies (lying predominantly at the edges of clusters) may produce a fake SZ effect (Tarter 1978).

Efficient heterodyne observations can be made in the frequency range 10 to 30 GHz for radio telescopes in the 10 to 100-m class. At higher frequencies the atmospheric signals become excessive, and the reduced beam-switching angle makes it hard to select clusters for observation. At lower frequencies the increased radio-source confusion is a problem. For all observations of this type, a good site, a sensitive, stable receiver, and long observing times (months) are essential for successful work. Further improvements in this technique are likely. The development of wider-bandwidth and higher-sensitivity receivers will allow better use to be made of the best weather. Focal plane arrays permit many adjacent sky regions to be observed simultaneously. The greatest single improvement, however, would come from the use of new telescopes with lower sidelobes, and hence reduced systematic errors from spillover.

## 2.2 Single-dish observations using bolometers

This type of observation is made at frequencies $\gtrsim 100\,\text{GHz}$, where radiometers tend to have high noise temperatures, and measures $\Delta I$, the change in energy flux of the background radiation, rather than $\Delta T_{\text{RJ}}$. At such high frequencies the contributions of background radio sources are not expected to be a problem, but rather there may be confusing signals from galactic gas and dust. The predominant difficulty with these observations is the atmospheric noise and transparency: sensitive data require work at an excellent site, and then good luck with the weather. Sky subtraction is achieved by beam-switching (Section 2.1) except that only a single beam is defined by the optics of the telescope. This single beam is switched quickly on the sky using, for example, a nutating subreflector. Position-switching and drift-scanning may also be used, to improve the precision of the sky subtraction. Limitations in the telescope optics, and the need to remove the sky noise as precisely as possible, force the use of small beam-throws and constrain the redshifts of observable clusters. An example is the use of UKIRT on Mauna Kea by Chase *et al.* (1987). These observations were performed with a beam-throw of only 3 arcmin, so that only clusters with X-ray core radii less than about 1 arcmin would be expected to produce strong signals. This restricts sensitive observations to redshifts $> 0.1$, and requires the use of a beam-size $\lesssim 1$ arcmin to avoid excessive beam dilution. In fact the FWHM of the Chase *et al.* beam was 1.9 arcmin.

A great advantage of this type of measurement, if the sky noise difficulty can be overcome (perhaps by satellite observations), is that the nominal sensitivity of the systems is excellent because of the extremely wide bandwidths that are available to bolometers. The large $\Delta I$ signal that is predicted by (3) corresponds to relatively bright sub-mm sources (particularly at high redshift), so that excellent measurements of the spectrum of the SZ effect may be possible.

## 2.3 Interferometric observations

A third method that can be used to observe the Sunyaev-Zel'dovich effect is interferometry, where the entire field of a cluster of galaxies is mapped in a single long observation using a number of antennas arranged to provide many baselines. The great advantage of this method is that it provides an *image* of the sky, showing both radio sources and the SZ effect. However, interferometers measure only a restricted range of Fourier components of the structure of the radio sky: in particular, they cannot observe baselines less than the antenna diameter. The missing short baselines are of critical importance in observations of large-scale structures, and for any configuration of antennas there is some largest angular scale to which the interferometer is sensitive. One way of quantifying this effect is to calculate the fraction of the integrated Sunyaev-Zel'dovich flux of a cluster that is seen as correlated signal on any baseline. Consider, for example, an SZ effect distributed according to equation (5), with a central decrement $-1$ mK, $\theta_{\text{cx}} = 1$ arcmin, and $\beta = 1$, which is observed at $\lambda 6$ cm by an interferometer with 25-m antennas. Then the total Sunyaev-Zel'dovich signal that would be seen

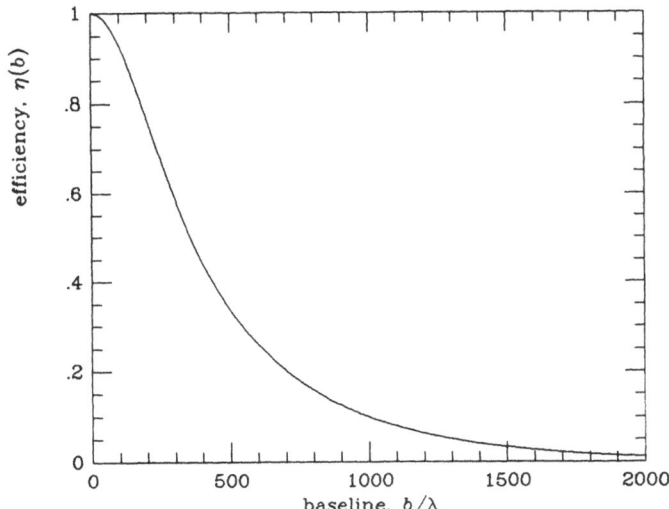

Figure 3. The fraction, $\eta(b)$, of the total SZ effect flux density of a cluster with $\beta = 1$ and $\theta_{cx} = 1$ arcmin that would be seen at $\lambda 6$ cm on baseline $b$ by an interferometer with 25-m antennas. On baselines greater than the diameter of an antenna, $\eta < 0.4$. Since most baselines of an interferometer are $\gg$ the antenna diameter, this interferometer would be inefficient at mapping the SZ effect.

---

by a single antenna is $-0.6$ mJy, less than the confusion from background radio sources in the 9 arcmin primary beam of the antennas. $\eta(b)$, the fraction of this flux density that would be seen by a pair of antennas separated by baseline $b$, is shown in Figure 3. Since the minimum possible separation of antennas is $420\lambda$, the antenna diameter, no baseline sees $> 40$ per cent of the SZ signal. Only a few baselines see even as much as 10 per cent of the signal: antenna pairs with longer baselines are sensitive to smaller-scale structures. Thus the data must be heavily tapered for the resulting map to display more than 60 $\mu$Jy of correlated flux from the SZ effect and very long observations are necessary.

Since interferometry relies on the correlated signals seen by a pair of antennas, it is less sensitive than single-dish observations to many types of systematic error, such as ground spillover signals, which are not coherent between different antennas. On the other hand, a new class of systematic errors is introduced by the correlators, by antenna-antenna crosstalk (particularly on the important short baselines), by imperfections in the local oscillator signal, etc. Because different systematic errors are important, and because a map of the SZ effect is produced, an interferometric strategy is attractive for confirming a single-dish detection of a cluster, even though the efficiency of interferometric observations may be low.

Only a single paper presenting results from this method has so far appeared (Partridge *et al.*, 1987), although several groups have attempted to use the VLA or other interferometers to do such work. It is likely that it will be the development of interferometers with small antennas, many short spacings, and sensitive receivers that will first map the SZ effect, rather than interferometers designed for other purposes (usually high resolution). It is certainly possible to redevelop suitable interferometers to add the ability to observe the microwave background radiation — this is currently being done for the 5-km telescope at Cambridge (Saunders 1988) — and small interferometers provide an exciting possibility for the study of all types of arcminute-scale structures.

## 3. OBSERVATIONS

Many attempts to detect the Sunyaev-Zel'dovich effect have been reported. It is clear that the earliest measurements suffered from the presence of systematic errors — from spillover signals, confusion with radio sources near the cluster centres and in the reference beam positions, errors in the gain of the telescope used, non-gaussian statistics for the sky noise, etc. In consequence, the earliest results are generally not in agreement with more recent work. The current state of the literature on the SZ effect is summarized in Table 1, which collects the *final* reports of each experiment (and excludes preliminary or progress reports, e.g. Gull & Northover 1976; Lake & Partridge 1977).

Many of the results presented in the papers in Table 1 are of little help in a study of the SZ effect because the errors are too large. In some cases this is because the paper reflects the development of a technique to observe the SZ effect; in others it represents short observing times, bad luck with the weather or the equipment, or the presence of large systematic errors. The best results from the papers of Table 1 (those with errors in $\Delta T_{RJ} < 0.5$ mK at the frequency of observation) are collected in Table 2. The results for any particular cluster in Table 2 cannot be compared directly because they arise from observations at slightly different positions, made with telescopes with different beam-widths and beam-throws, at different frequencies, and in different observing modes. A rough correction for this has been made by assuming that $\Delta T_{RJ}(\theta)$ follows equation (5) with $\beta = 0.8$ and a value of $\theta_{cx}$ appropriate to the redshift of the cluster, assuming $r_{cx} = 500$ kpc, and the deconvolved, central, SZ effect at zero frequency, $\Delta T(0)$, is given in the last column of Table 2.

It is notable that all the observations contained in Table 2 were made using single-dish radiometry: the other techniques have not yet achieved the same sensitivity. As this implies, the quality of the spectral data on even the best-observed cluster is poor, and no useful constraints on the spectrum of the microwave background radiation can be found.

The results for those clusters for which several observations have been reported are generally in poor agreement. Thus for Abell 2218, the minimum-$\chi^2$ estimate for $\Delta T(0)$ is $-0.54 \pm 0.05$ mK, but the value of $\chi^2 = 39.5$ (the probability of a greater value of $\chi^2$ appearing by chance is $< 1$ per cent). This suggests that residual systematic errors at the level of the random errors lurk in the data of Schallwich (1979), Lake & Partridge (1980), and Birkinshaw *et al.* (1981), which make the largest contributions to $\chi^2$.

On the basis of the simple estimate for the Coma cluster (Section 1), it might be expected that every cluster in Table 2 should display an SZ effect at a level $\approx -0.5$ mK. In general, the deconvolved results for $\Delta T(0)$ do not exclude this possibility. Where positive signals are seen, they are likely to arise from strong,

Table 1. Observations of the Sunyaev-Zel'dovich effect

| Reference | frequency (GHz) | beamsize (arcmin) | beamthrow (arcmin) | Notes |
|---|---|---|---|---|
| Parijskij 1972 | 7.5 | 1.3 × 40 | — | 1B, DS, RA |
| Rudnick 1978 | 15.0 | 2.2 | 17.4 | 2B, DS, RA |
| Perrenod & Lada 1979 | 31.4 | 3.5 | 8 | 2B, PS1, RA |
| Schallwich 1979 | 10.7 | 1.2 | 8.2 | 2B, DS, RA |
| Lake & Partridge 1980 | 31.4 | 3.6 | 9 | 1B, PS1, RA |
| Birkinshaw *et al.* 1981 | 10.6 | 4.5 | 15 | 2B, PS1, RA |
| Andernach *at al.* 1983 | 10.7 | 1.2 | 3.2, 8.2 | 3B, DS, RA |
| Lasenby & Davies 1983 | 5.0 | 8 × 10 | 30 | 2B, PS1, RA |
| Meyer *et al.* 1983 | 90 – 300 | 5.0 | 5.0 | 1B, PS2, BO |
| Birkinshaw & Gull 1984 | 10.7 | 3.3 | 14.4 | 2B, PS1, RA |
| *ditto* | 20.3 | 1.8 | 7.2 | 2B, PS1, RA |
| Birkinshaw *et al.* 1984 | 20.3 | 1.8 | 7.2 | 2B, PS1, RA |
| Uson 1985 | 19.5 | 1.8 | 8.0 | 2B, PS2, RA |
| Andernach *at al.* 1986 | 10.7 | 1.2 | 3.2, 8.2 | 3B, DS, RA |
| Radford *et al.* 1986 | 89.6 | 1.3 | 4.0 | 1B, DS, RA |
| *ditto* | 89.6 | 1.2 | 4.2 | 1B, PS1, RA |
| *ditto* | 105 | 1.7 | 19 | 1B, PS1, RA |
| Birkinshaw 1987 | 20.3 | 1.8 | 7.2 | 2B, PS1, RA |
| Chase *et al.* 1987 | 261 | 1.9 | 3.0 | 1B, PS2, BO |
| Partridge *et al.* 1987 | 4.9 | 0.3 – 1.4 | — | IN(4 arcmin) |
| Uson 1987 | 19.5 | 1.8 | 8.0 | 2B, PS2, RA |

Notes: 
$n$B — detector system providing $n$ beams
BO — bolometer detector
DS — drift or driven scans
PS1 — position switching by fixed azimuth, elevation offsets
PS2 — position switching to fixed $\alpha, \delta$
IN($\theta_{max}$) — interferometer, sensitive to scales $< \theta_{max}$
RA — radiometer

extended, central radio sources which mask the SZ signals at low frequencies (e.g. for Abell 376). Significant corrections for such central sources, or for sources near the reference arcs, have been made for several of the $\Delta T_{RJ}$ values in Table 2, but consistent corrections have not been applied to all the results. Variations in the temperatures, central gas densities, and sizes of the clusters will cause significant variations in the expected values of $\Delta T(0)$: a rough guess might be that the distribution of central SZ effects should extend from $-1.5$ to $0$ mK. Since the clusters in Table 2 were (usually) selected on the basis of X-ray properties, and the SZ effect and the X-ray luminosity depend on the properties of the atmosphere in different ways, some variation in the central decrements corresponding to variations in the structures of the cluster atmospheres, is to be anticipated.

## 4. RECENT RESULTS AT 20.3 GHZ FROM OVRO

As a more detailed example of observations of the Sunyaev-Zel'dovich effect,

Table 2. Measurements with $\Delta T_{\mathrm{RJ}}$ error $< 0.5$ mK

| Cluster | $\theta_{\mathrm{cx}}$ (arcmin) | $\Delta T_{\mathrm{RJ}}$ (mK) | Reference | $\Delta T(0)$ (mK) |
|---|---|---|---|---|
| 0016+16 | 1.0 | $-0.72 \pm 0.18$ | Birkinshaw & Gull 1984 | $-1.83 \pm 0.46$ |
| | | $-0.37 \pm 0.16$ | Birkinshaw & Gull 1984 | $-0.71 \pm 0.31$ |
| | | $-0.64 \pm 0.08$ | Birkinshaw *et al.* 1984 | $-1.10 \pm 0.34$ |
| | | $-0.48 \pm 0.12$ | Uson 1987 | $-0.78 \pm 0.20$ |
| | | $-0.41 \pm 0.08$ | Birkinshaw 1987 | $-0.70 \pm 0.14$ |
| PHL 957 | 1.0 | $-0.60 \pm 0.15$ | Andernach *et al.* 1986 | $-0.91 \pm 0.23$ |
| Abell 347 | 16.6 | $+0.34 \pm 0.29$ | Birkinshaw *et al.* 1981 | $+1.05 \pm 0.90$ |
| Abell 376 | 6.3 | $+1.22 \pm 0.35$ | Birkinshaw *et al.* 1981 | $+1.94 \pm 0.56$ |
| Abell 401 | 4.4 | $-0.64 \pm 0.18$ | Uson 1987 | $-1.09 \pm 0.31$ |
| Abell 478 | 3.7 | $-0.71 \pm 0.47$ | Birkinshaw *et al.* 1981 | $-1.11 \pm 0.73$ |
| | | $+0.44 \pm 0.32$ | Birkinshaw & Gull 1984 | $+0.77 \pm 0.56$ |
| Abell 545 | 2.3 | $+1.64 \pm 0.44$ | Lake & Partridge 1980 | $+3.07 \pm 0.82$ |
| | | $+0.51 \pm 0.43$ | Uson 1985 | $+0.70 \pm 0.59$ |
| Abell 576 | 7.9 | $-1.26 \pm 0.27$ | Lake & Partridge 1980 | $-3.35 \pm 0.72$ |
| | | $-1.12 \pm 0.17$ | Birkinshaw *et al.* 1981 | $-1.92 \pm 0.29$ |
| | | $+1.10 \pm 0.44$ | Lasenby & Davies 1983 | $+1.82 \pm 0.73$ |
| | | $-0.14 \pm 0.29$ | Birkinshaw & Gull 1984 | $-0.37 \pm 0.77$ |
| Abell 586 | 2.1 | $-0.09 \pm 0.38$ | Birkinshaw & Gull 1984 | $-0.16 \pm 0.67$ |
| Abell 665 | 2.0 | $-0.53 \pm 0.22$ | Birkinshaw *et al.* 1981 | $-1.02 \pm 0.42$ |
| | | $+0.03 \pm 0.25$ | Birkinshaw & Gull 1984 | $+0.05 \pm 0.45$ |
| | | $-0.55 \pm 0.13$ | Birkinshaw & Gull 1984 | $-1.01 \pm 0.24$ |
| | | $-0.34 \pm 0.05$ | Birkinshaw *et al.* 1984 | $-0.50 \pm 0.07$ |
| | | $-0.37 \pm 0.14$ | Uson 1987 | $-0.51 \pm 0.19$ |
| | | $-0.47 \pm 0.08$ | Birkinshaw 1987 | $-0.69 \pm 0.12$ |
| Abell 669 | 1.3 | $+0.38 \pm 0.24$ | Birkinshaw & Gull 1984 | $+0.82 \pm 0.52$ |

this Section contains a description of work undertaken using the 40-m telescope at the Owens Valley Radio Observatory. Since 1982 the telescope has been equipped with a tunable K-band maser operating with a bandwidth of about 350 MHz about a centre frequency $\approx 20.3$ GHz. The receiver package illuminates the dish to provide two 1.8-arcmin primary beams separated by 7.2 arcmin on the sky. The gains in these beams are almost equal, and the system noise temperature is about 40 K. The first use of this system, in the 1982–3 winter season, provided about 100 hr of test data, and about 3000 hours of useful data have been taken since 1983. The 1989 data have not yet been adequately studied, so that only the data from 1983-8 are used in the discussion that follows.

Observations were made using beam-switching and position-switching in azimuth (see Fig. 2), which results in the point under study being compared with points on surrounding reference arcs. Each position-switching cycle lasts about 75 sec, made up of 10 sec with the 'main' beam on source, then $2 \times 10$ sec with the 'reference' beam on source, and a final 10 sec with the 'main' beam on source again. The remainder of the time is spent moving and steadying the telescope. During each 'on' or 'off' segment of the cycle, the difference between the brightnesses in the two beams is recorded, and the mean brightness difference is stored at the end of each cycle.

Although this method of observing is efficient at subtracting atmospheric

Table 2. *Continued.*

| Cluster | $\theta_{cx}$ (arcmin) | $\Delta T_{RJ}$ (mK) | Reference | $\Delta T(0)$ (mK) |
|---|---|---|---|---|
| Abell 777 | 1.7 | $-0.22 \pm 0.44$ | Lake & Partridge 1980 | $-0.46 \pm 0.91$ |
| Abell 1656 | 12.8 | $-0.19 \pm 0.22$ | Lake & Partridge 1980 | $-0.87 \pm 1.00$ |
| 1305+29 | 0.8 | $-0.28 \pm 0.22$ | Birkinshaw & Gull 1984 | $-0.84 \pm 0.66$ |
| Abell 1689 | 2.1 | $+0.24 \pm 0.38$ | Birkinshaw & Gull 1984 | $+0.43 \pm 0.67$ |
| Abell 1763 | 2.0 | $-0.36 \pm 0.25$ | Uson 1985 | $-0.50 \pm 0.34$ |
| Abell 1904 | 4.5 | $+0.55 \pm 0.40$ | Birkinshaw et al. 1981 | $+0.84 \pm 0.61$ |
| Abell 2079 | 4.8 | $-0.05 \pm 0.25$ | Lake & Partridge 1980 | $-0.10 \pm 0.49$ |
| Abell 2125 | 1.6 | $+0.77 \pm 0.44$ | Lake & Partridge 1980 | $+1.65 \pm 0.94$ |
|  |  | $-0.39 \pm 0.22$ | Birkinshaw et al. 1981 | $-0.73 \pm 0.41$ |
|  |  | $-0.31 \pm 0.39$ | Birkinshaw & Gull 1984 | $-0.60 \pm 0.76$ |
| Abell 2218 | 2.1 | $-1.04 \pm 0.48$ | Perrenod & Lada 1979 | $-2.00 \pm 0.92$ |
|  |  | $-1.22 \pm 0.25$ | Schallwich 1979 | $-1.58 \pm 0.32$ |
|  |  | $+0.71 \pm 0.38$ | Lake & Partridge 1980 | $+1.36 \pm 0.73$ |
|  |  | $-1.05 \pm 0.21$ | Birkinshaw et al. 1981 | $-1.97 \pm 0.39$ |
|  |  | $-0.38 \pm 0.19$ | Birkinshaw & Gull 1984 | $-0.74 \pm 0.37$ |
|  |  | $-0.31 \pm 0.13$ | Birkinshaw & Gull 1984 | $-0.57 \pm 0.24$ |
|  |  | $-0.34 \pm 0.05$ | Birkinshaw et al. 1984 | $-0.50 \pm 0.07$ |
|  |  | $-0.29 \pm 0.24$ | Uson 1985 | $-0.40 \pm 0.33$ |
|  |  | $+0.16 \pm 0.43$ | Radford et al. 1986 | $+0.40 \pm 1.08$ |
|  |  | $+0.41 \pm 0.32$ | Radford et al. 1986 | $+0.89 \pm 0.69$ |
|  |  | $-0.33 \pm 0.07$ | Birkinshaw 1987 | $-0.48 \pm 0.10$ |
| Abell 2319 | 5.9 | $-0.30 \pm 0.19$ | Lake & Partridge 1980 | $-0.64 \pm 0.41$ |
|  |  | $-0.40 \pm 0.29$ | Birkinshaw et al. 1981 | $-0.63 \pm 0.46$ |
| Abell 2666 | 11.3 | $+0.60 \pm 0.30$ | Lake & Partridge 1980 | $+2.33 \pm 1.16$ |
|  |  | $+0.34 \pm 0.29$ | Birkinshaw et al. 1981 | $+0.72 \pm 0.62$ |

(and many systematic) signals, the overall system noise temperature is significantly increased in poor weather. Checks that the increased noise does not co-exist with systematic offsets have been made by subdividing the data, and further controls on systematic offsets are made by observing blank sky regions near the points of interest at approximately the same times that the cluster observations are made.

A second advantage of this observing method is that the modulation of the signals contributed by sources in the reference arcs can be searched for and used to locate those sources. An example is shown in Fig. 4, where the data for a point near Abell 665 are displayed binned by parallactic angle, and then compared with the signals predicted on the basis of an interferometric survey of the field (Moffet & Birkinshaw 1989). For full accuracy, it is necessary that good source surveys be conducted in each field that is observed in the SZ effect.

The process of analysing the data involves (1) removing data taken under periods of poor weather, when the noise temperature was excessive, or when problems were noted with the telescope, the receiver, the computer, or the operator; (2) rejecting data that are seriously contaminated by radio sources in the reference arcs; and (3) applying source corrections to the remaining data. The (much smaller) data-set is then combined to produce an estimated brightness temperature at each point observed, but note that step (2) means that the results are not a brightness temperature *uniformly* compared with points on a reference arc.

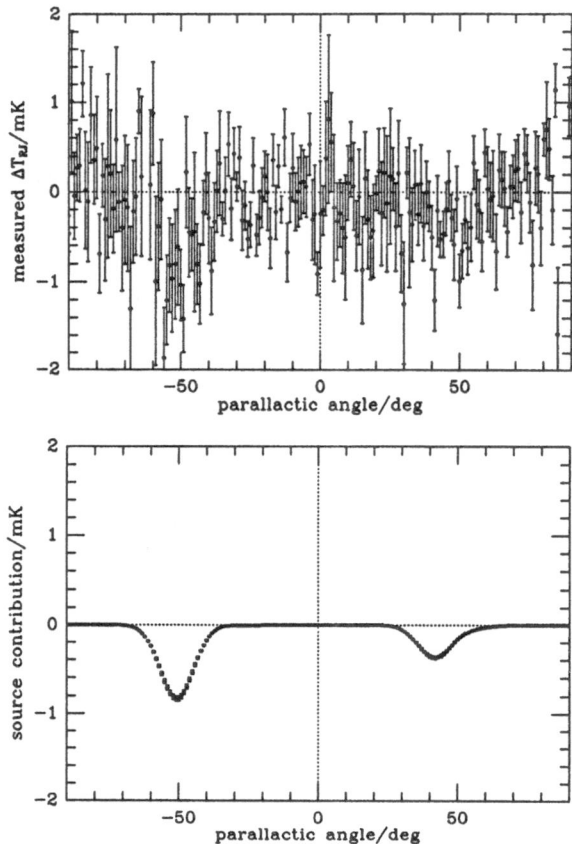

Figure 4.  The upper panel shows the SZ data taken at a point 7 arcmin from the nominal centre of Abell 665 binned by parallactic angle.  Significant deviations of the 1-degree means can be seen near parallactic angle −50° and +45° (note that since the beams occupy an angle of about 15° on the reference arcs, only groups of about 15 points are independent samples of the sky although all points contain wholly independent data).  The lower panel shows the predicted parallactic angle scan, based on the multi-frequency VLA survey of Abell 665 by Moffet & Birkinshaw (1989). The locations and brightnesses of the two main sources agree well with the the values found directly from the SZ data.

The errors are calculated in two stages. In the first step the random, sampling, error is calculated using the selected data. The second source of error arises from systematic problems with the observations, the telescope, and the data-processing. Such errors can only be estimated, although control observations of blank sky

Table 3. Reference sky results

| name | measured $\Delta T_{\mathrm{RJ}}$ ($\mu$K) | maximum systematic signal ($\mu$K) |
|---|---|---|
| Ref 1 | $+ 94 \pm 49$ | $-260$ to $+320$ |
| Ref 3 | $+150 \pm 74$ | $-330$ to $+330$ |
| Ref 5 | $- 26 \pm 44$ | $-320$ to $+230$ |

Table 4. Cluster centre results

| name | measured $\Delta T_{\mathrm{RJ}}$ ($\mu$K) | maximum systematic signal ($\mu$K) |
|---|---|---|
| 0016+16 | $-444 \pm 65$ | $- 30$ to $+120$ |
| 0302+17 | $-442 \pm 109$ | $-330$ to $+330$ |
| Abell 665 | $-301 \pm 49$ | $-130$ to $+ 50$ |
| 1358+62 | $-183 \pm 114$ | $-250$ to $+250$ |
| Abell 2218 | $-354 \pm 43$ | $-120$ to $+ 30$ |

regions limit their size. The sources of systematic error that have been considered here are (a) errors in the telescope pointing; (b) errors in the radio source list used for source corrections; (c) errors in the zero level of brightness temperature; and (d) the discordance in the data from year to year in excess of the random errors. (d) corrects for any underestimation in the random errors through a factor that increases the sizes of the error bars. The results tabulated here incorporate factor (d) into the errors, and specify maximum ranges for residual systematic errors resulting from (a – c). The strictest control on systematic errors *must* always be the comparison of results from independent measurements by different observers on different telescopes using alternative techniques.

Consider first the results for three regions of blank sky, observed as controls on the experiment (Table 3). It is clear that the results are close to zero, and that there is no large negative offset in the measurements (a 'fake' SZ effect caused by systematic errors). The maximum possible systematic errors on these results are large because no useful radio source surveys have been done in these three fields. These are distant controls on the performance of the system — the three fields lie far from any cluster. Closer controls on the systematic errors are also important, since the magnitude and sign of any systematic errors due to spillover may be a sensitive function of the sampling in azimuth and elevation.

A number of clusters of galaxies were observed using this system: the results at the nominal centres of the clusters are collected in Table 4 and supercede the measurements by Birkinshaw (1987) in Table 2. Statistically-significant detections of the SZ effect have been achieved in clusters 0016+16, 0302+17, Abell 665, and Abell 2218, if only the random errors are considered. However, adequate radio source survey data are only available for 0016+16, Abell 665, and Abell 2218, and the systematic error limits are rather large in the other cases. It is important to note that Uson (1985, 1987), working with the NRAO 130-ft telescope (which has a similar beam-width) confirms the detections of 0016+16, Abell 665, and Abell 2218 (Table 2): this provides a further, and stronger, check on the systematic errors than the blank sky data of Table 3.

Although these cluster centre results are of interest, and display highly-significant detections of the SZ effect towards 0016+16, Abell 665, and Abell 2218, the main emphasis of these observations has been on the measurement of the angular structure of the SZ effect. The measurements were made by observing seven points on a NS line through the nominal centre of each of these three clusters: the scan results are displayed in Figure 5. For each of the clusters there is a significant dip in the temperature of the microwave background radiation towards the cluster centre. For Abell 665, this dip is offset $\approx$ 2 arcmin from the cluster centre — a similar offset in the X-ray emission was found independently, after the SZ offset was discovered. The close coincidences between the centres of the X-ray structures and the SZ profiles constitutes strong evidence for the reality of the SZ effect.

## 5. IMPLICATIONS OF THESE RESULTS

### 5.1 The microwave background radiation

The existence of a well-detected SZ effect towards 0016+16, which lies at a redshift 0.541, immediately proves that the microwave background radiation originates at greater redshifts. This is a satisfying confirmation of a result that was already known, on the basis of the lack of a suitable local source for the radiation field, and the absence of small-scale structure.

From Table 2 it appears that not all clusters of galaxies display SZ effects $\Delta T(0) < -1$ mK, but from the detection of three clusters with $\Delta T(0) \approx -0.5$ mK it seems likely that all rich clusters should display the SZ effect at a level $\Delta T(0) \approx -0.2$ mK. The integrated effect from these clusters should produce an SZ confusion level $\approx$ 30 $\mu$K on arcminute angular scales (Rephaeli 1981). This confusion level may complicate the search for arcminute-scale primordial anisotropies in the background radiation. By the same token, the integrated SZ effect of clusters will appear in spectral observations of the background radiation at a level $y \approx 10^{-5}$.

Much has been made at this Workshop of the difficulty of calibrating primordial searches for anisotropy on scales $0°.5$ and greater. Although strong calibrators are available, a reliable low-signal calibrator on the sky would be useful. If the SZ effect exists at a level $\approx -1$ mK in nearby clusters, these clusters can provide such a calibration. The Coma cluster (Sec. 1), although confused by strong radio sources at cm-wavelengths, should appear only as an SZ object with an FWHM about 43 arcmin at mm wavelengths, making it a possible low-signal calibrator for degree-scale experiments.

### 5.2 The intracluster medium

The results for each of the clusters in Table 2 may be studied in terms of the intracluster medium, with the assistance of X-ray data from the *Einstein*, *EXOSAT*, and *Ginga* satellites. In most cases, however, the quality of the information gained is poor — the fractional errors in the SZ data are large, and they contribute little that is new. This is not the case for the three scanned clusters of Fig. 5, where the structural information as well as the central SZ decrement may be used to constrain models of the cluster atmospheres. As a first step, consider fitting the data of Fig. 5 to $\beta$-models centered on the centres of X-ray emission. Since the SZ data effectively constrain only the half-width of the SZ profile, $\beta$ and $\theta_{cx}$ cannot be estimated independently. The fit results in Table 5 therefore give the fitted full-width at half maximum of the SZ decrements, $\theta_{1/2}$, and the corresponding depths of the decrements, $\Delta T(0)$, if $\beta = 0.8$. The errors in Table 5 include an allowance

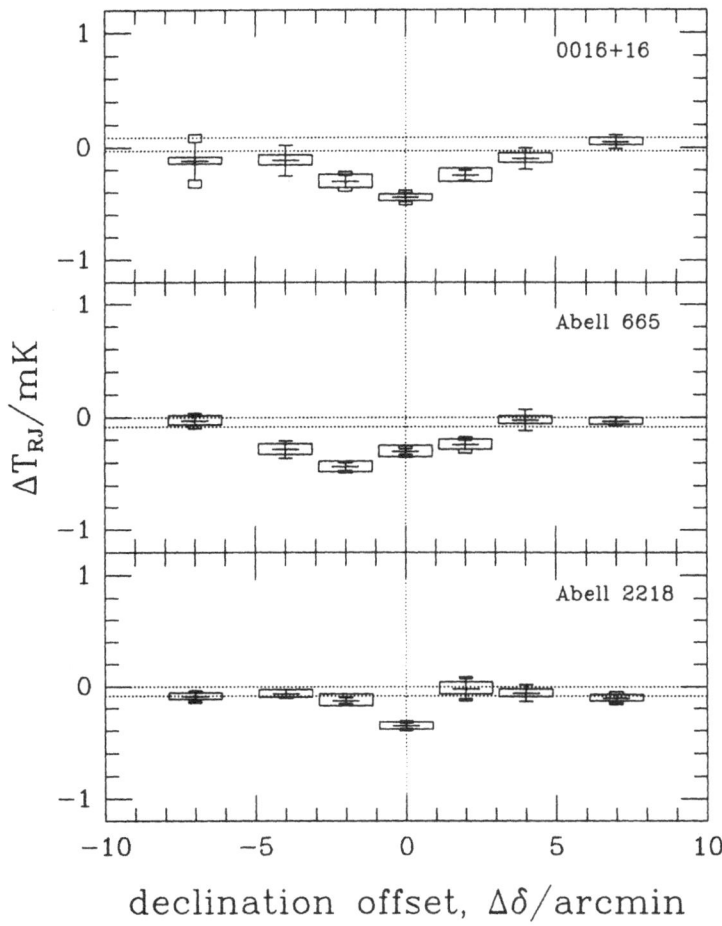

Figure 5. The scan results for 0016+16, Abell 665, and Abell 2218.
The declination offsets of each point are measured with respect to a
nominal centre for the cluster, and the cross with error bars respresents
the scan result with $\pm 1\sigma$ errors. The box around each point indicates
the *maximum* estimated additive systematic error, and the small boxes
at the ends of the error bars indicate the increase in the error suggested
by the year-to-year discordance in the data. The horizontal dotted lines
indicate the systematic error on the overall zero level.

Table 5. Derived parameters for the scanned clusters

| cluster | fitted ($\beta = 0.8$) | | inferred |
| | $\theta_{1/2}$/arcmin | $\Delta T(0)$/mK | $T_{\mathrm{e}}$/keV |
|---|---|---|---|
| 0016+16 | $5 \pm 2$ | $-0.61 \pm 0.14$ | 10 |
| Abell 665 | $9 \pm 3$ | $-0.79 \pm 0.14$ | 9 |
| Abell 2218 | $< 3$ | $-0.55 \pm 0.12$ | 7 |

for residual systematic errors.

Note first that the fitted values of $\theta_{1/2}$ in Table 5 display no simple relationship with the redshift of the cluster. Abell 665 and Abell 2218 have similar redshifts ($\approx 0.18$), and are closer than 0016+16 ($z = 0.54$), but the angular sizes of the SZ effects in Abell 665 and Abell 2218 are distinctly different, and the SZ effect in 0016+16 is of intermediate angular size. This suggests that these three clusters have different gas density or temperature distributions.

Using the fitted values of $\Delta T(0)$ in Table 5, and the X-ray images, weighted mean temperatures of the cluster gas can be obtained. If the Hubble constant is $50 \ \mathrm{km\,s^{-1}\,Mpc^{-1}}$, then the temperatures given in the last column of Table 5 are obtained. These may be compared with the values $T_{\mathrm{e}} > 6$ keV, $T_{\mathrm{e}} = 10$ keV, and $T_{\mathrm{e}} = 6.7$ keV deduced from the X-ray data for 0016+16, Abell 665, and Abell 2218, respectively (White, Silk & Henry, 1981; Hughes & Tanaka, 1989; McHardy et al., 1989).

The accuracy of the SZ effect data must be improved if further information on the properties of the cluster gas is to be obtained. This will be difficult, because the remaining major sources of error are systematic. A better strategy than simply increasing the observing time, and adding more points in scans with the 40-m telescope, would be to image the clusters using a small interferometer. A first attempt should hope to produce maps of the clusters with central signal/noises similar to the original *Einstein* X-ray images, which requires a sensitivity $\approx 30 \ \mu$K in a 20-arcsec diameter synthesized beam. This corresponds to a flux density sensitivity of about 9 $\mu$Jy at 1 cm, which requires long exposures with an array of small ($< 4$ m diameter) antennas.

5.3 The Hubble constant

For each of the well-detected clusters for which good X-ray imaging and spectroscopy exist, the value of the Hubble constant can be deduced by comparing the SZ and X-ray data. A detailed discussion of this process for Abell 665 is given by Birkinshaw et al. (1989). For this cluster the SZ and X-ray structures are consistent with an isothermal $\beta$-model with $\beta = 0.66$ and $\theta_{\mathrm{cx}} = 1.6$ arcmin (the best fit to the X-ray image). The central SZ effect is then $\Delta T(0) = -0.92 \pm 0.16$ mK (including an allowance for systematic errors) and the temperature of the cluster gas, from a *Ginga* spectrum, is $10 \pm 1$ keV (Hughes & Tanaka 1989).

In calculating the Hubble constant, the normalizations of the X-ray and SZ fits are compared (as suggested by equation 3). The central value of the Hubble constant values that is implied by this comparison is about 60 $\mathrm{km\,s^{-1}\,Mpc^{-1}}$, although the permitted range of models, and the residual systematic errors in the SZ data imply errors of about 30 per cent on this result. Values of $H_0$ as large as $100 \ \mathrm{km\,s^{-1}\,Mpc^{-1}}$ are formally permitted only if the SZ and X-ray data contain the maximum estimated systematic errors.

This result depends on the SZ signal and the X-ray flux arising entirely from processes involving an unclumped intracluster medium, and the cluster being at rest in the Hubble flow. Any contribution to the X-ray flux from inverse Compton emission, or to the SZ signal from radio emission or cluster motion, will have a direct systematic effect on this result. The only control on the external precision of this result for $H_0$ is to repeat the calculation for the other well-detected clusters: indeed the agreement between the inferred gas temperature (Table 5) and the measured temperature for Abell 2218 suggests a value of $H_0$ closer to $50\,\mathrm{km\,s^{-1}\,Mpc^{-1}}$ (but see McHardy et al. 1989), whilst only an uninteresting lower limit to $H_0$ can be found from the data for 0016+16.

The accuracy of this method is limited by the assumptions involved in determining the constant of proportionality in equation (8), although a major contribution to the error budget arises from the difficulty of measuring $\Delta T_{\mathrm{RJ}}$ accurately, and with strong constraints on the systematic errors. The problem is that while the X-ray data are sensitive to the highest-density, central, gas in the cluster, the SZ data are sensitive to the low-density, outer, gas. The constant of proportionality in (8) relates the properties of the gas in these regions through a simple model (equation 1), but there is no reason to suppose that such a simple model is an adequate representation of the global properties of the cluster atmosphere. A detailed understanding of cluster atmospheres is necessary if this method of determining $H_0$ is to be useful.

## 6. REFERENCES

Abramopoulos, F. & Ku, W.H.-M., 1983. *Astrophys. J.*, **271**, 446.
Andernach, H., Schallwich, D., Sholomitski, G.B. & Wielebinski, R., 1983. *Astr. Astrophys.*, **124**, 326.
Andernach, H., Schlickeiser, R., Sholomitski, G.B. & Wielebinski, R., 1986. *Astr. Astrophys.*, **169**, 78.
Birkinshaw, M., 1987. NRAO GreenBank Workshop **16**, 261; eds O'Dea, C. & Uson, J.; NRAO GreenBank, WV.
Birkinshaw, M., 1979. *Mon. Not. R. astr. Soc.*, **187**, 847.
Birkinshaw, M. & Gull, S.F., 1984. *Mon. Not. R. astr. Soc.*, **206**, 359.
Birkinshaw, M., Gull, S.F. & Hardebeck, H., 1984. *Nature*, **309**, 34.
Birkinshaw, M., Gull, S.F. & Northover, K.J.E., 1981. *Mon. Not. R. astr. Soc.*, **197**, 571.
Birkinshaw, M., Hughes, J.P & Arnaud, K.A., 1989. In preparation.
Cavaliere, A., Danese, L. & De Zotti, G., 1979. *Astr. Astrophys.*, **75**, 322.
Cavaliere, A. & Fusco-Femiano, R., 1976. *Astr. Astrophys.*, **49**, 137.
Cavaliere, A. & Fusco-Femiano, R., 1978. *Astr. Astrophys.*, **70**, 667.
Chase, S.T., Joseph, R.D., Robertson, N.A. & Ade, P.A.R., 1987. *Mon. Not. R. astr. Soc.*, **225**, 171.
Gull, S.F. & Northover, K.J.E., 1976. *Nature*, **263**, 572.
Gunn, J.E., 1978. In *Observational Cosmology*, 1; eds Maeder, A., Martinet, L. & Tammann, G.; Geneva Obs., Sauverny, Switzerland.
Hughes, J.P., Yamashita, K., Okumura, Y., Tsunemi, H. & Matsuoka, M., 1988. *Astrophys. J.*, **327**, 615.
Hughes, J.P. & Tanaka, Y., 1989. In preparation.
Kogut, A., Bersanelli, M., De Amici, G., Friedmann, S.D., Griffith, M., Grossman, B., Levin, S., Smoot, G.F. & Witebsky, C., 1988. *Astrophys. J.*, **325**, 1 and

94

erratum **332**, 1092.

Lake, G. & Partridge, R.B., 1977. *Nature*, **270**, 502.

Lake, G. & Partridge, R.B., 1980. *Astrophys. J.*, **237**, 378.

Lasenby, A.N. & Davies, R.D., 1983. *Mon. Not. R. astr. Soc.*, **203**, 1137.

Matsumoto, T., Hayakawa, S., Matsuo, H., Murakami, H., Sato, S., Lange, A.E. & Richards, P.L., 1988. *Astrophys. J.*, **329**, 567.

McHardy, I.M., Stewart, G.C., Edge, A.C., Cooke, B.A., Yamashita, K. & Hatsukade, I., 1989. *Mon. Not. R. astr. Soc.*, in press.

Meyer, S.S., Jeffries, A.D. & Weiss, R., 1983. *Astrophys. J.*, **271**, L1.

Moffet, A.T. & Birkinshaw, M., 1989. *Astr. J.*, in press.

Mushotzky, R.F. & Smith, B.W., 1980. *Highl. Astr.*, **5**, 735.

Parijskij, Yu.N., 1972. *Astr. Zhurn.*, **49**, 1322.

Partridge, R.B., Perley, R.A., Mandolesi, N. & Delpino, F., 1987. *Astrophys. J.*, **317**, 112.

Perrenod, S.C. & Lada, C.J., 1979. *Astrophys. J.*, **234**, L173.

Radford, S.J.E., Boynton, P.E., Ulich, B.L., Partridge, R.B., Schommer, R.A., Stark, A.A., Wilson, R.W. & Murray, S.S., 1986. *Astrophys. J.*, **300**, 159.

Rephaeli, Y., 1981. *Astrophys. J.*, **245**, 351.

Rudnick, L., 1978. *Astrophys. J.*, **223**, 37.

Saunders, R.D.E., 1988. In *The Post-Recombination Universe*, 187; eds. Kaiser, N. & Lasenby, A.N.; Kluwer, Dordrecht.

Schallwich, D., 1979. Poster presented at IAU Symposium **97**, *Extragalactic Radio Sources*, Albuquerque, New Mexico.

Silk, J.I. & White, S.D.M., 1978. *Astrophys. J.*, **226**, L103.

Sunyaev, R.A. & Zel'dovich Ya.B., 1972. *Comm. Astrophys. Sp. Phys.*, **4**, 173.

Tarter, J.C., 1978. *Astrophys. J.*, **220**, 749.

Uson, J., 1985. *Observational and Theoretical Aspects of Relativistic Astrophysics and Cosmology*, 269; eds Sanz, J.L. & Goicoechea, L.J.; World Scientific Publ. Co.

Uson, J., 1987. NRAO GreenBank Workshop **16**, 255; eds O'Dea, C. & Uson, J.; NRAO GreenBank, WV.

White, S.D.M., Silk, J.I. & Henry, J.P., 1981. *Astrophys. J.*, **251**, L65.

# EXPERIMENTAL METHODS FOR THE INVESTIGATION OF THE LARGE SCALE CBR ANISOTROPY

I.A. Strukov
Space Research Institute of Academy of Science USSR
Moscow

We review shortly the recent state of large scale anisotropy searches; the advantages of satellite-borne experiment for studying the large-scale anisotropy in comparison with ground-based and balloon experiments; the state of RELICT-2 project.

## 1. INTRODUCTION

One of the most outstanding problems of modern cosmology is the understanding of basic physical processes occurring in the Universe at high redshifts ($z > 5$). Recent investigations by means of large optical and radiotelescopes allowed studying the universe structure at $z$=0.01-3. Up to now there is no evidence of presence of light matter in such form as stars, galaxies, quasars at $z > 5$. The detecting of the cosmic background radiation by Penzias and Wilson in 1965 gave us new possibility to study the early universe. Cosmic background radiation gives us an information about processes that have occurred in 700,000 years after the Big Bang. Fluctuations of the microwave background are the unique tool for studying the matter distribution and gravitational fields from the epoch of recombination ($z \sim 1,000$) up to now. Searches for fluctuations have the status of a fundamental experiment

95

*N. Mandolesi and N. Vittorio (eds.), The Cosmic Microwave Background: 25 Years Later, 95–113.*
© 1990 *Kluwer Academic Publishers.*

in cosmology because the can provide us with the information on: a) very early stages of the universe evolution; b) properties of the matter filling the universe (including the dark matter); c) the epoch of galaxy formation.

It is generally accepted to divide angular fluctuations of the microwave background into three ranges: small scales (arc minutes), intermediate scales (30'-5°) and large scales (larger than 5°). The division is due to difference in physical problems which could be solved by measurements carried out in a particular range. Observations of the large scale anisotropy of cosmic background radiation are important for number of other reasons as well:

1) Recently observed CMB isotropy on a scale larger than about 10° (Fixen et al., 1983; Lubin et al., 1983; Strukov and Skulachev, 1988) is the right confirmation of existence of inflation stage in evolution of universe, and the principle justification for the acceptance of the Robertson-Walker metric and friedmannian world models. This observation together with small-scale anisotropy, discovery of superclusters and voids (Gregory and Thompson, 1978; Tarenghi et al., 1979; Kirshner et al., 1981) and indication of large scale streaming motions has spurred great activity in the theoretical searches history of universe (Rubin et al., 1976; Collins et al., 1986; Burstein et al., 1986; Wilson and Silk, 1981; Bond and Szalay, 1983; Starobinski , 1983; Wilson, 1983; Abbott and Wise, 1984; Bond and Efstathiou, 1984; Doroshkevich and Kholopov, 1984; Vittorio and Silk, 1984; Sazhin, 1985; Kofman and Linder, 1986; Kofman et al., 1986; Bond and Efstathiou, 1987; Juszkiewicz et al., 1987; Peebles, 1987; White et al., 1987).

2) In case of gravitational waves Sachs and Wolfe effect initiates CMB-anisotropy with the amplitude that is proportional to the gravitational wave amplitude (Starobinski, 1985 and reference therein).

A source of gravitational waves are quantum fluctuations and current limits on the amplitude of large scale anisotropy of CMB have allowed to set significant constraint of the energy density of false vacuum. This study is a direct link between the physical processes that gave rise to the fluctuations and the large scale structure observed today.

Fluctuations due to unresolved non-uniform space distributed discrete sources may seriously hinder search for anisotropies on the intermediate and small scale.

This problem was first discussed by Longair and Sunyaev, 1971. A more recent discussion of this problem have been presented by Danese et al., 1983, Korolev and Sazhin, 1984, Franceschini et al., 1989, de Zotti et al., 1989.

Only within a relatively narrow frequency range from 20 GHz to 200 GHz the contribution of sources to anisotropies on scale $< 1°$ is expected to be $\Delta T/T \leq 10^{-5}$. Effect of clustering may strongly enhance fluctuations at least on same angular scales.

Of course our galaxy emission may seriously hinder search for the large scale anisotropy, but our preliminary calculations give us the hope that the many frequency experiment with very high-sensitivity equipment allow to receive the constraints for all recent cosmological theoretical models.

We review shortly the recent state of large scale anisotropy searches; the advantages of satellite-borne experiment for studying the large-scale anisotropy in comparison with ground-based and balloon experiments; the state of RELICT-2 project.

## 2. MEASUREMENTS OF THE DIPOLE COMPONENT

In spite of great efforts and many years searches for anisotropy of CMB today is well established only the dipole component. That is caused by Doppler-shift due to the motion of the solar system with respect to a frame in which the background would appear isotropic. Peebles was first who predicted that phenomenon. The first measurements of the large-scale anisotropy were undertaken by Berkeley and Princeton groups. Results of measurements of dipole component in different frequency bands are listed in the Table 1.

The dipole moment has weighted mean $T_{therm.}=3.25\pm0.07$ mK and direction $\alpha=11.32^h \pm 0.09^h$, $\delta = -6.4° \pm 0.9°$. Weighted mean for components are marked by $*$.

All recent results including brilliant observations made by groups of american scientists by means of balloons at 19 GHz and represented at "The Cosmic Microwave Background: 25 years later", 1989, L'Aquila, are in excellent agreement with each other.

Table 1:

| | f GHz | T$_{CMB}$ mK thermo. | R.A. hr | $\delta$ S° |
|---|---|---|---|---|
| Partridge, Wilkinson, 1967 | 9.37 | 0.9±2.5 | - | - |
| Conclin, 1969 | 8.0 | 1.6±0.8 | - | - |
| Boughn, Fram, Partridge, 1971 | 34.9 | 7.5±11 | - | - |
| Smoot, Lubin, 1979 | 33.0 | 3.1±0.4* | 11.4±0.4* | 9.6±6 |
| Boughn, Cheng Wilkinson, 1981 | 19.0 | 3±0.8* | 12±1.3 | -18±18 |
| " | 24.8 | 4±0.3 | 11.6±0.3* | -2±3* |
| " | 31.6 | 3.7±0.3 | 11±0.3* | -8±4* |
| " | 46.0 | 3.9±1 | 11.6±0.9 | -12±14 |
| Melchiorri, 1982 | 300 | 3.5±0.5* | - | - |
| Fixsen, Cheng, Wilkinson, 1983 | 24.6 | 3.18±0.17* | 11.2±0.1* | -8±2* |
| Lubin, Epstein, Smoot, 1983 | 90.0 | 3.46±0.17* | 11.3±0.1* | -6±1.4* |
| Strukov, Skulachev, 1984 | 37 | 3.16±0.12* | 11.3±0.16* | -7.5±2.5* |
| Halpern et al., 1988 | 40-240 | 3.4±0.42* | 12.1±0.24* | -23±5 |
| " | 150-540 | 4.7±1.4 | $9,9^{+1.7}_{-1.1}$ | -38±21 |
| Bernstein et al., 1989 | 75-500 | 2.6±0.46* | - | - |
| Weighted mean: | | 3.25±0.07 | 11.32±0.09 | -6.4±0.9 |
| Boughn et al. | 19 | 3.28 | 11.3 | -6.6 |

## 3. CONSTRAINTS ON ANISOTROPY

Consider the possibility of improving of methods of data reduction in satellite-based experiment. The satellite-borne experiment has following advantages in comparison with ground-based and balloon experiments:

1) The possibility to determine the level of instrument noise with a high accuracy and hence to use the method of variance analysis that gives the most stringent constraints on multipole components.

2) The possibility to use the radiation cooling.

3) The possibility to remove systematical errors due to the thermal emission of the Earth and its atmosphere and the Moon's thermal emission.

4) Essential increase of the total time to actual measurements.

5) Possibility of determining of variation of radiometer noise temperature with very high accuracy.

There are two possibilities in determination of variation of the radiometer noise temperature:

a) periodical sample - when temporal interval between two measurements is equal one period of satellite rotation. In this case in each measurement the horn looks at the same point. So sample gives information about initial variance of noise temperature.

b) sample ortogonalization methods. In this case the ortogonal basis is formed. In this basis some of samples of full scan receive the positive sign and the other ones receive the negative sign. If number of positive and negative signs are equal then this basis forms the population of samples without signal. If number of measurements of the brightness temperature of the same region on the sky is more 10, then the difference between sample variance mean and population variance will be less than 1%.

For constraining the amplitude of primordial density fluctuations usually are adopted the following methods: spectral analysis, correlation analysis and variance analysis.

Spectral and correlation analysis were the most employed up to now (Fixsen *et al.*, 1983; Lubin *et al.*, 1983; Lubin and Villela, 1986). But the possibility of determining the variance of instrumental noise $\sigma^2$ with high accuracy gives the advantage to variance analysis (Strukov and Skulachev, 1988).

It is necessary to stress that the capability of any method increases if we from the beginning use a priori knowledge of spectral response of primordial fluctuations. The model independent estimations always give the less stringent constraints of amplitude of cosmic background fluctuations.

In any analysis method for improving the accuracy it should be used an information about the spectral response of one cosmological model, because in comparing experimental and theoretical data always is used the concrete model.

Most stringent constraints on the quadrupole component have been provided in balloon experiments by Berkeley and Princeton groups (Lubin et al., 1983; Lubin, 1986; Fixsen et al., 1983). The quadrupole limits of both data sets are $7 \cdot 10^{-5}$ at 90% confidence level upper limit.

Fig. 1, and Tables 2,3 show the advantages of the variance analysis in comparison with the spectral one. For the variance analysis it is necessary to pay attention to

a) difference between spectrum responses of primordial density fluctuations and of instrument noise. In fact if we have the model with inflation spectrum then (Abbott and Wise, 1984)

$$< Q_l^2 > = \frac{2\pi^2 \varepsilon_H^2 (2l + 1)}{l(l + 1)}$$

and

$$Var < Q_l^2 > = \frac{2 < Q_l^2 >^2}{2l + 1}$$

where $\varepsilon_H$ - amplitude of density fluctuations at the horizon crossing. Then for N spherical harmonics the sample size (number of degrees of freedom) is

$$n_1 = \frac{\sum_{l=2}^{N} < Q_l^2 >^2}{\sum < Q_l^2 >^2 / (2l + 1)} = \frac{[\sum(2l + 1)/l(l + 1)]^2}{\sum(2l + 1)/l^2(l + 1)^2}$$

Then for system with angular resolution 7° $n_1$=68 and size of sample for instrumental noise is equal 1,000;

b) observational sample may be not uniform: different sample units have different statistical weight;

c) cosmological signal samples and (sometimes) instrumental noise have non-gaussian statistic.

For the correct determination of upper limit on (or value of) cosmological signal, it is necessary to keep in mind the above-mentioned features. If we have the uniform observational sample with the Gaussian statistic then

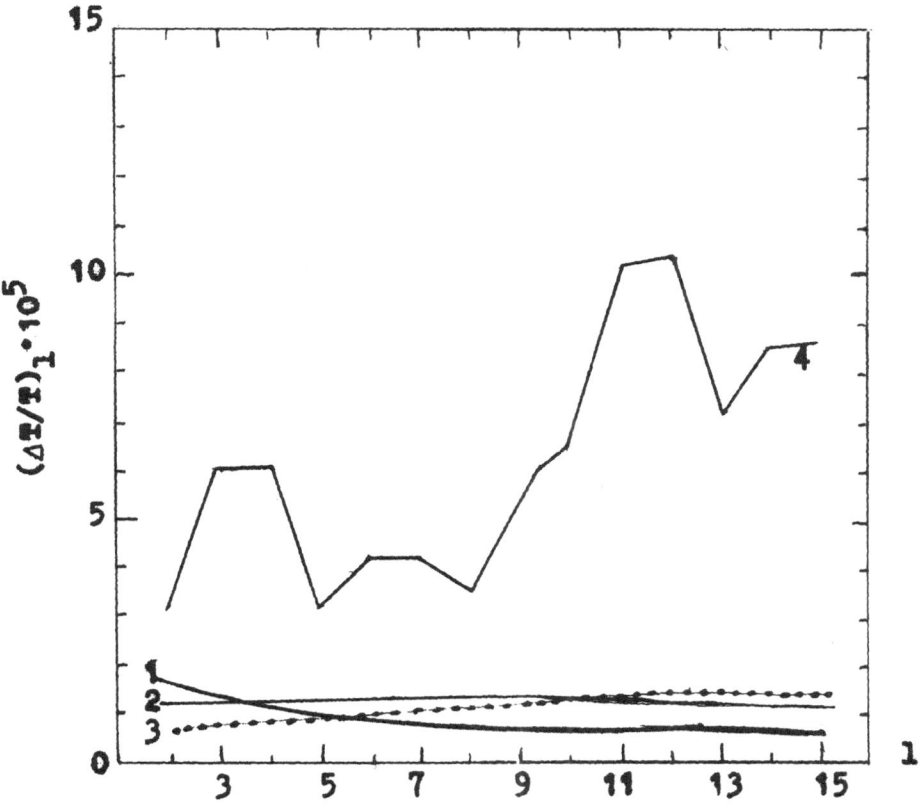

*Fig. 1*: Upper limit on the spectral components. 1, 2, 3 - variance analysis; 4 - spectral analyisis.

Table 2:

| $(\Delta T/T)_l^2$ | $(\Delta\rho/\rho)_k^2$ | $(\Delta T/T)_2$ | $(\sum_2^{25}(\Delta T/T)_l^2)^{1/2}$ | Adv. variance over spectral |
|---|---|---|---|---|
| (1) $\frac{2l+1}{l(l+1)}$ | $\sim K^1$ | $< 1.6 \cdot 10^{-5}$ | $< 4 \cdot 10^{-5}$ | 6.4 |
| (2) const | $\sim K^2$ | $< 1.2 \cdot 10^{-5}$ | $< 5.7 \cdot 10^{-5}$ | 5.0 |
| (3) $2l+1$ | $\sim K^4$ | $< 0.6 \cdot 10^{-5}$ | $< 7 \cdot 10^{-5}$ | 5.4 |

usual form for 95% confidence level is

$$\sigma_{u.l.} = 1.5\sigma n^{-1/4}$$

where $n$-sample size; $\sigma$-r.m.s. of instrumental noise.

In case sample size for cosmological signal $(n_1)$ is not equal to sample size of instrument noise then we have

$$\sigma_{u.l.} = 1.5\sigma n^{-1/4}(n/n_1)^{1/2} = 1.5\sigma n^{1/4}n_1^{1/2}$$

For the non-uniform sample in the points with larger statistical weight there is some increasing of cosmological signal. So we have some changes of cosmological signal response and for the correct estimation of the upper limit it is necessary to know transmission response $K(l)$ that depends from the weighting process

$$< Q_l^2 >_{\text{output}} = K_l^2 < Q_l^2 >_{\text{input}}$$

Most precise solution is given by Monte-Carlo simulations. If we know population variance of instrumental noise accuracy, then by increasing the amplitude of cosmological signal at system input we obtain for 95 cases over 100 that amplitude of the output signal is larger than the measured value.

For the rather large population size (about 1,000 theoretical samples) we obtain constraints on the population variance of the cosmological signal, hence constraints on its spectral component, see Fig. 2. In Table 4 are listed constraints on amplitude $\varepsilon_H$ for different methods. The most stringent constraint we obtain for 24°x24° weighted averaging:

$$S^2 = \frac{\sum \Delta T_i^2 N_i^2}{\sum N_i^2},$$

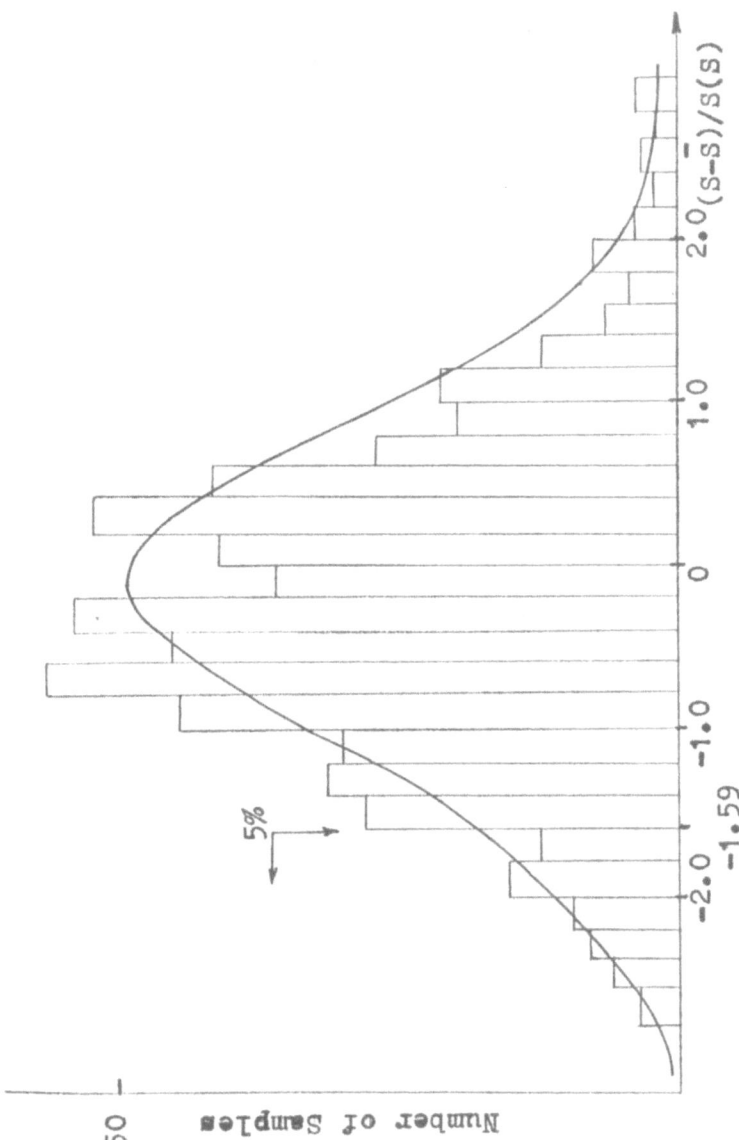

*Fig. 2:* Histogram of the probability density. R.m.s. receiver noise, S - for 621 samples of noise. The superimposed curve is Gaussian. $\chi^2(25)=29.84$. Sample mean $\bar{S} = 1.008 \cdot 10^{-4}$ K; $S(S) = 1.4 \cdot 10^{-5}$ K; $S_{obs} = 9.2 \cdot 10^{-5}$ K.

Table 3:

| Number of harmonic | 2 | 3 | 4 | 5 | 6 | 7 | 8 | 9 | 10 |
|---|---|---|---|---|---|---|---|---|---|
| Measured $(\Delta T/T)_l \cdot 10^5$ | 1 | 3 | 3.1 | 2 | 2.9 | 2.5 | 2.4 | 1.8 | 2.7 |
| Upper limit $(\Delta T/T)_l \cdot 10^5$ | 3 | 7 | 5.9 | 4.4 | 5.8 | 3.6 | 5.1 | 5.5 | 5.9 |
| Upper limit $\varepsilon_H \cdot 10^4$ | 0.7 | 2.0 | 1.9 | 1.6 | 2.2 | 1.5 | 2.3 | 2.6 | 2.9 |
| Upper limit from variance analysis $(\Delta T/T)_l \cdot 10^5$ | 1.6 | 1.12 | .86 | .7 | .59 | .51 | .45 | .4 | .37 |

where $N_i$ - number of measurement included in $i$-square.

Method of correlation analysis for detection of large scale anisotropy was more developed by Scaramella and Vittorio, 1988. Lubin and Villela, 1986, unified the results of both balloon experiments. An autocorrelation of the residual map shows no obvious structure from 10° to 180° giving r.m.s. $V$ 0.01 mK$^2$.

We also obtained a limit on the autocorrelation function for the anisotropy $< \Delta T_1 \Delta T_2 > \leq 0.005$ mK$^2$.

For the correct estimation of the amplitude of cosmological fluctuations (in case of non-uniform observational samples) both in correlation analysis and in variance analysis it is necessary to keep in mind transmission response $K(l)$.

COBE-mission (launched in November 1989) will give new possibilities in study of large scale anisotropy. The results of data analysis will be the best sky maps ever made at frequencies 31, 52.6 and 91 GHz with an uncertainty of only 0.0003 K at 9.6 mm and 0.00015 K at 5.7 and 3.3 mm, for each 7° pixel.

Table 4:

| $<\sigma>$ Population standard deviation (mK) | $S_1$ observed standard deviation (mK) | S(S) (mK) | $n_1$ | $N$ Number of simul. samples | Dimen. of square | 95% upper limit $\varepsilon_H \cdot 10^5$ | Type of weight *) |
|---|---|---|---|---|---|---|---|
| .214 | .2 | .015 | 102 | 102 | 6x6 | 3.96 | 1 |
| .38 | .39 | .024 | 468 | 102 | 6x6 | 7.08 | 2 |
| .11 | .9 | .014 | 26 | 621 | 24x24 | 3.48 | 1 |
| .14 | .15 | .014 | 52 | 621 | 24x24 | 6.3 | 2 |
| .188 | .181 | .011 | 146 | 200 | 7°beam | 3.77 | 2 |
| .17 | .17 | .011 | 119 | 600 | 7°beam | 5.1 | 2 |
| .294 | .288 | .0118 | 310 | 800 | | 4.5 | 2 |

*) 1: $S^2 = \frac{\sum \Delta T_i^2 N_i^2}{\sum N_i^2}$

*) 2: $S^2 = \frac{\sum \Delta T_i^2 N_i^2}{\sum N_i}$

## 4. THE STATE OF RELICT-2 PROJECT

From the treatment of results of recent observations of large-scale cosmic background radiation we conclude that the main difficulties in the study of CMB anisotropy are the following:      1) In spite of the high apogee of the orbit of the RELICT-1 satellite and of the rather low level of antenna side-lobes, we observed the existence of large contributions fromm the Moon and Earth thermal emission to the antenna temperature. This gives the reduction of number of measurements fit for analysis two times. So we must change the satellite orbit for successful experiment in which we use 10 times more sensitive equipment than it was used in our first experiment.

2) Intense emission of galaxy was observed between latitudes $\pm 5$ and longitudes $50°$ and $250°$. This radiation increases the variance in the background radiation by 0.5 mK. The main part of this radiation may be explained by the emission from giant HII regions (Strukov and Skulachev, 1988). Besides Halpern *et al.*, 1988 reported measurements of the dust emission in the two frequency bands from 1.25 to 8 .3 mm and from 0.5 to 2.0 mm at high galaxy latitudes.

That causes the necessity for separating the cosmic background anisotropy from the anisotropy produced by galaxy emission. This is the main task at last stage of experiment.

For decreasing the contamination of the thermal Moon's emission in RELICT-2 project will be used special orbit near Lagrange-2 point as shown in Fig. 3. Maximal distance between L2 and the satellite is about 500,000 *km*. But american scientists (NASA) proposed to use an interaction between Moon and satellite for decreasing that distance. Success in this direction will give the most optimal orbit for study of large scale CMB anisotropy.

We propose the 5-frequency band (from 22 to 189 GHz) experiment for separation of anisotropies of different nature. Instrumental complex RELICT-2 project consists of 5 different radiometers. Each receiver will have two antennae with $7°$ beam width. One horn is pointed along the satellite's axis of rotation, away from the Sun, and the other one - the measuring antenna - collects the incident radiation from direction perpendicular to the satellite's axis of rotation.

Three radiometers will be cooled by radiational loss of heat into space.

The 22 GHz radiometer is designed (in collaboration with Bulgarian scientists) with HEMT-amplifier at the input. Its sensitivity is 10 mK$\cdot$s$^{-1/2}$.

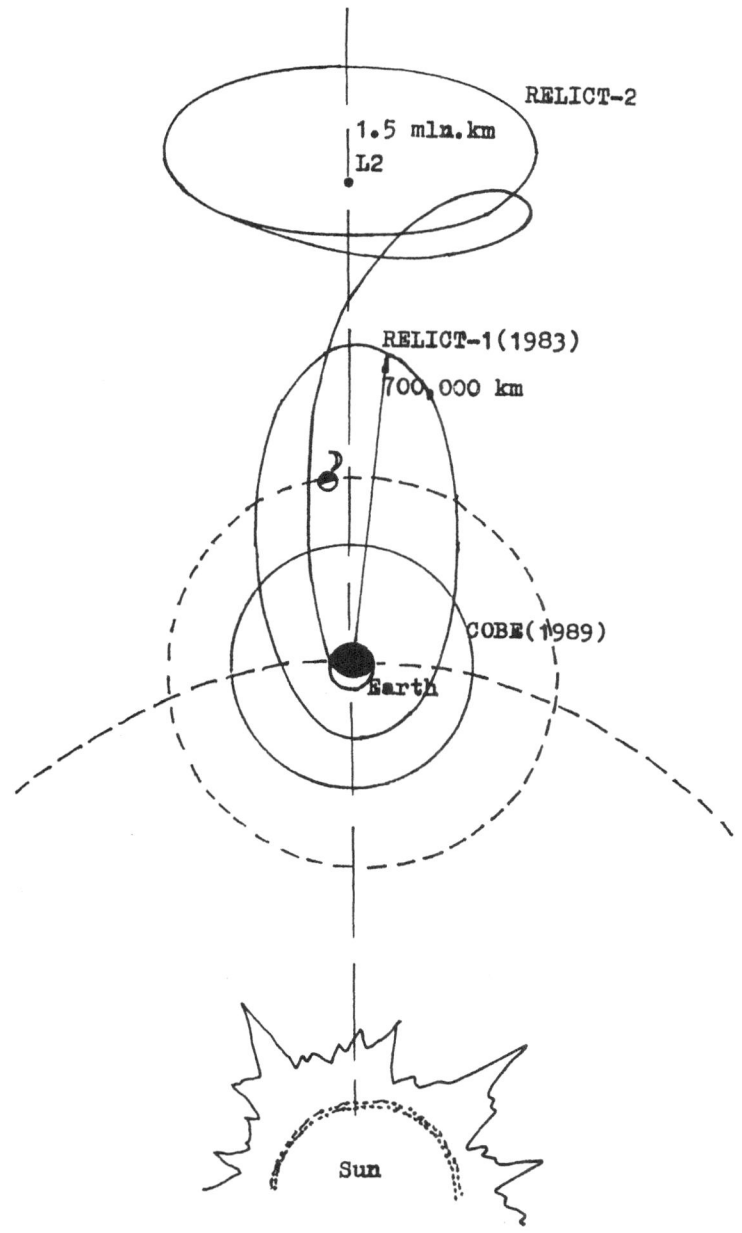

*Fig. 3:* RELICT-2 orbit.

At 34 GHz will be used a Dike-radiometer with degenerate parametric amplifier at the input. The scheme of the 34 GHz radiometer is shown in Fig. 4.

At present our group has designed and tested these radiometers. The radiometer cooled to 77 K has an equivalent noise temperature of 60 K (double side band averaged over 1.7 GHz). The measured $\Delta T$ is less than 3 mK$\cdot$s$^{-1/2}$.

By means of laboratory piece of room temperature radiometer were made measurements of intermediate scale anisotropy with an angular resolution 3.5° and sensitivity 10 mK$\cdot$s$^{-1/2}$ (Bryuchanov et al., 1990, in press). The scheme of the radiometers with its parameters are shown in Fig. 4.

Our preliminary calculations showed the importance of improving the receiver sensitivity at 60 GHz. At present we have designed and tested several types of radiometers. Schematics are shown in Fig. 5,6. In the radiometer shown in Fig. 5,6 as a switch we use a magnetic rotary polarizer, that brings in additional losses and constraints the frequency band. That scheme is not promising for improving the sensitivity. A two-input radiometer is more promising, see Fig. 7. This room-temperature radiometer has sensitivity 15 mK$\cdot$s$^{-1/2}$ and we hope that by means of this type of radiometer will be possible to realize sensitivity 3-5 mK$\cdot$s$^{-1/2}$ by using of HEMT intermediate frequency amplifier. This scheme of radiometer was tested at 3 mm.

At present is performed the vibration testing of the thermal model of RELICT-2 complex

## 5. CONCLUSIONS

In the 25 years since the discovery of the microwave background radiation no intrinsic structure has been convincingly measured. At present only the dipole anisotropy has been measured sufficiently precise.

The separation of anisotropy caused by galaxy emission and cosmological background anisotropy is the most important task in last stage of experiment.

Technical possibilities promise near order of magnitude increasing in system sensitivity and realizing of multi-frequency satellite-borne mission allows to receive the constraints for all recent cosmological theoretical models.

I should like to thank N. Mandolesi and N. Vittorio, organizers of that school, and the other participants in that school, J. Bond, D. Wilkinson, G.

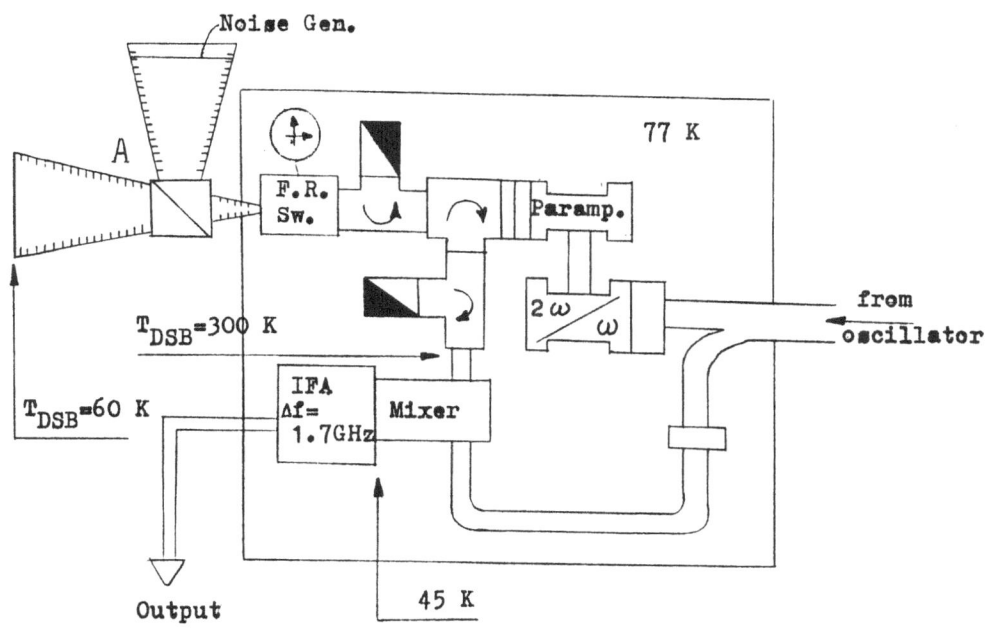

Noise Gen.

A

77 K

F.R.
Sw.

Paramp.

$T_{DSB}=300$ K

$2\omega$ / $\omega$

from
oscillator

$T_{DSB}=60$ K

IFA
$\Delta f=$
1.7GHz

Mixer

Output

45 K

| | |
|---|---|
| $T_o = 77$ K | $T_o = 293$ K |
| $f_o = 36$ GHz | $f_o = 36$ GHz |
| $\Delta f = 1.7$ GHz | $\Delta f = 1.7$ GHz |
| $T_{DSB\,\Sigma} = 60$ K | $T_{DSB\,\Sigma} = 200$ K |
| $L_{A+F.r.sw.} = 0.5$ dB | $L_{A+F.r.sw.} = 0.7$ dB |
| $L_{circ.} = 0.3\text{x}2$ dB | $L_{circ.} = 0.25\text{x}2$ dB |
| $T_{mix.+IFA} = 300$ K (DSB) | $T_{mix.+IFA} = 700$ K (DSB) |
| R.m.s. $\angle$ 3 mK$\cdot$s$^{1/2}$ | R.m.s. $\approx$ 8 mK$\cdot$s$^{1/2}$ |

Fig. 4: Schematic of 34 GHz radiometer.

110

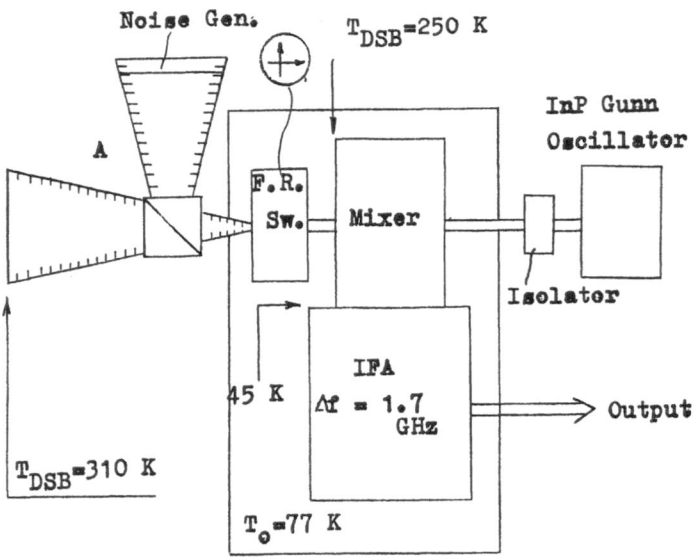

Fig. 5 : Schematics of 60 GHz radiometer.

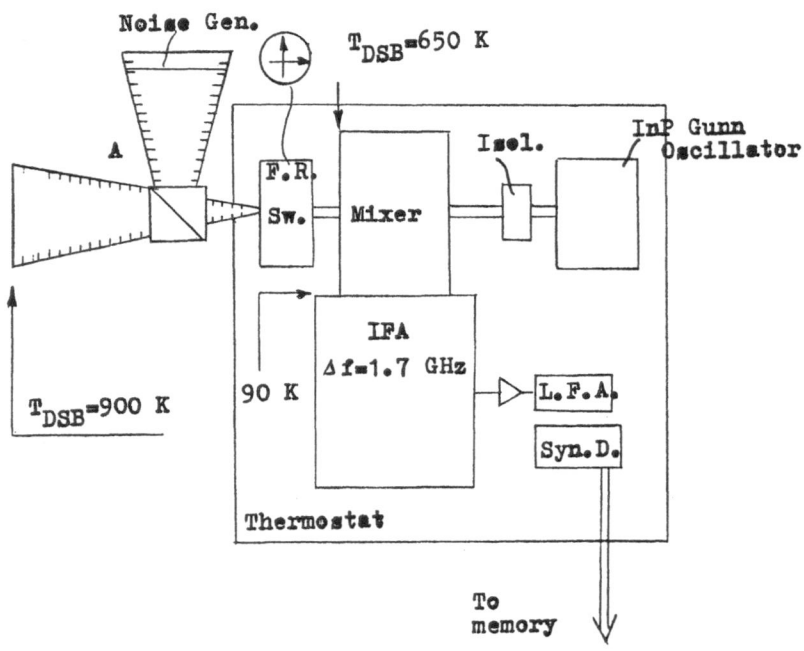

Fig. 6: Schematics of 60 GHz radiometer.

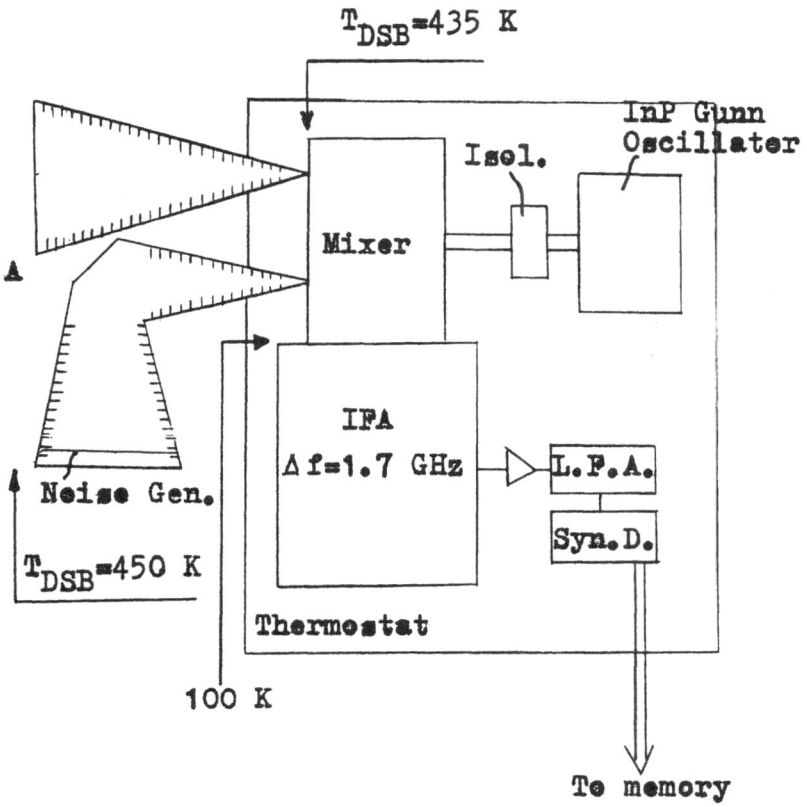

$T_{DSB}$=435 K

InP Gunn Oscillater

Isol.

Mixer

A

IFA
Δf=1.7 GHz

L.P.A.

Syn.D.

Noise Gen.

$T_{DSB}$=450 K

100 K

Thermostat

To memory

Fig. 7: Schematic of two-input radiometer.

de Zotti, P. Lubin, G. Smoot, de Bernardis, S. Boughn, F. Melchiorri, R. Scaramella, for many useful discussions about the microwave background radiation. This work would not have been possible without the help of number of my collegues in developing the microwave receivers and theirs components and in calculations, particularly A. Kosob, V. Korogod, N. Budilovich, D. Skulachev, Yu. Nemlicher, A. Rukavicin, A. Klypin, A. Tkachev and other. I should like especially R. Syunyaev and A. Starobinski for the permanent support of RELICT-2 project.

## 6. REFERENCES

Abbott, L.F. and Wise, M.B. 1984 *Astrophys. J. Letters* **282** L47.

Abbott, L.F. and Wise, M.B. 1984 *Phys. Letters* **135B** 279.

Bond, J.R. and Szalay, A.S. 1983 *Astrophys. J.* **276** 443.

Bond, J.R. and Efstathiou, G. 1984 *Astrophys. J. Letters* **285** L45.

Bond, J.R. and Efstathiou, G. 1987 *M.N.R.A.S.* **226** 655.

Bryukhanov, A. *et al.,* 1990 in press.

Burstein, D., Davies, R.L., Pressler, A., Faber, S., Terlevich, R.J. and Wenger, G., 1986, in *Galaxy Distances and Deviations from Universal Expansion*, B.F. Madore and R.B. Tully eds. (Dordrecht: Reidel), p.123.

Collins, C.A., Joseph, R.D. and Robertson, N.A., 1986 *Nature* **320** 506.

Danese, L.G., de Zotti, G. and Mandolesi, N., 1983 *Astron. Astrophys.* **121** 114.

De Zotti, G., Danese, L., Toffolatti, L. and Franceschini, A. 1989, in *Highlights of Astronomy* **8** in press.

De Zotti, G. and Toffolatti, L. 1989 in *Highlights of Astronomy* **8** in press.

Doroshkevich, A.G. and Khlopov, M.Yu. 1984 *M.N.R.A.S.* **211** 277.

Fixsen, D.J., Cheng, E.S. and Wilkinson, D.T. 1983 *Physics Rev. Lett.* **50** 620.

Gregory, S.A. and Thompson, L.A. 1978, *Astrophys. J.* **222** 784.

Halpern, M., Benford, R., Meyer, S., Muehlner, D. and Weiss, R., 1988 *Astrophys. J.* **332** 596.

Juszkiewicz, R., Gorski, K. and Silk, J. 1987 *Astrophys. J. Letters* **323** L1.

Kirshner, R.P., Oemler, A., Schechter, P.L. and Schechtman, S.A. 1981, *Astrophys. J. Letters* **248** L57.

Kofman, L.A. and Linde, A.D. 1986, *Phys. Letters* **174** 400.

Kofman, L.A., Pogosyan, D.Yu. and Starobinski, A.A. 1986 *Sov. Astron. Lett.* **12** No. 2, 419.

Korolev, V.A. and Sazhin, M.V. 1984, *Astron. Zh.* **61** 1054 (*Sov. Astron.*, **28** No. 6, 616).

Longair, M.S. and Syunyaev, R.A. *Nature* **233** 719.

Lubin, P.M., Epstein, G.L. and Smoot, G.F. 1983 *Physics Rev. Letters* **50** 616.

Lubin, P. and Villela, T. 1986 in *Galaxy Distances and deviations from Universal Expansion* B.F. Madore and R.B. Tully eds. (Dordrecht: Reidel), p.169.

Peebles, P.J.E. 1987 *Nature* **327** 210.

Rubin, V.G., Thonnard, N., Ford, W.K. and Robert, M.S. 1976 *Astrophys. J.* **81** 719.

Sazhin, M.V. 1985 *M.N.R.A.S.* **216** 25.

Scaramella, R. and Vittorio, N. 1988 *Astrophys. J. Letters* **331** L53.

Starobinski, A.A. 1983, *Sov. Astron.* **9** No.5 302.

Starobinski, A.A. 1985, *Sov. Astron. Letters* **11** 133.

Strukov, I.A. and Skulachev, D.P. 1988, SSR/S.E., *Astrophys. Spa. Physics Rev.* **6** part 2. Cosmology.

Tarenghi, M., Tifft, V.G., Chincarini, B., Rood, H.J. and Thompson, L.A. 1979 *Astrophys. J.* **234** 793.

Vittorio, N. and Silk, J. 1984 *Astrophys. J. Letters* **285** L39.

White, S.D.M., Frenk, C.S., Davies, M. and Efstathiou, G. 1987, *Astrophys. J.* **313** 505.

Wilson, M.L. and Silk,J. 1981 *Astrophys. J.* **243** 14.

Wilson, M.L., 1983 *Astrophys. J.* **273** 2.

Franceschini, A., Toffolatti, L., Danese, L., and de Zotti, G. 1989 *Astrophys. J.* in press.

# MEDIUM SCALE MEASUREMENTS OF THE COSMIC MICROWAVE BACKGROUND AT 3.3 mm

Philip M. Lubin
Peter R. Meinhold †
Alfredo O. Chingcuanco ‡

Physics Department
University of California
Santa Barbara, CA 93106

We have developed a system for making measurements of spatial fluctuations in the Cosmic Microwave Background Radiation at 3 mm wavelength, on an angular scale of .5 to 5 degrees. The system includes a telescope with a Gaussian beam with full width at half maximum (FWHM) of 20 to 50 arc-minutes, a Superconductor-Insulator-Superconductor (SIS) coherent receiver operating at 90 GHz, and for balloon flights, a pointing system capable of 1 arc-minute RMS stabilization. We report on results from the first flight of the stabilized platform, as well as results from ground based measurements made from the South Pole station in December, 1988.

## INTRODUCTION

Searches for structure in the spatial distribution of the Cosmic Background Radiation (CBR) are one of the few experimental tests of cosmological models. Currently no definitive detections of anisotropy have been made except for the dipole term, and limits of 20 to 200 parts per million have been established from 10 arc-seconds to 90 degrees angular scale (see figure 1). In the region from 1 to 10 degrees few experiments have been done with sufficient sensitivity to seriously constrain cosmological models, galaxy formation scenarios in particular. Recent reports of detection in this region are suggestive but may suffer from systematic and galactic emission subtraction problems.

At the largest scale (180 degrees) measurements of the doppler shift dipole anisotropy produced by our peculiar motion relative to the frame of the CBR are

---

† also, Physics Dept., UC Berkeley
‡ also, M.E. Dept., UC Berkeley

*N. Mandolesi and N. Vittorio (eds.), The Cosmic Microwave Background: 25 Years Later, 115–139.*
© 1990 *Kluwer Academic Publishers.*

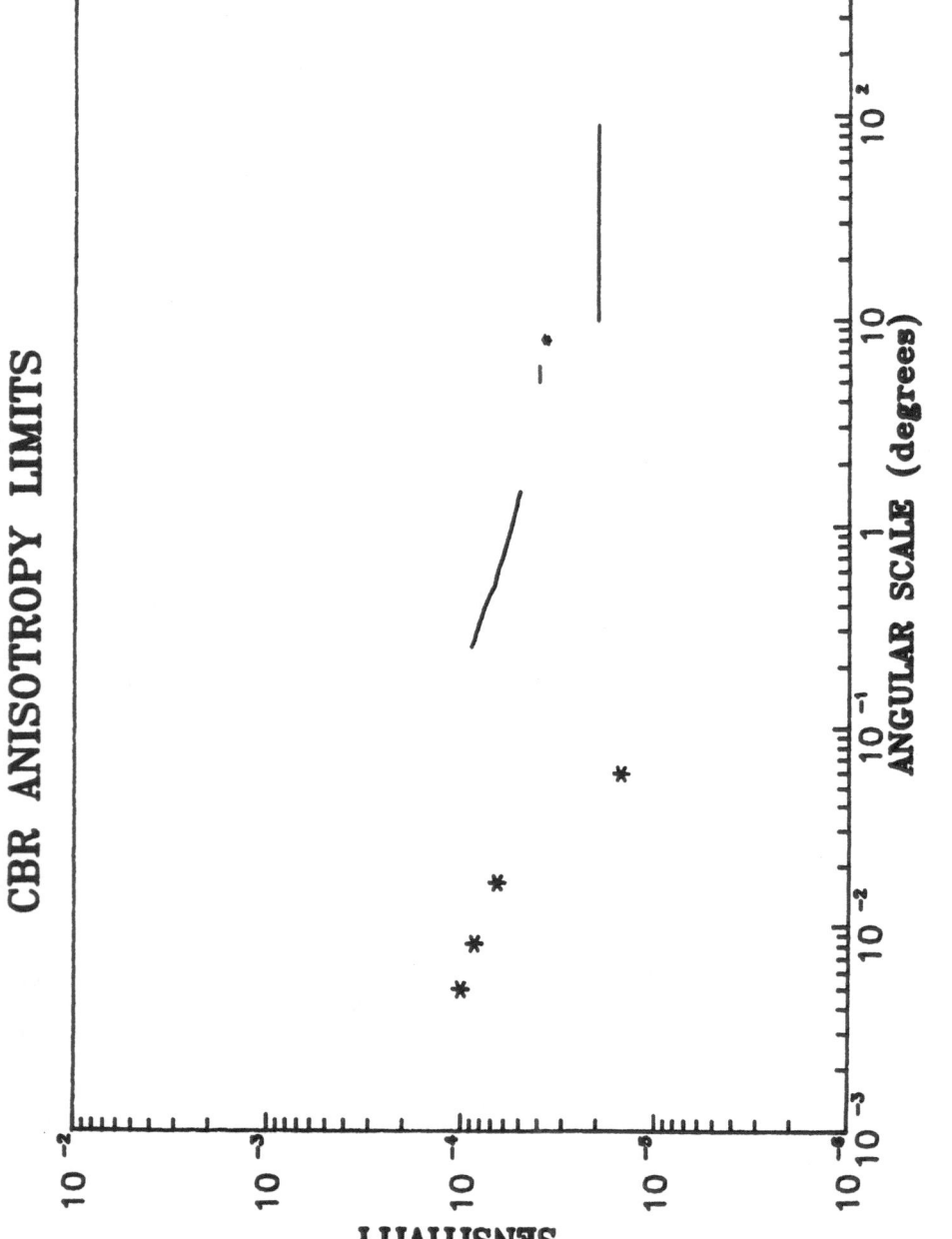

Figure 1

limited not by statistics but by calibration errors and galactic contamination. Measurements at this scale have also been used to place upper limits at angular scales down to about 10 degrees. (Strukov et al., (1988)) To date, all measurements on these scales (greater than about 10-20 degrees) have been done from space or high altitude balloon payloads in order to avoid contributions due to atmospheric fluctuations at large angles. Because of the large solid angle of these measurements, point source contamination is generally not a problem, whereas diffuse galactic emission from Bremsstrahlung, synchrotron, and dust are, in addition to off axis sidelobe contamination by the earth, sun, and moon.

Measurements on very small angular scales (arc-seconds to a few arc-minutes) have been performed from large, ground based, single dish and synthesized aperture telescopes. For these scales, source confusion begins to be a problem at the $\frac{\Delta T}{T} \sim 10^{-5} - 10^{-6}$ level (Franchesini et al., (1988)). In addition, current ideas about the generation of structure in the CBR include smoothing on scales of order 10 arc-minutes due to the finite "thickness" of the surface of last scattering. This tends to lower the level at which upper limits on $\frac{\Delta T}{T}$ put constraints on theories of galaxy evolution.

The Sachs-Wolfe (SW) effect (Sachs, Wolfe, (1967)), gravitational doppler shifting of photons moving through evolving gravitational potentials, is the primary theoretical mechanism for temperature fluctuations at angular scales greater than a few degrees. The most sensitive tests for theories of galaxy evolution are expected to be on scales of about 1 degree, above the scale where recombination and reheating effects are important, and below that where correlations due to the Sachs Wolfe effect begin to dominate. Some Cold Dark Matter scenarios for galaxy formation predict $\frac{\Delta T}{T}$ on .5 to 5 degree scales to be larger than the SW fluctuations at larger angles (Vittorio et al.(1988)). For these reasons, interest in experiments in the .5 to 10 degree range has risen in the past few years. The two primary systematic difficulties with doing sensitive experiments in this angular range are the atmosphere, which has time varying structure, and galactic dust contamination, which must be modelled and possibly subtracted.

## Our Experiment

We have chosen to work at 3 mm, where emission from the galaxy is low. Figure 2 shows the pole value of a cosecant fit in galactic coordinates, to several different large scale data sets, as a function of frequency. The plot shows that 3 mm is near the minimum.

While this choice of frequency reduces the problem of galactic contamination, problems with atmospheric emission are increased. Figure 3 shows the antenna temperature due to the atmosphere as a function of frequency at sea level, 3.6 km, and balloon altitudes. The plot is based on an atmospheric model with a standard temperature and pressure versus altitude profile, using water, oxygen and ozone absorption lines. It is evident that in order to work at 3 mm, one requires either a very stable atmosphere or a high enough altitude that the emission lines are not saturated and the measurement can be done between molecular transitions. For example, at sea level, the atmospheric emission is more than 6 orders of magnitude higher than a desired sensitivity of $\frac{\Delta T}{T} = 10^{-5}$.

We have built a system to make measurements on .5 to 5 degree scales, and have carried out experiments at balloon altitude and at the South Pole Station. Our gondola flew at about 30 km, where the precipitable water is approximately 3 x $10^{-4}$ mm. We chose the South Pole as a ground observation site because of the

## GALACTIC EMISSION

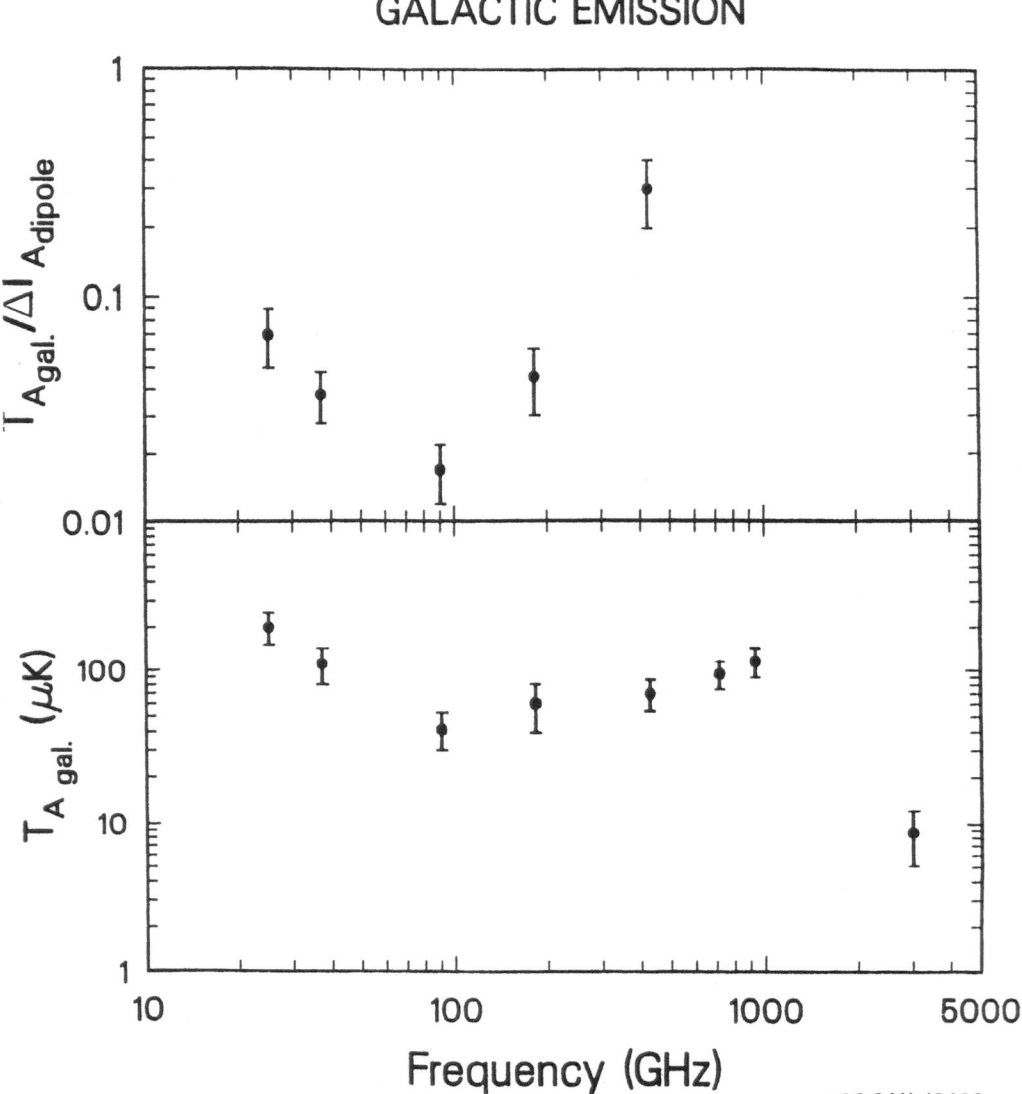

XCG 8411-13463

*Figure 2*

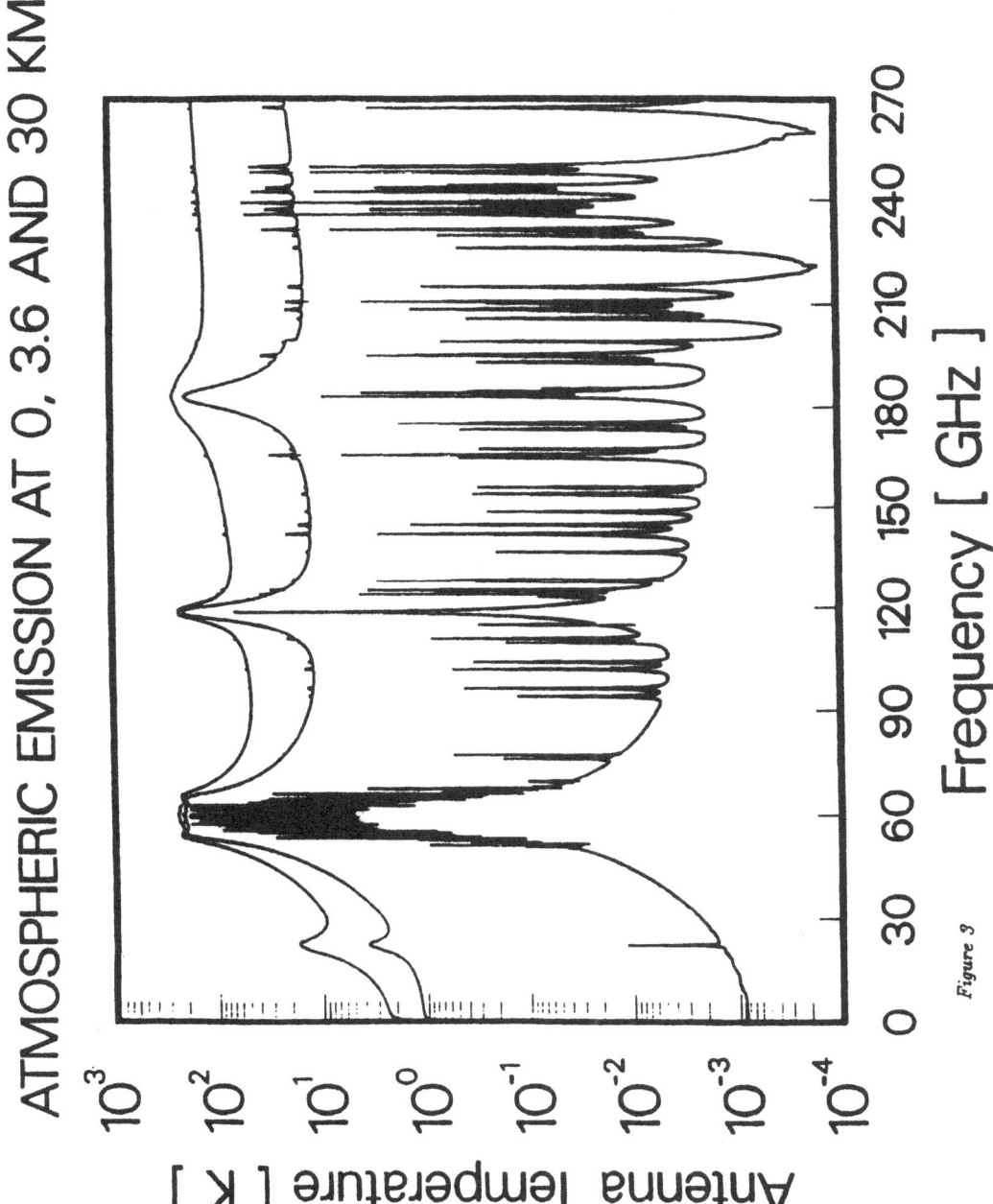

Figure 3

low water content and previously reported high stability of the atmosphere there. Figure 4 shows precipitable water for the time we were observing. Following is a brief description of the balloon payload, flight performance, and details of the South Pole expedition and results.

## Balloon Payload

In order to get useful integration time, we need a balloon gondola capable of stabilizing to a fraction of a beam width, which is 15 arc-minutes for our system. This requires an active stabilization system and some absolute pointing reference. Figures 5 and 6 show a diagram and schematic for our pointing platform. The primary elements in the pointing system are: an inertial guidance system consisting of 3 axis gyros, accelerometers, and a navigation processor; a CCD Star camera, for real time verification of absolute pointing accuracy and stability performance; A reaction wheel for azimuth angle stabilization and control; and an active triple race bearing system which serves to decouple the rotations of the balloon from the gondola, in addition to providing a controlled way to dump angular momentum accumulated in the reaction wheel from external perturbations on the gondola. The servo control and data taking are implemented with an on-board computer, with real-time interaction via telemetry from the ground. This package was first flown from the National Scientific Balloon Facility in August, 1988.

## Optical System

Our optical system is an off axis Gregorian telescope, consisting of a 6.5 degree (FWHM) corrugated scalar feed, a 1 meter diameter, 1 meter focal length primary, with a confocal elliptical secondary mirror. The resulting beam can have a FWHM of 20 to 50 arc-minutes, depending on the secondary mirror used (our results are for a FWHM of 36 arc-minutes). Rotation of the secondary about the axis of the feed horn throws the beam horizontally on the sky. We chop the beam by a physical angle of 1 degree on the sky at 10 Hz to make a first difference measurement of temperature fluctuations. Our primary reason for using this configuration is the very low sidelobe response of such an antenna. For the central lobe, the beam is well approximated by a Gaussian of $\sigma = 15$ arc-minutes. $P(\Omega) = e^{-\theta^2/2\sigma^2}$ With a FWHM of 36 arc-minutes, the ratio of solid angle available for contamination to that in the beam puts stringent limits on the allowable sidelobe response. We measured our sidelobes down to -85 dB, without ground shields. In addition a ground shield was attached during data taking both during the balloon flight and at the South Pole.

## SIS Receiver

A schematic of our detection system is shown in figure 7. We use a Niobium SIS (Superconductor-Insulator-Superconductor) based coherent radiometer, operating at 90 GHz. Our mixer, HEMT IF amplifier (spot noise about 1 K), and cooled RF section enable us to achieve a system spot noise of about 33 Kelvin at a mixer physical temperature of 3.5 Kelvin. Receiver performance is shown in figure 7. During data taking at the South Pole, our full band (0.6 GHz) noise was approximately 40 K, providing a theoretical system sensitivity (before chopping) of $\Delta T = 1.6 \frac{mK}{\sqrt{Hz}}$.

## Flight Results

Our flight from Palestine Texas in August, 1988 was a highly successful test

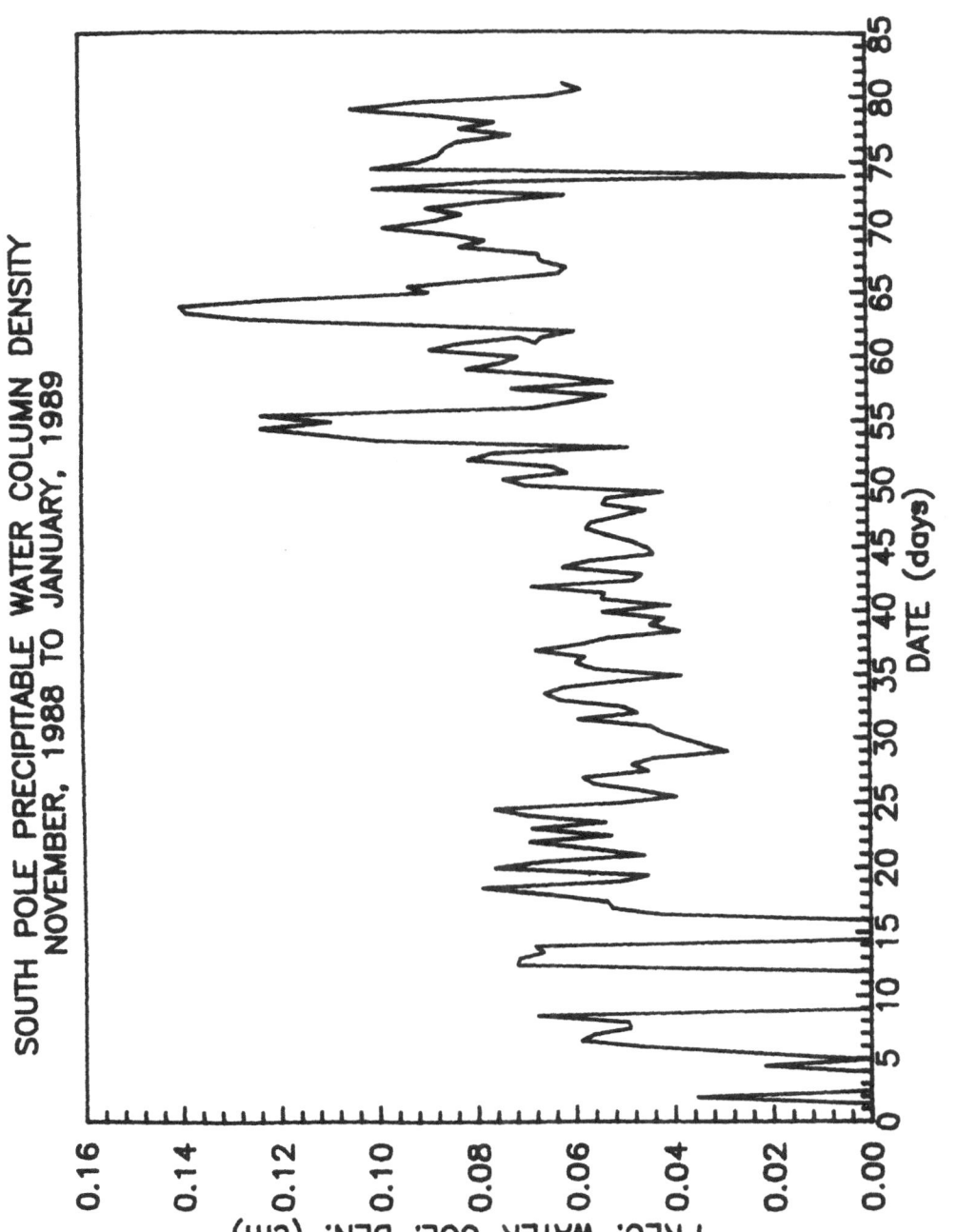

Figure 4

122

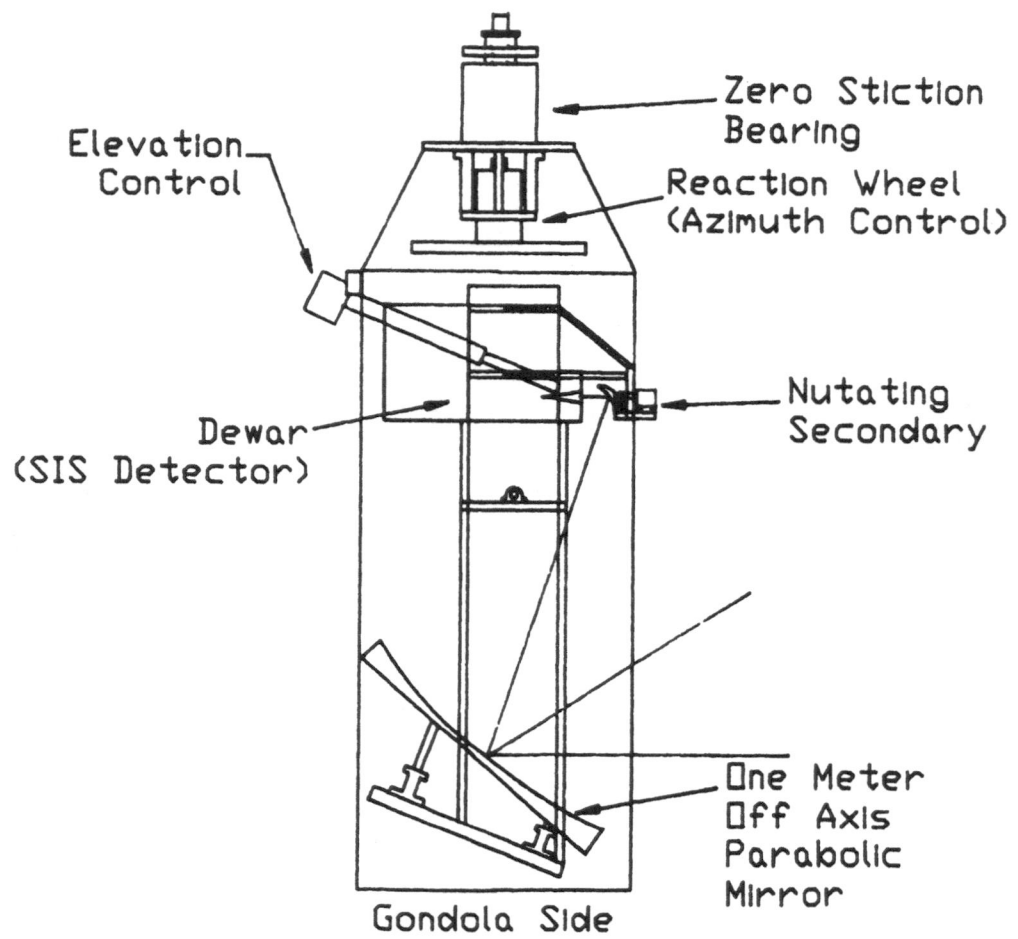

Zero Stiction
Bearing

Elevation
Control

Reaction Wheel
(Azimuth Control)

Nutating
Secondary

Dewar
(SIS Detector)

One Meter
Off Axis
Parabolic
Mirror

Gondola Side

*Figure 5*

# Pointing Control System

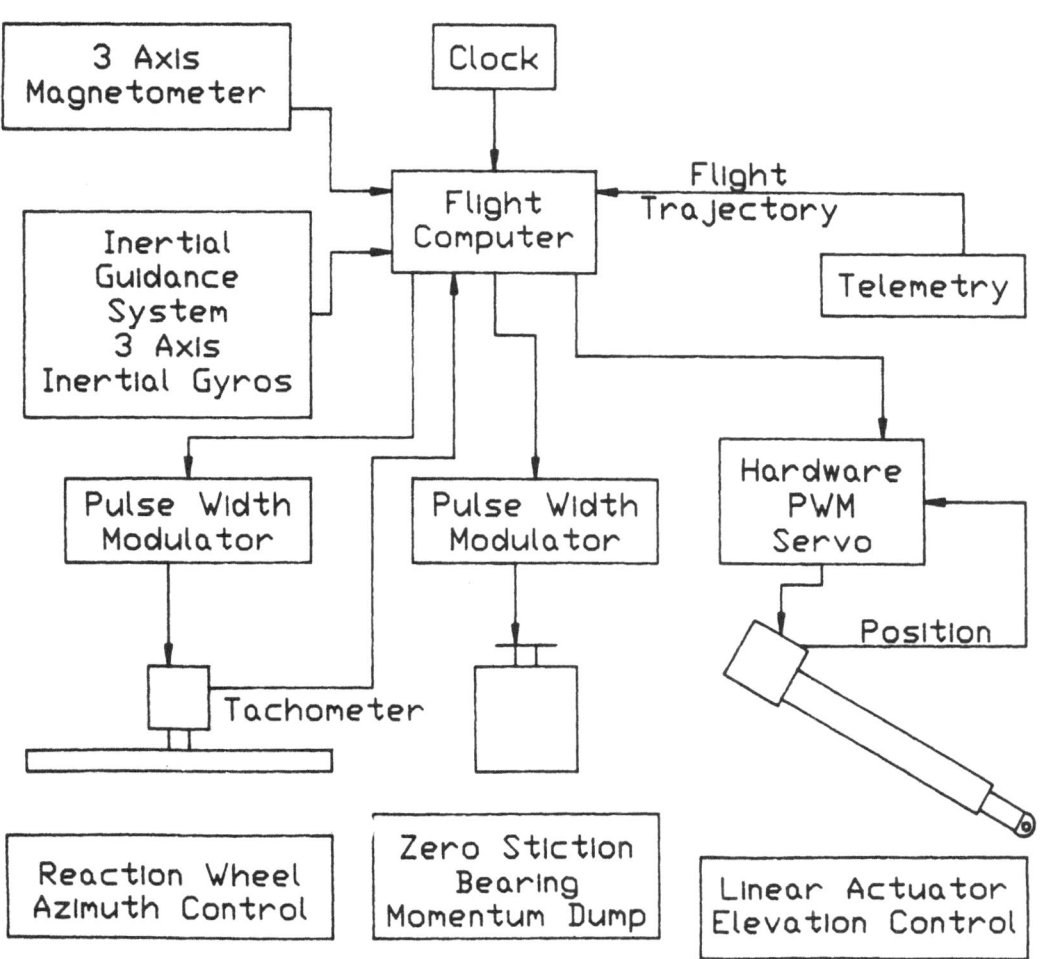

*Figure 6*

124

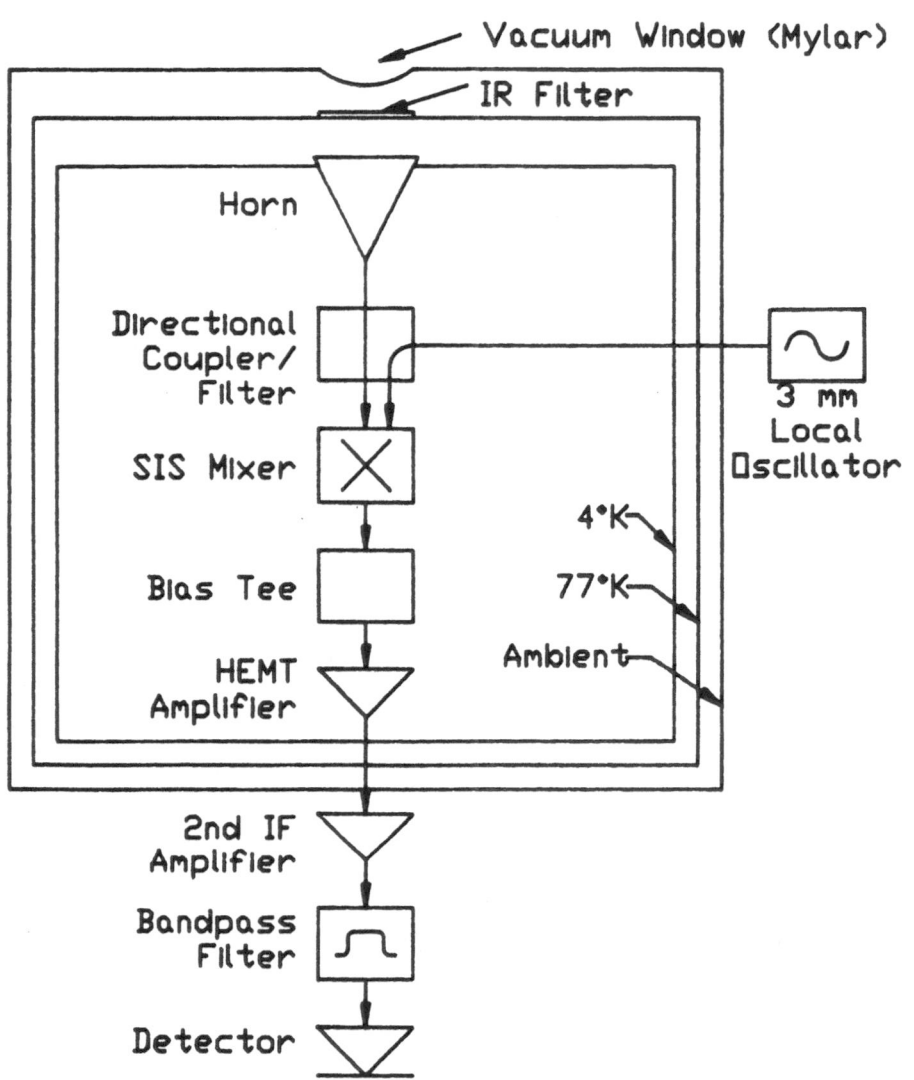

*Figure 7a*

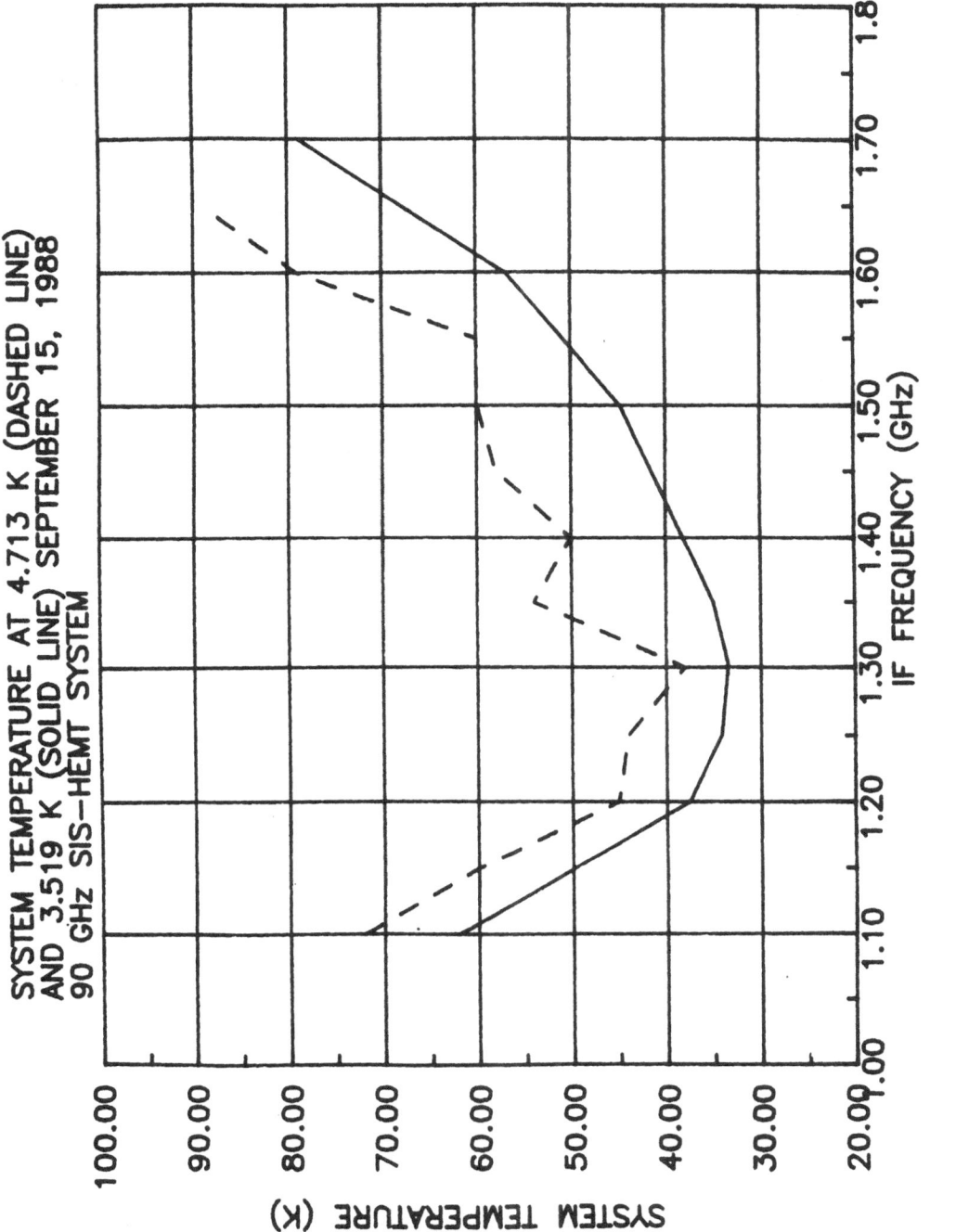

Figure 7b

of the stabilization and detector systems. The package got about eight hours at a float altitude between 95,000 and 100,000 ft., with no major damage on landing. The gondola achieved a pointing stability better than 1 arc-minute (RMS), and we were able to perform several important system tests during the flight.

Figure 8 shows actual azimuth angle as a function of time, showing our three point scan trajectory superimposed on the tracking for an az-el mount. Figure 9 is a calibration scan of Jupiter done during flight as a check of calibration factor and beam profile. Figure 10 shows a real-time scan of the galactic center, which we use to scale IRAS 100 micron data to our frequency for subtraction (this is discussed in detail later in this work). Unfortunately, a telemetry problem prevented us from obtaining enough integration time to get useful data on CBR fluctuations.

## South Pole Results

From late November, 1988 to early January, 1989, we made measurements of CBR fluctuations and galactic emission from the South Pole station, replacing the azimuth stabilization with a servoed rotation table. Figure 11 shows a calibration scan of the moon, along with an approximate theoretical curve, based on a model for moon emission as a function of moon phase.

Since galactic dust emission is a probable cause of error, we need to determine the scaling between short wavelength data to 3 mm. Comparing a simple cosecant fit to the IRAS 100 micron data and the cosecant amplitude for 90 GHz from our earlier large scale anisotropy flights (see figure 2) gives approximately $10 \frac{\mu K}{MJy/Sr}$, which is consistent with the number obtained by comparing the IRAS data to the flight scan of the galactic center. Figure 12 shows our South Pole plane crossing scan with error bars, as well as a first difference scan of the IRAS 100 micron map, scaled by the above number. Since these three comparisons are consistent, we can estimate the contribution of dust emission to our data. We implicitly must assume here that the dust emissivity scaling is the same over the sky. We chose to measure in a region around RA=21.5, DEC=-73 ($l^{II} = -40.6, b^{II} = -37.43$), where the IRAS 100 micron map shows a total intensity minimum of about 4-10 MJy/Sr, and first differences only of order 1-2 MJy/Sr (see figure 13a,b). Using the galaxy data described above, this would be about 10-20 microKelvins in our first difference data, which is small though not completely negligible compared to our errors (about 40 to 60 microKelvin per data point). Figure 13 shows two views of the IRAS 100 micron data in celestial coordinates, with an expanded view of the measurement region.

The total dust intensity contribution in this region of 4-10 MJy/Sr could then be about 40 to 100 microKelvins, clearly of concern for future measurements. We are currently at the point in sensitivity where even in the best parts of the sky, dust emission at 3 mm wavelength is predicted to be near our detection limit. To do an order of magnitude more sensitive measurement will almost certainly require galactic subtraction preferably by multiple wavelength measurements.

## CBR Data

We observed 9 points with 1 degree physical chop angle on the sky, in a strip, spaced so that one beam from each point coincided with one beam from the next point. Several strips were measured to different sensitivities. This gives us a powerful test for systematic errors, as well as providing information on a variety of angular scales, from the beam sigma of 15 arc-minutes up to approximately 5 degrees. After time lost due to setting up, equipment problems and bad weather,

127

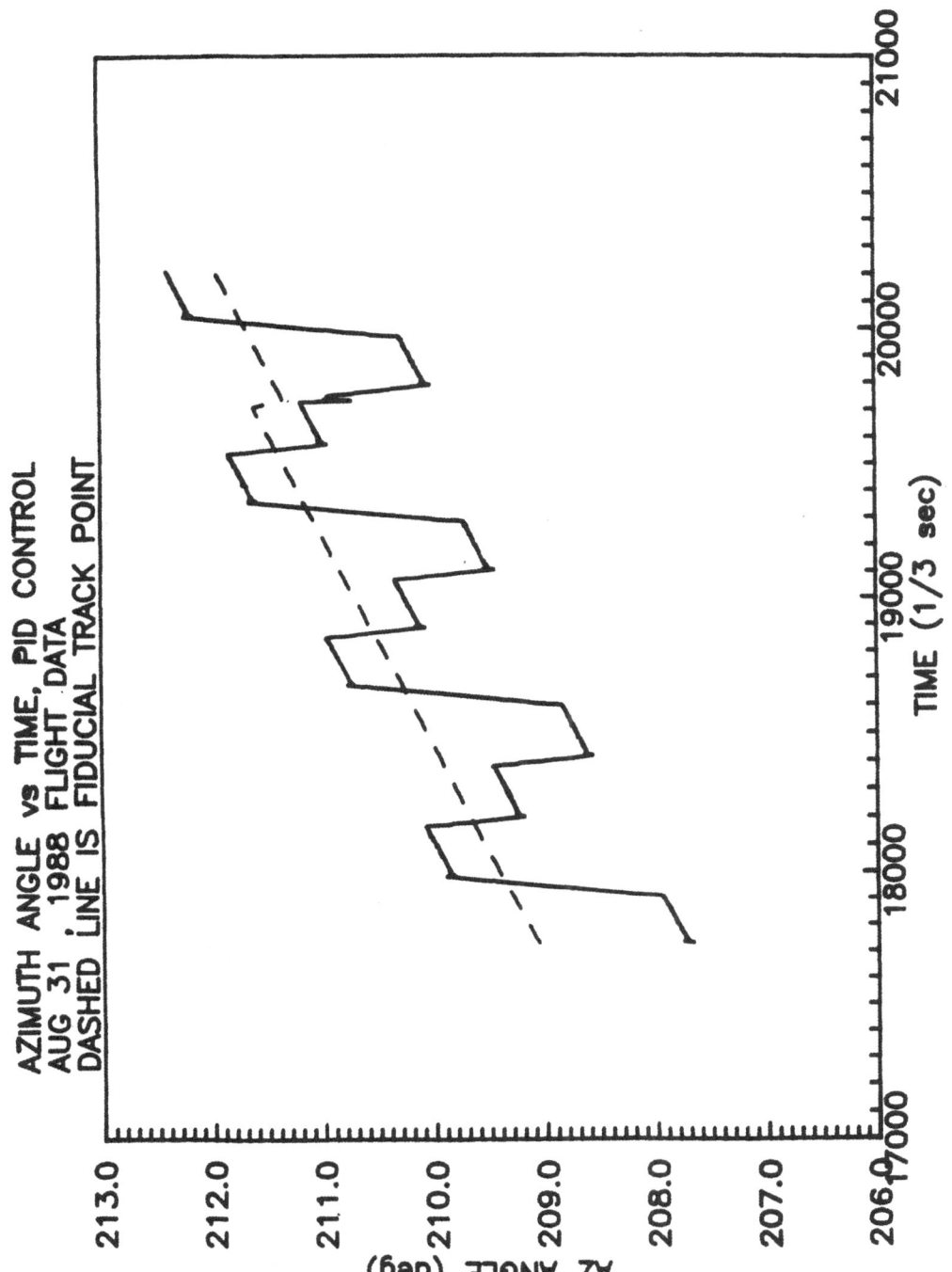

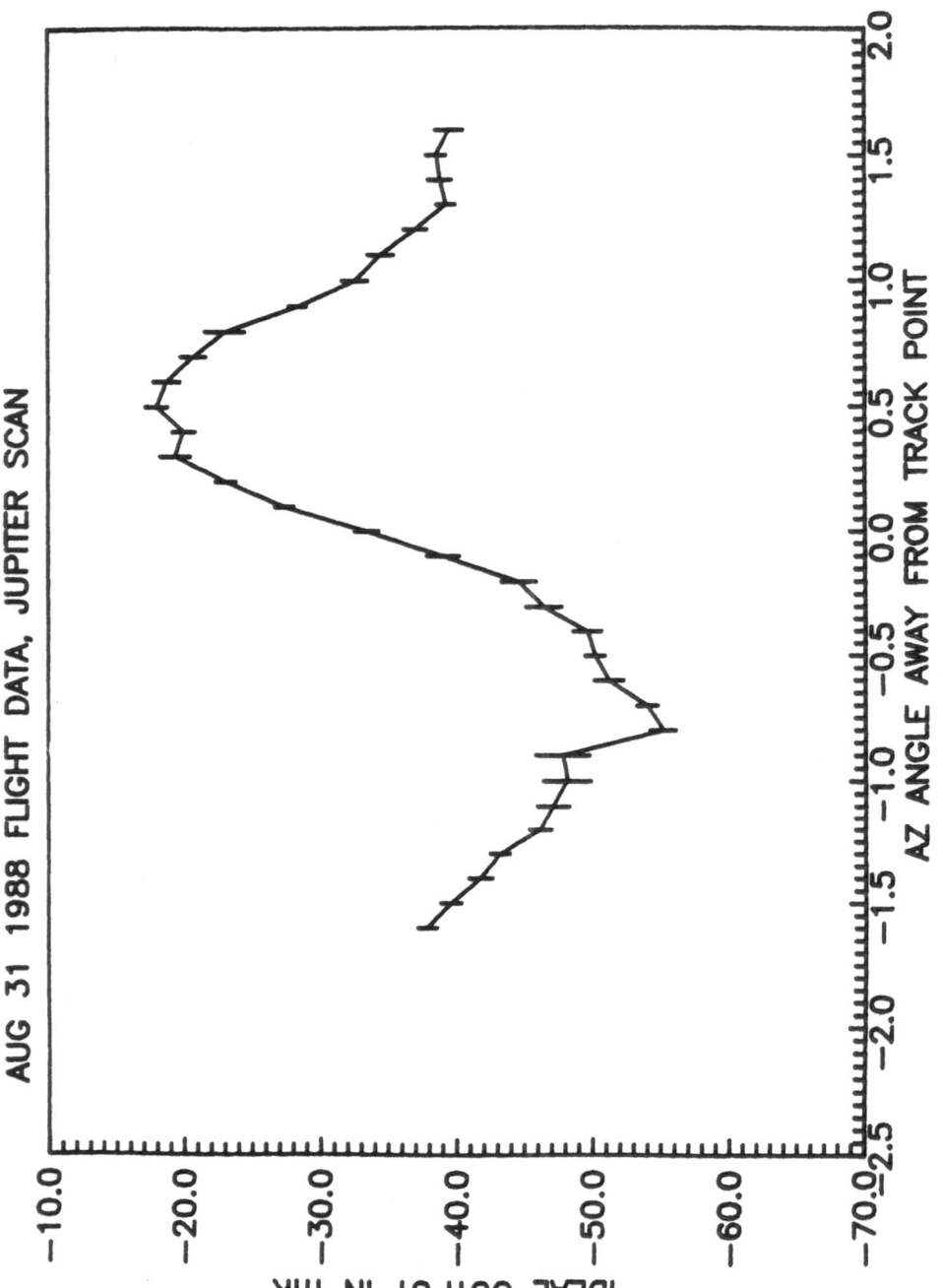

Figure 9

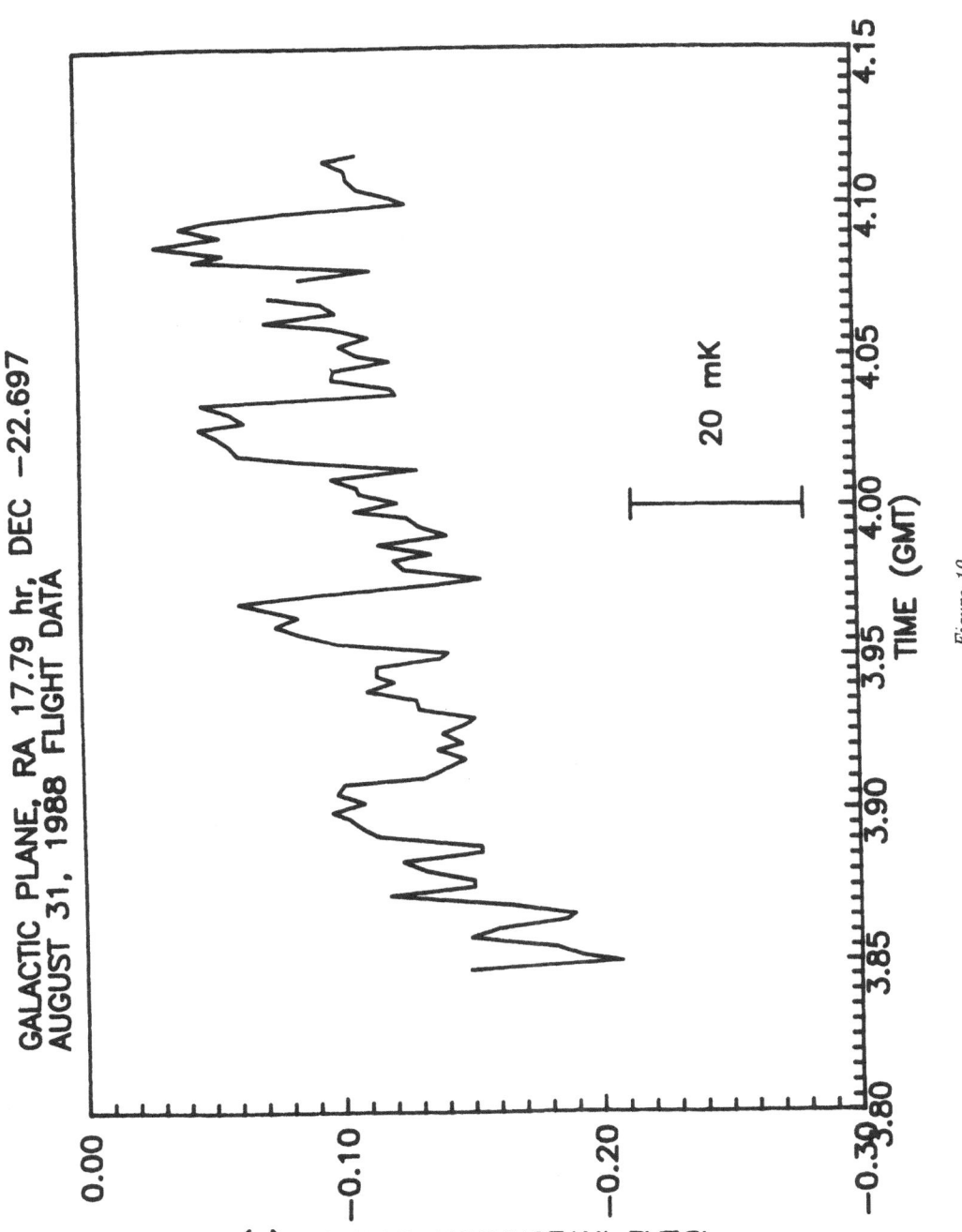

*Figure 10*

130

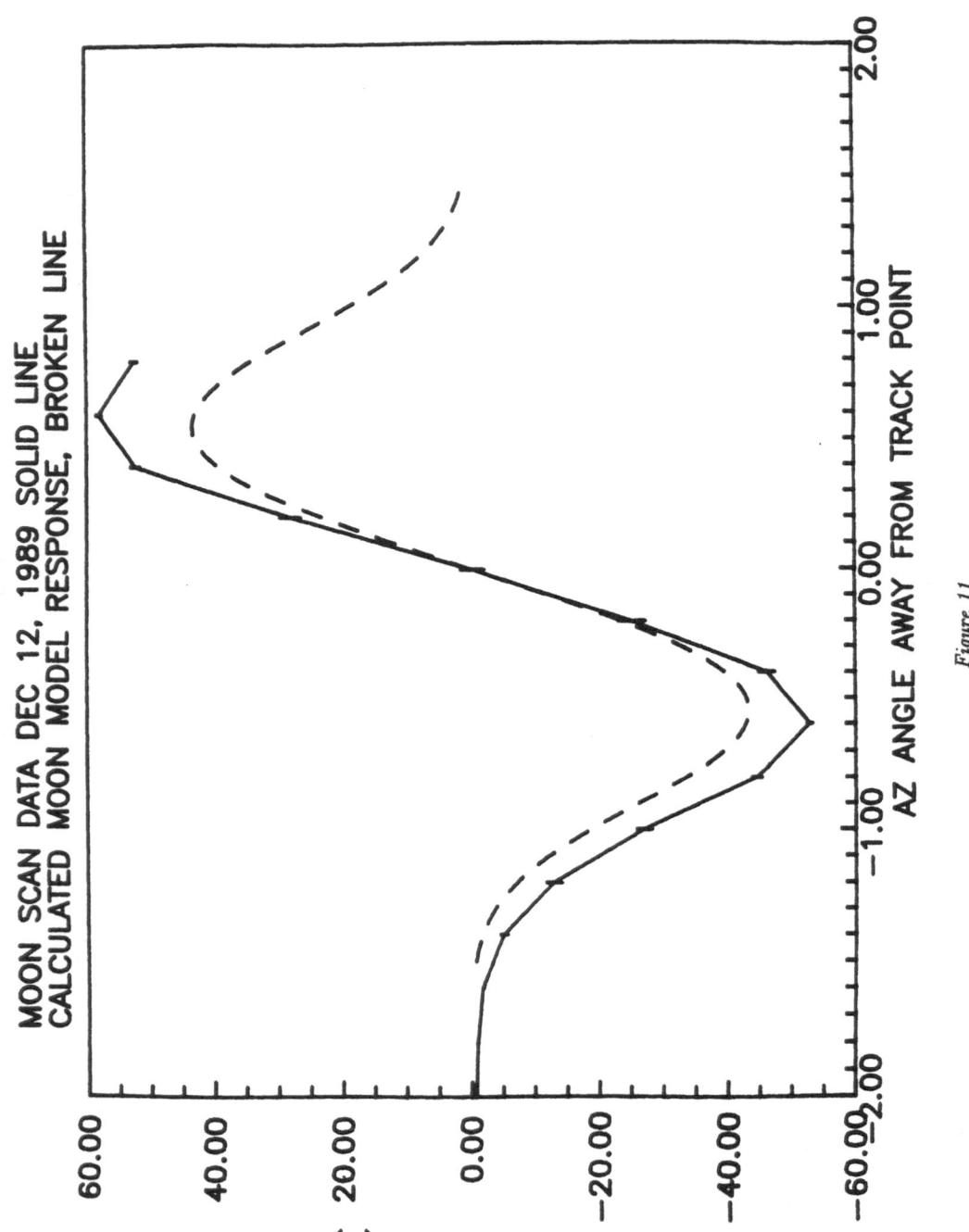

*Figure 11*

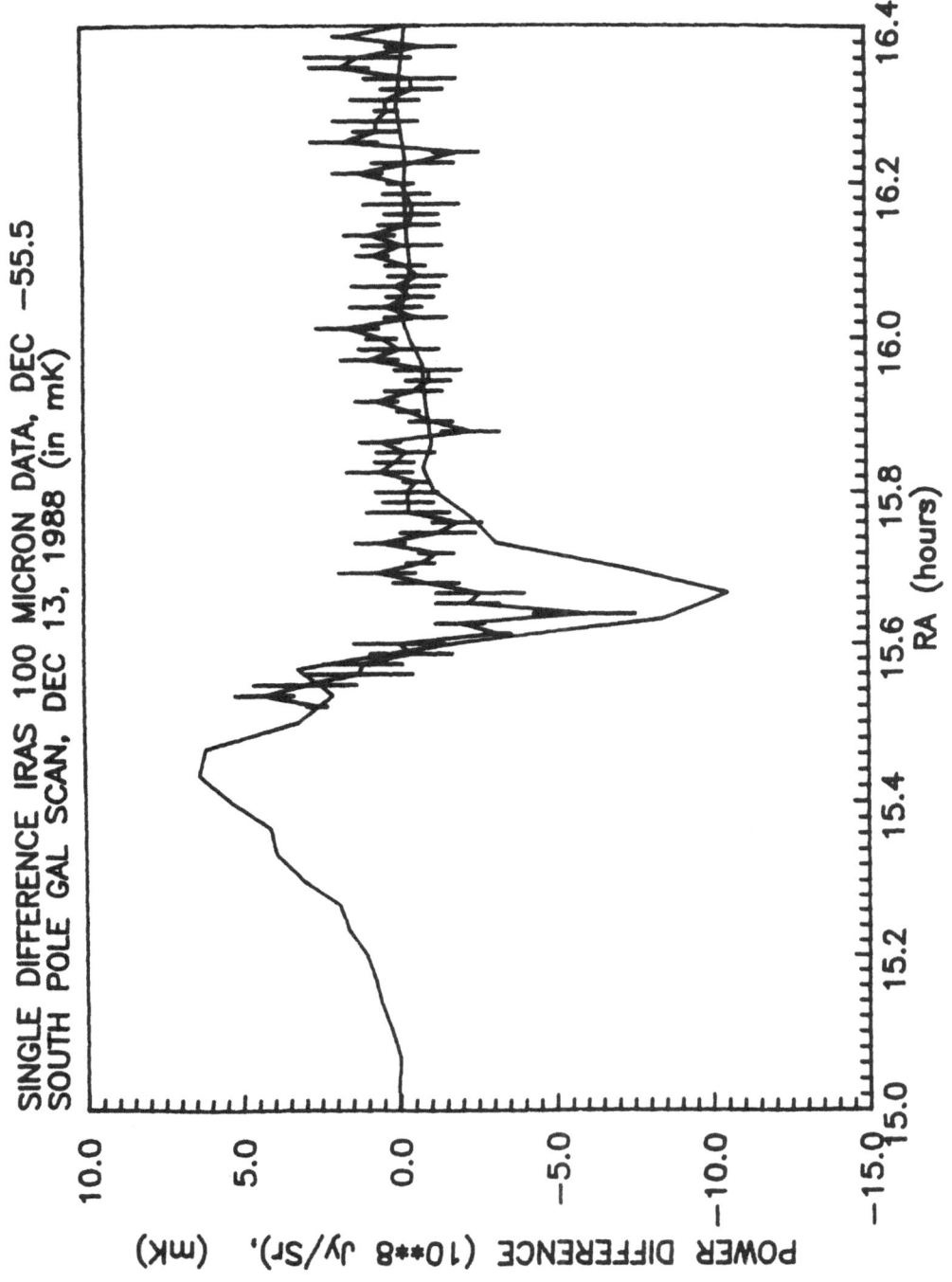

IPAS 100 micron 2 x 2 DEGREE MAP. CELESTIAL COORDINATES

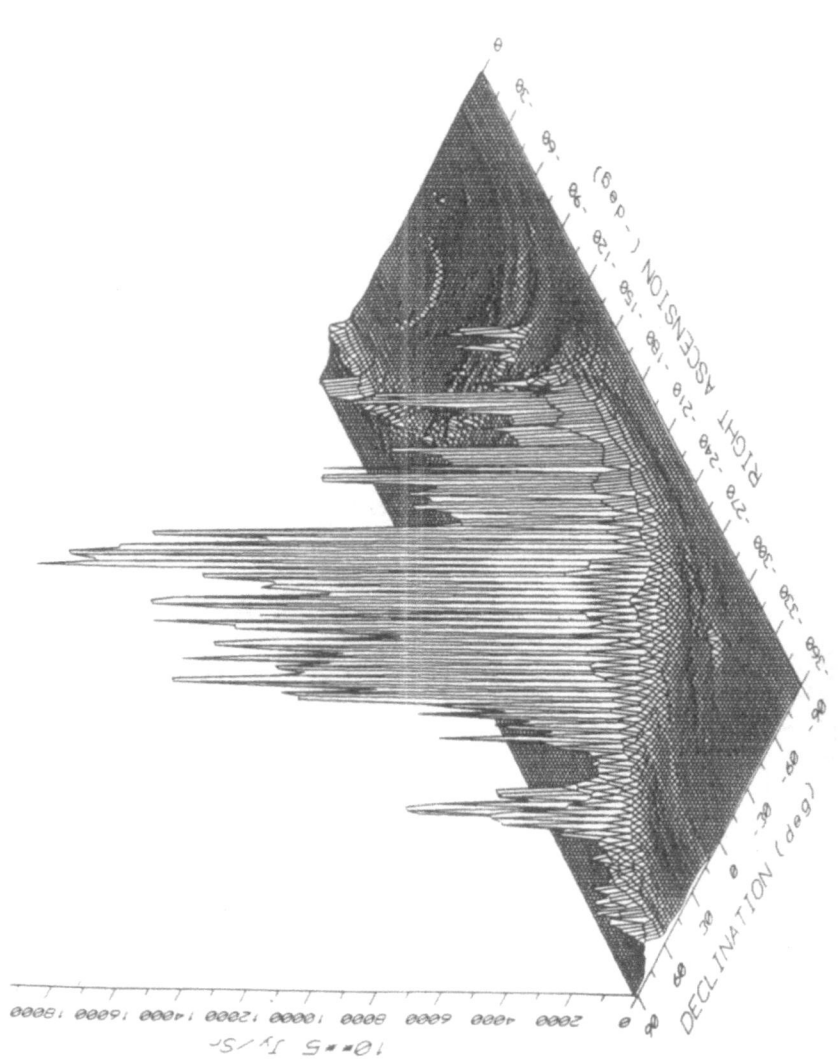

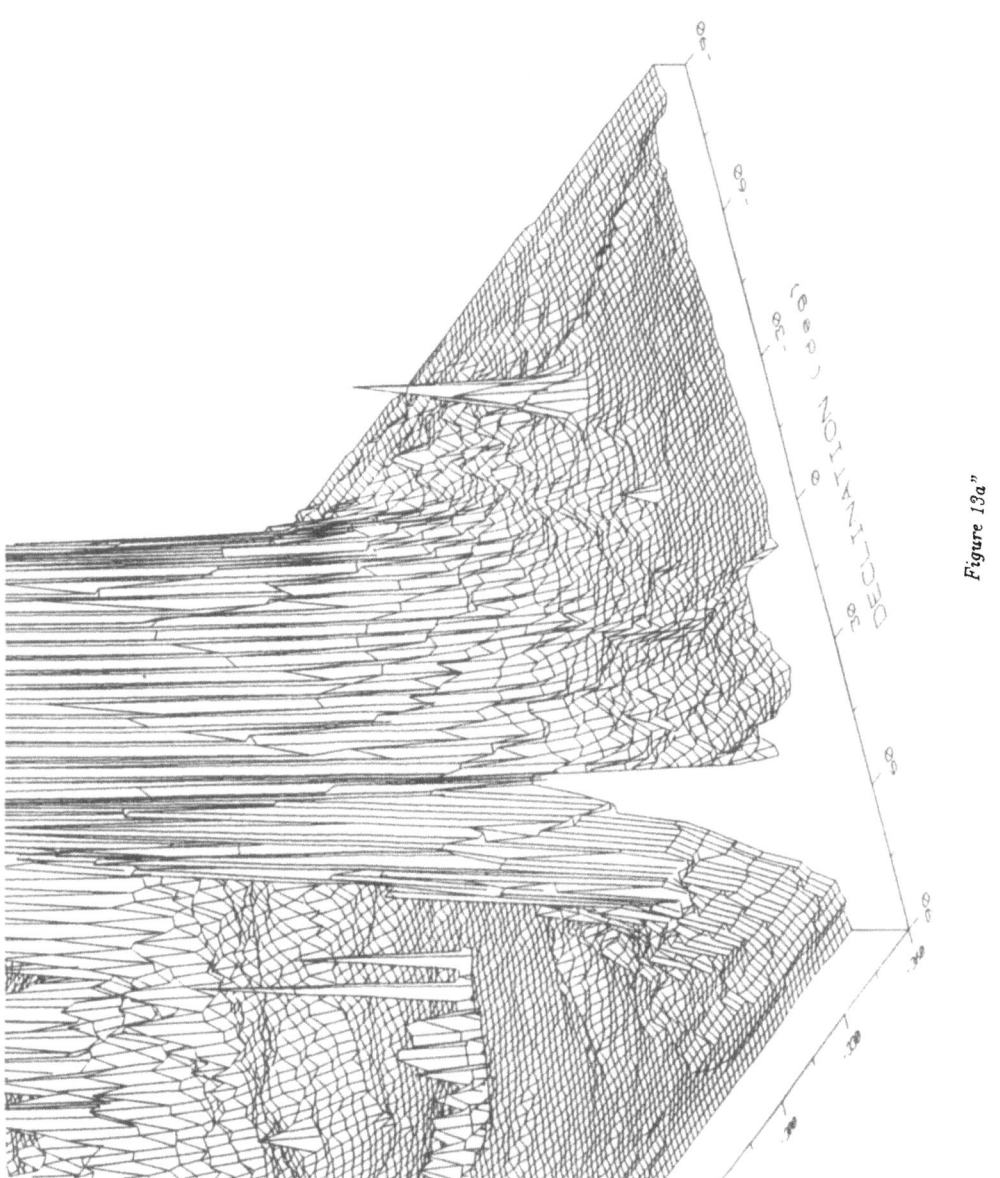

*Figure 13a"*

134

IRAS 100 micron .5 x .5 DEG MAP. MEASUREMENT REGION

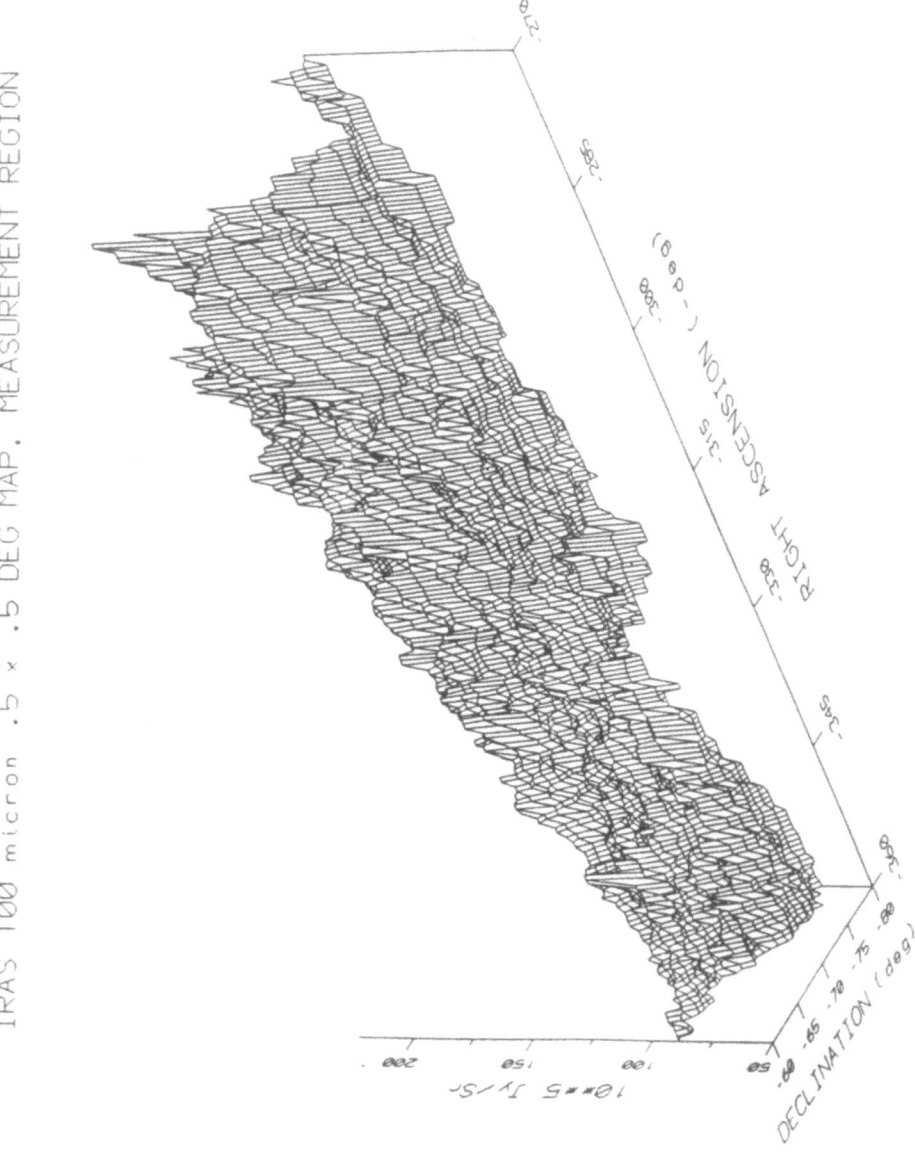

*Figure 13b*

we obtained about 80 hours of data, which reduced to about 70 hours after editing out radio interference and bad sky data. Our scan system gave us an efficiency (time spent on the measurement points) of only 60 percent, reducing the real data further to about 43 hours.

With a calculated statistical system sensitivity (on the sky) of 3 $\frac{mK}{\sqrt{Hz}}$, or 4 $\frac{mK}{\sqrt{Hz}}$ with sky shot noise included, we measured approximately 6 $\frac{mK}{\sqrt{Hz}}$ (RMS) on the sky for short time scales. Figure 14 is a histogram of the RMS in 100 second bins, showing the stability of the short term sky noise over time (for 'good' days). Several runs were made of just atmospheric noise and are being investigated to help understand the nature of the sky noise.

**Raw Data Fitting**

In order to work with the data, we have found it necessary to remove slow drifts in offset, which can be attributed to long term sky variations, changing electrical offsets, and temperature gradients on the primary. Our observing technique allows a natural way to remove such non- intrinsic shifts. Since we scan from one side of the strip to the other and then back in a period of about 30 minutes, linear variations on time scales long compared to 30 minutes can be removed without removing CBR structure. The results plotted in figure 15 are the summed data for each point, with statistical error bars, where the raw data have been edited and piecewise linear fit in time, over times of approximately 3 hours. The results for a truncated Fourier fit subtraction, constructed to fit only structure longer than 3 scans, as well as a Legendre polynomial fit, are consistent with the linear fit presented. The error bars on this data set are consistent with the short term RMS fluctuations.

**Data Analysis**

Looking at the data set in figure 15, a linear trend is evident across the points. Although this could be taken as an indication of intrinsic structure in the background radiation, we are unwilling to rule out some systematic effect to produce this. As an example, the sun was at RA of about 18 hours during our data taking, and contributions from this on the 100 $\mu K$ level are not out of the question. We choose to remove the linear component from the data and consider the result to be our final set, which is shown in figure 16. This set with error bars shown has a reduced chisquare of 1.53, corresponding to approximately 20 percent probability of being consistent with the null hypothesis. We are currently analyzing the data to test for various cosmological models, such as the cold dark matter galaxy formation model, scale invariant Gaussian fluctuations, etc.. These calculations will be presented in a forthcoming paper.

3. CONCLUSIONS

**Acknowledgements**

This work was supported by the National Aeronautics and Space Administration, the National Science Foundation, California Space Institute, the University of California, and the U.S. Army. This work would not have been possible without the support and encouragement of Nancy Boggess, Buford Price, and John Lynch. We wish to especially thank Anthony Kerr and S.K. Pan of NRAO for supplying the exceptional SIS mixer. The $Nb/Al - Al_2O_3/Nb$ junctions used were supplied by

136

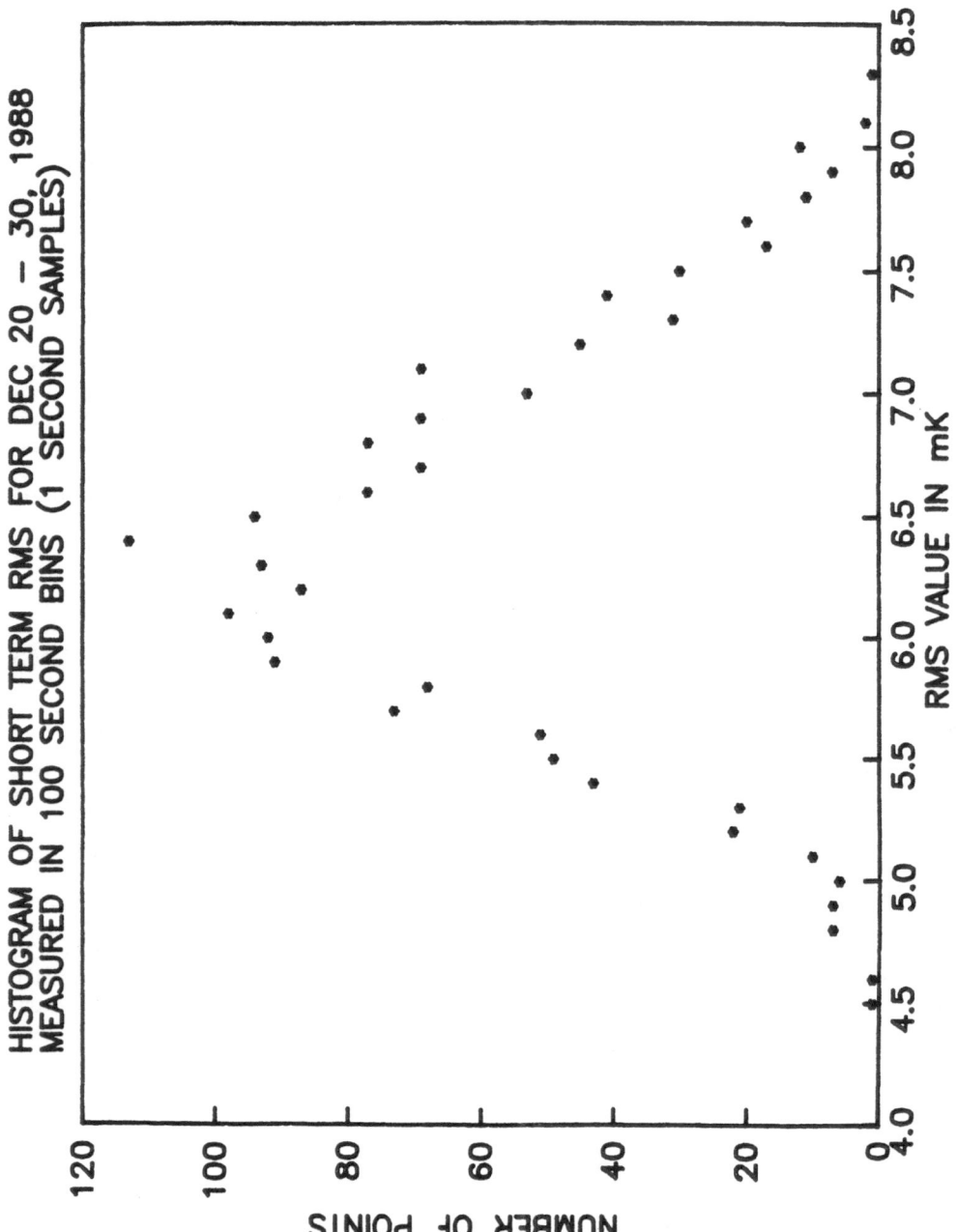

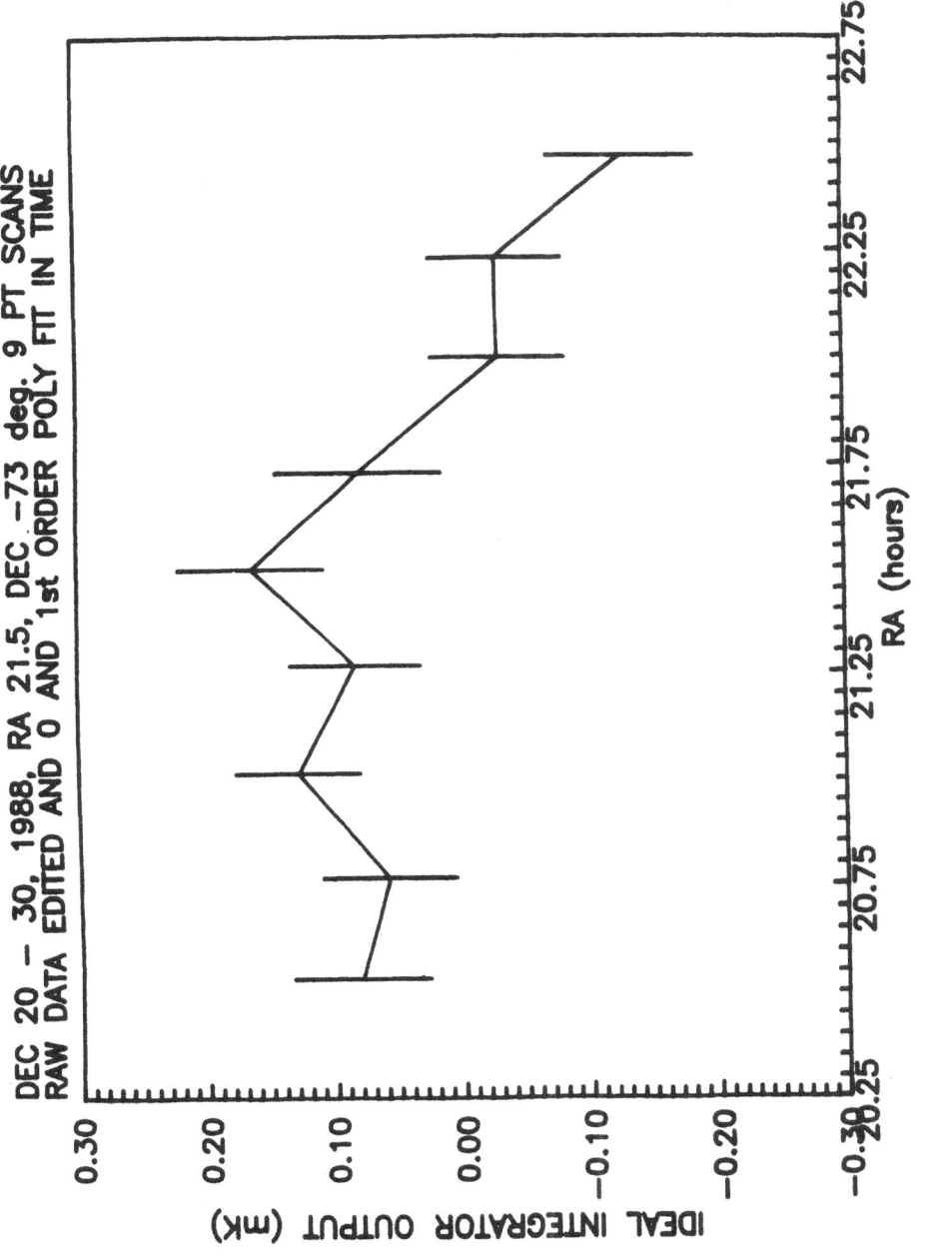

*Figure 15*

138

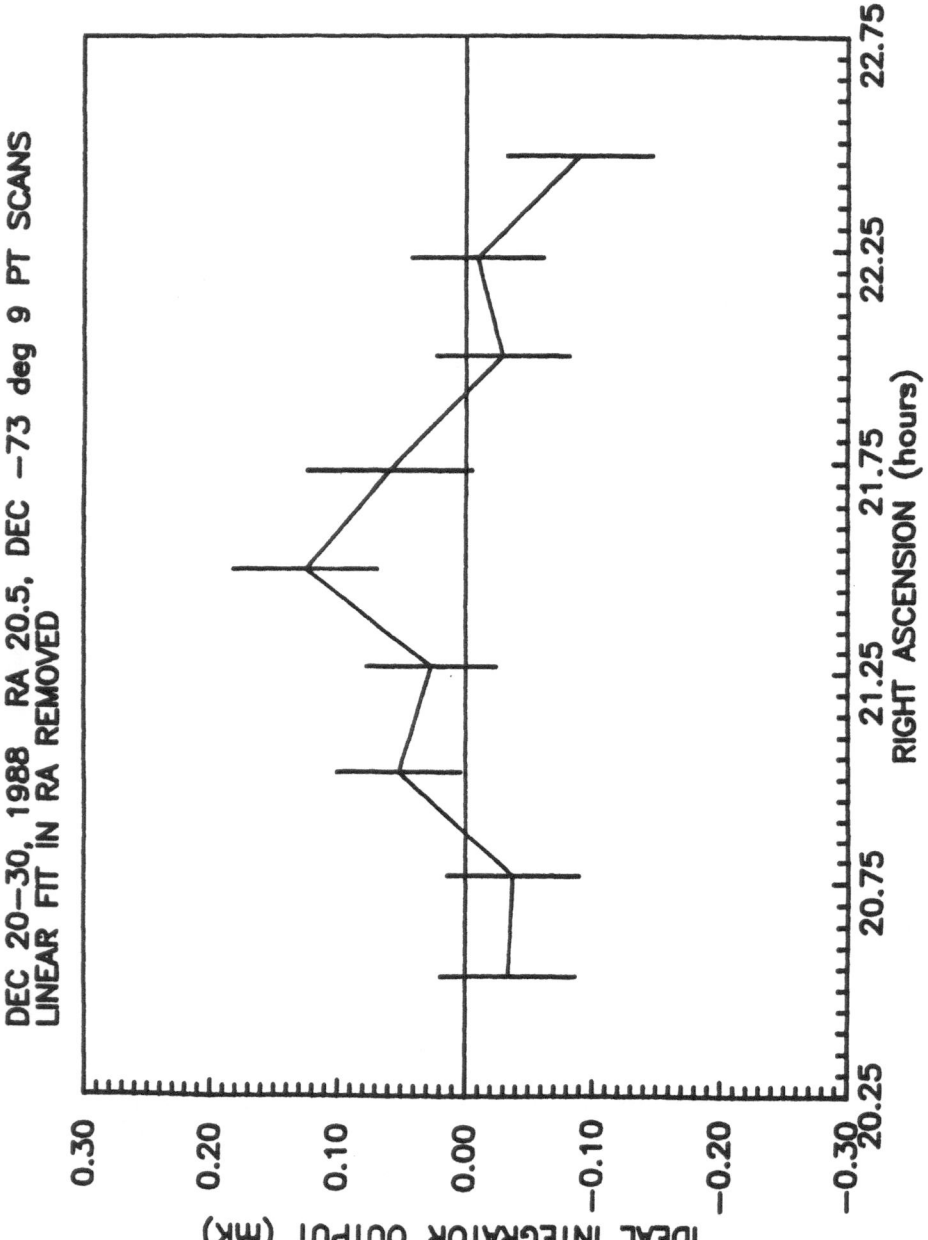

Figure 16

Hypres Corporation. Special thanks to Robert Wilson, Anthony Stark, Joe Stack, and Paul Moyer at Bell Labs for assistance in machining the primary and secondary mirrors. We gratefully acknowledge the support of Donald Morris. In particular we would like to thank Bill Coughran, and rest of the South Pole ANS support staff for the 1988-89 polar summer.

## 4. REFERENCES

R.D. Davies, A.N. Lasenby, R.A. Watson, E.J. Daintree, J. Hopkins, J. Beckman, J. Sanches-Almeida, and R. Rebolo, *Nature*, **326**, 462, (1987).

E.B. Fomalont, K.I. Kellerman, M.C. Anderson, D. Weistrop, J.V. Wall, R.A. Windhorst, and J.A. Kristian, *Ap. J.*, submitted (1988).

A.N. Lasenby, Ph.d. Thesis (1981).

F. Melchiorri, B.O. Melchiorri, C. Ceccarelli, and L. Pietranera, *Ap. J.*, **250**, L1, (1981).

S.K. Pan, M.J. Feldman, A.R. Kerr, and P. Timbie, *Applied Physics Letters*, **43**. 8, (1983).

A.C.S. Readhead, C.R. Lawrence, S.T. Myers, W.L.W. Sargent, H.E. Hardebeck, and A.T. Moffet, *Ap. J.*, submitted (1989).

R. Sachs and A. Wolfe, *Ap. J.*, **147**, 73, (1967).

I.A. Strukov, D.P. Skulachev, A.A. Klypin, *Large Scale Structure of the Universe*, J. Audouze, et al (eds.), IAU #130.

# OBSERVATIONS OF THE CMB SPECTRUM

P. L. RICHARDS
Department of Physics, University of California,
Berkeley, California 94720, U.S.A.

A review is given of the status of direct measurements of the spectrum of the CMB prior to COBE.

## 1. INTRODUCTION

The spectrum of the cosmic microwave background (CMB) is one of the few cosmological observables which can, in principle, give information about the conditions of the early universe during the era of decoupling. Spectral measurements have been made over a factor $10^3$ in frequency from ~600 mHz to ~600 GHz. The experimental techniques and the factors which limit experimental accuracy vary markedly over this broad frequency range. This review will describe the measurements in order of increasing frequency.

## 2. MICROWAVE MEASUREMENTS

There is strong interest in measurements of the spectrum of the CMB in the Rayleigh-Jeans region because one class of cosmological model predicts a non-zero chemical potential for photons. This would give a frequency dependent antenna temperature, rather than the constant value given by a Planck spectrum. There exists a well developed heterodyne receiver technology below ~90 GHz with low enough receiver noise temperatures that the measurement uncertainty $\Delta T$ due to receiver noise is small compared with 2.7 K. For example, a very modest receiver with noise temperature $T=10^3$ K, intermediate frequency bandwidth $B=10^8$ Hz and integration time $\tau=100$ sec gives $\Delta T=T_R(B\tau)^{-1/2}=10^{-2}$ K. The radiation from 300 K objects is small enough in this spectral range that ambient temperature receivers can be used to observe the difference in signal between the sky, a cold load, or blackbody cooled with liquid helium, and a second higher temperature load.

141

*N. Mandolesi and N. Vittorio (eds.), The Cosmic Microwave Background: 25 Years Later, 141–152.*
© 1990 *Kluwer Academic Publishers.*

Measurements of the spectrum of the CMB at the high frequency end of this range $\gtrsim 10$ GHz encounter significant signals from the atmosphere. This atmospheric contribution is usually minimized by observing from dry high altitude sites in "windows" between atmospheric emission lines. The remaining contribution is calibrated by measuring the temperature of the sky as a function of zenith angle. The requirement for the maximum permissible sidelobe response of the antenna is set by the antenna temperature of the horizon which is $\sim 10^2$ times that of the CMB.

At frequencies $\lesssim 1$ GHz, galactic radiation becomes troublesome. Antennas many wavelengths across should be used to minimize confusion from compact sources. Such large antennas, however, are not compatible with straightforward cold load techniques. In practice, the galactic contribution is identified from its spectral index and subtracted. The intermediate frequency range $1 \lesssim f \lesssim 10$ GHz is relatively free of special problems, but measurements are still limited by the achievable quality of the cold load and by the thermal and mechanical stability of the receiver over the periods of observation and calibration.

Despite the difficulty of working at high dry sites, including the South Pole, it has been possible to check and repeat experiments on the time scale of about one year. Many experiments have converged to a temperature of the CMB between 2.6 and 2.8 K. Frequent updates of this data set are provided by Smoot and collaborators (Kogut et al., 1990). Recently, the experiment of Johnson and Wilkinson (1986) at 24 GHz has broken with tradition by the use of a balloon platform to avoid atmospheric emission and a LHe-cooled antenna and receiver to reduce emission and to improve thermal stability. This experiment was the first to report error limits less than one percent of $T_{CMB}$. Related techniques should permit improved experiments at other frequencies near the high end of the microwave range.

Because of the low noise in available receivers, all of these experiments are limited by systematic errors. There are inherent difficulties in interpreting such experiments, and especially in estimating the error limits. Despite these problems, the history of microwave measurements of the spectrum of the CMB has been one of reasonably orderly progress. Recent measurements are illustrated in Fig. 1, which was adapted from Kogut et al., (1990).

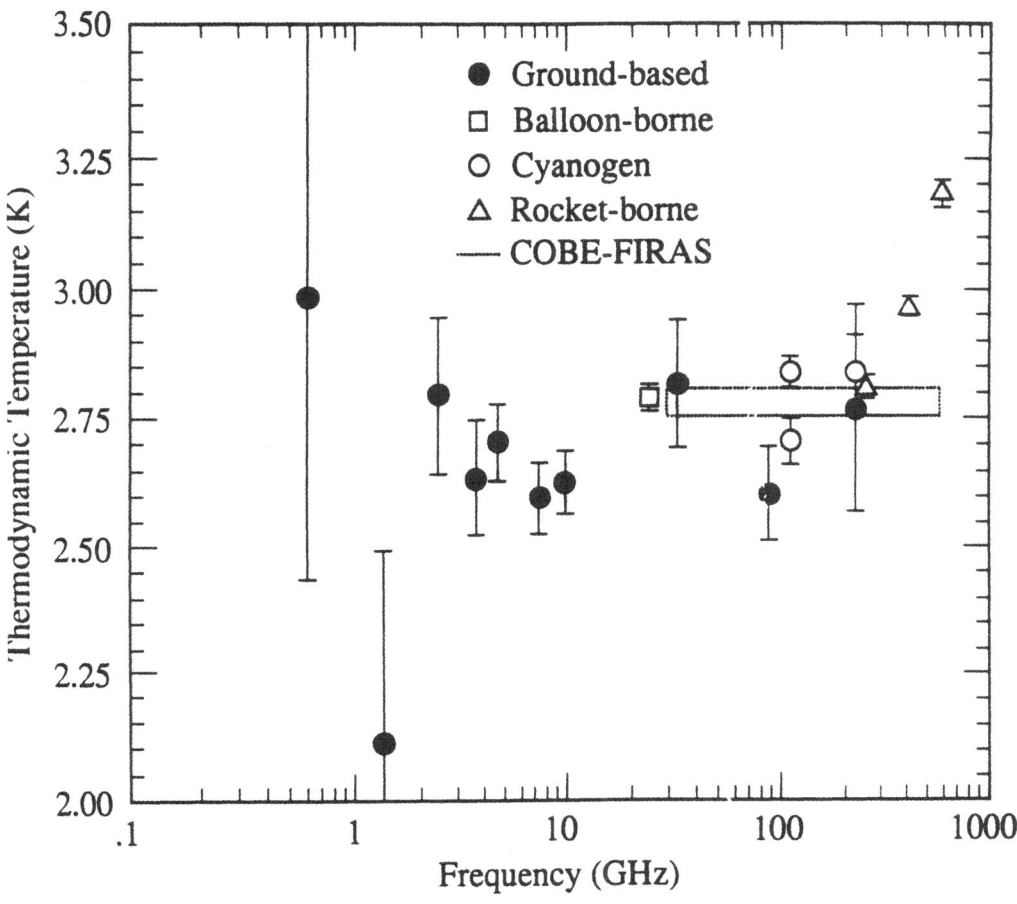

Figure 1. A collection of recent measurements of the spectrum of the CMB prepared by Kogut et al., with the preliminary result from COBE added.

## 3. NEAR-MILLIMETER WAVE EXPERIMENTS

Measurements of the spectrum of the CMB near the peak of the Planck curve at ~160 GHz are of special interest because they include most of the photon energy in the universe. As the frequency is increased above 90 GHz, however, spectral measurements become more difficult extremely rapidly. This is due to a combination of factors including the decrease of the antenna temperature of the CMB, the relative absence of developed receiver technology, and the rapid increase of the brightness of interfering emission from the apparatus, the horizon and the atmosphere.

Historically, a change has been made from coherent heterodyne receivers to incoherent bolometric photometers for frequencies above 90 GHz. These photometers are more sensitive in principle, but less well developed. Although bolometers can be operated as single-mode receivers, they are often used multimode to help compensate for the exponential decrease of the antenna temperature of the CMB with frequency. Such photometers cannot make use of the high performance scalar horn antennas developed for microwave frequencies and do not have the precise frequency selectivity of the heterodyne receiver. Fourier transform spectrometers and band-pass filters that are related to Fabry-Perot etalons have been used to provide frequency selectivity. Steady progress in the sensitivity of coherent heterodyne receivers at higher frequencies now makes heterodyne measurements possible well above 90 GHz. Similar progress in bolometer sensitivity makes single mode bolometric receivers more attractive. As a consequence, there are opportunities for a new generation of high frequency experiments which make use of the superior performance of antennas and filters available for single mode receivers.

The brightness (or antenna temperature) of a 300 K blackbody is ~600 times that of the CMB at the peak and nearly $10^6$ times at 600 GHz. Consequently, it is often necessary to cool the entire receiver with LHe, not just the cold load. Serious problems are encountered when a window is used to separate the thermal vacuum surrounding a cooled receiver from the outside atmosphere. Since such windows must operate near ambient temperature, emission from them can be substantial and difficult to measure. Reflection of warm radiation into the beam can also be troublesome. Because of the relative brightness of the horizon, the requirements on antenna sidelobe response are so severe that laboratory verification of the performance of incoherent antennas becomes extremely difficult. Because the strength and density of the atmospheric lines of $O_2$, $O_3$ and $H_2O$ increase with frequency, the problem of emission from the ~300 K atmosphere becomes rapidly very severe. In general, experiments with moderate frequency resolution can be done from balloons up to ~360 GHz. Higher frequencies are possible only from rockets and spacecraft (or possibly by balloon observations between emission lines with very narrow band heterodyne receivers). The difficulties of space cryogenics experiments are partially offset by the ability to operate a LHe cooled receiver with no window. Some success has been achieved at balloon altitudes by flowing cold He gas out of the antenna aperture. This requirement for balloon or space platforms has been the most severe limitation on spectral measurements at or above the peak. The resources available to groups doing balloon and rocket experiments did not increase in proportion to the complexity of the task. It frequently took 3 to 5 years before an experiment could be tested for systematic errors. Only the development program for the COBE satellite was able to escape these restrictions.

Despite the vigorous efforts of many groups, progress in our knowledge of the spectrum of the CMB near the peak have not been orderly. Most early measurements are known in retrospect to have contained serious errors, but many played a very important role in developing the technology used in later work. It also took time for the community to develop a consensus on how to evaluate the relative quality of different experiments. An experiment is judged favorably if its design adequately addresses all known sources of systematic error and if a wide variety of diagnostic tests show that it performed properly. Rightly or wrongly, the acceptance of a result has also depended on whether it is broadly consistent with theoretical expectations. This set of criteria has proved to be flawed, as it does not help to identify experiments that contain unanticipated and undetected systematic errors.

The modern history of spectral measurements near the peak at 160 GHz began with the work of the MIT group that carried out a series of balloon measurements that identified many of the technical difficulties and provided solutions for several of them. These experiments used a cooled horn antenna, an incoherent photometer with band-pass filters, and zenith angle scans to subtract the atmospheric contribution. It was shown that the CMB spectrum did not continue to rise beyond 160 GHz, but the results were not sufficiently accurate to reveal the existence of a peak (Muehlner and Weiss, 1973).

Following the MIT work, groups at Queen Mary College and Berkeley flew second-generation balloon experiments. Both experiments used a bolometric detector and a polarizing Michelson interferometer as a Fourier transform spectrometer to identify the atmospheric contribution from its spectrum. This type of spectrometer was later used in the FIRAS experiment on COBE. The QMC experiment reported evidence for a peak (Robson et al., 1974), but this report was not judged to be reliable by the community. The experiment was criticized for a lack of absolute calibration. Also an analysis of the contrast of the observed atmospheric lines suggested some unidentified malfunction.

The Berkeley experiment also reported evidence for a peak which became unmistakable after a second flight with an improved $^3$He-cooled bolometric detector. This Woody-Richards (1981) experiment was of historical importance because the reported spectrum gave the best evidence for the blackbody character of the CMB and thus for standard big bang cosmology. Two features of the experiment, however, were surprising. The data did not fit a blackbody curve perfectly and the best fit blackbody temperature was 2.96 K compared with an average of 2.77 K for the available microwave frequency data. Conventional, but unreliable, statistical interpretations of known systematic errors in the experiment suggested that this excess was a 3σ effect and therefore likely to be real. Many theoretical attempts were made to find a cosmological source for this Woody-Richards excess. A general consensus evolved, however, that the shape of the excess was very difficult to explain.

Stimulated by this intense theoretical interest, the Berkeley group with their collaborators mounted two subsequent balloon experiments designed to search for unknown, undetected systematic errors in the Woody-Richards experiment. The approach used was to change as many features of the experiment as possible and see if the results were the same. The Fourier spectrometer was replaced with a filter wheel and zenith angle scans, rather than spectral fits, were used to evaluate and subtract the atmospheric emission. The antenna and bolometric detector remained the same. The instrument was calibrated during the flight by means of a transfer standard. Peterson et al. , (1985) observed the sky in five passbands centered from 86 to 299 GHz. The sky signal was observed to decrease in time, even though the signals from the transfer calibrator were stable. Making generous allowances for this decrease, values of CMB temperature were deduced which were lower than those of Woody and Richards and generally consistent with no excess. The situation remained unsatisfactory, however, because the source of the error was not identified and the accuracy of the measurement was therefore severely compromised.

Bernstein et al. , (1990) re-flew the experiment with more diagnostics and discovered that the time varying contribution to the signal arose from the top edge of the antenna which cooled slowly during the flight. Although it was not possible to make an accurate measurement of this contribution, it was possible to place upper limits on $T_{CBR}$ at four frequencies. These confirm that an excess of the magnitude reported by Woody and Richards does not exist. Since the same antenna was used in all of the Berkeley balloon measurements, it probably contributed an excess flux corresponding to about 10% of the peak CMB flux to all experiments.

It is useful to inquire how this effect could have escaped detection for so long in a program that was attempting to proceed with great care. The antenna used in the Berkeley experiments was a Winston light concentrator, capped with a straight flared section designed to apodize the diffraction sidelobes. The antenna pattern was measured in the laboratory and agreed with the pattern calculated using the geometrical theory of diffraction. Since the horn is a non-imaging device, the rays diffracting around the upper edge of the horn (in the time reversed sense) were assumed to have random phases in this calculation. Measurements of the emissivity of the horn combined with these calculations gave a very low estimate for the product of throughput times emissivity reaching the detector from the flare. In flight, the horn and flare were cooled with cold gas from the boiling LHe. The temperature of the top of the horn was 3-6 K. There was a rapid temperature gradient across the flare; its top edge was in good thermal contact with the ambient atmosphere at ~230 K. The possibility of emission from the antenna was tested by heating the antenna in flight. A signal was seen from the warm antenna which was in good agreement with predictions from the measured temperature distribution and the calculations (Woody and Richards, 1981). The model predicted that the emission would be negligible at the lower temperatures used for data collection. This test was not sensitive to emission from the top lip of the flare, however, because its temperature did not change appreciably when the lower part of the flare and the top of the horn were heated. Also, a very low upper limit on antenna emission at ~600 GHz was obtained from the Woody Richards measurement which gave a very small value for the sky brightness.

Faced with experimental evidence for antenna emission near the peak of the CMB, Bernstein et al., (1990) reexamined the diffraction analysis and concluded that the assumption of random phases might be a reasonable approximation near 600 GHz, but not from 90 to 300 GHz where the number of modes excited in the antenna was too small for the assumed phase cancellation to take place. Estimates from a more complete theory gave a large enough throughput-emissivity product from the top of the flare to the detector to produce the observed excess. The experimental tests described above failed to reveal the source of the problem; they either measured the wrong part of the antenna or the right part at the wrong frequency.

Several lessons can be learned from this experience. The most important is that the geometrical theory of diffraction with the approximation of random phases does not adequately describe flared horns in which only a few modes are excited. The usefulness of such horns is doubtful in balloon experiments where the lip of the flare cannot be cooled. They can, however, be used with good effect in the vacuum of space where the entire antenna can be isothermal as was the case with the Berkeley-Nagoya rocket experiments and with COBE. Another lesson is that speculations from members of the community about sources of systematic error in a complicated experiment are not likely to be correct. Even though all of the information necessary to identify the antenna problem was published, comments from the community focused on possibilities such as the incorrect interpretation of known systematic errors, incorrect atmospheric subtraction and a postulated shift in preamplifier gain during flight.

## 4. SUBMILLIMETER MEASUREMENTS

The comments made above about the difficulty of near-millimeter wave experiments are even more valid at submillimeter wavelengths. The spectrum of a 2.74 K blackbody falls exponentially with frequency in the submillimeter spectral range. Measurements in this spectral range are of particular interest, however, because there is a remarkably deep window in the known backgrounds. It appears possible that submillimeter measurements will be sensitive to distortions of the spectrum due to Compton scattering from hot electrons. Also, some theories of early star formation suggest that starlight absorbed by dust and re-radiated in the infrared would be redshifted to submillimeter wavelengths.

Some of the earliest high frequency spectral measurements of the CMB were made from sounding rockets and were designed to observe submillimeter wavelengths. Perhaps because of the lack of early success, sounding rockets were relatively neglected, despite the inherent advantage that a cooled receiver could be opened to space. Gush (1981) at UBC reported a rocket experiment that used a rapid scan Fourier spectrometer and a $^3$He cooled bolometric detector. Unfortunately, the flight was marred by the drift of the rocket motor into the field of view of the instrument. Also, the calibration data appeared to show some leakage of high frequency radiation. The analyzed data showed less radiation than expected at the peak of the CMB and a substantial excess at higher frequencies. Despite many excellent features of this apparatus, the results were not judged to be reliable. Gush and Halperin flew a much improved experiment in 1990, but results are not yet available.

In 1988, Matsumoto et al., (1988) reported a sounding rocket experiment developed by a collaboration between Berkeley and Nagoya. The six-channel dichroic photometer (Lange et al., 1987) with four bolometric and two photoconductive detectors, the chopper, and the Winston horn antenna with curved apodizing flare were all cooled to 1.1 K. This temperature is low enough that radiation from the instrument was negligible in all bands. The instrument was calibrated with a LHe cooled variable-temperature load in the laboratory and checked during flight with a transfer standard. The 7.6° FOV of the antenna precess around a 20° half-angle circle on the sky. The fluxes observed in the six bands revealed the expected deep minimum in the background radiation between the Wien side of the CMB and the Rayleigh-Jeans side of the interstellar dust emission. The lowest frequency channel at 258 GHz gave a temperature of $T = 2.799 \pm 0.018$, which was consistent with the most accurate spectral measurements at lower frequencies. The balloon measurement of Johnson and Wilkinson (1986) gave $T_{CMB} = 2.783 \pm 0.025$ at 25 GHz and Crane et al., (1989) obtained 2.796 (+0.014 -0.039) from optical measurements of interstellar cyanogen. Channels 2 and 3 at 423 and 624 GHz gave $T = 2.955 \pm 0.017$ and $3.175 \pm 0.027$, respectively. The measured sky brightness fell with frequency, but not nearly as rapidly as for a 2.74 K blackbody. The minimum in the observed brightness was in channel 4 at 1.15 THz (262 μm). Along with channel 4, the brightnesses in channels 5 and 6 at 2.19 THz (137 μm) and at 2.95 THz (102 μm) were fit to a model of the interstellar dust emission with T=20 K, and an emissivity varying as $\nu^2$. The interpretation of these signals as being primarily due to interstellar dust was confirmed by the fact that the dependence of sky signal with precession angle was found to be strongly correlated with the distribution of HI. A complete analysis of the results from bands 4-6 have been submitted for publication (Lange et al., 1990).

Community interest in the Berkeley Nagoya experiment focused on the excess (above the Planck spectrum) in channels 2 and 3, which was much larger than any known experimental errors. The shape of the excess fit reasonably well both to a Planck function distorted by Compton scattering from non-relativistic electrons and by dust emission from an early generation of star formation. Other mechanism considered included the decay of cosmic strings or exotic particles left over from the big bang. It is difficult to imagine a spectral shape that would attract more theoretical excitement. By the time of the L'Aquila conference, however, this enthusiasm was beginning to fade because it was judged difficult to provide the large energy release required to produce the submillimeter excess by either process.

During the conference there was much discussion of possible systematic errors that could produce the apparent submillimeter excess. None of the diagnostics carried on this rocket experiment showed any sign of malfunction. Suggested possibilities that might have escaped detection included the idea that electronic pickup synchronous with the chopper was observed in channels 2 and 3, the idea that goo from rocket exhaust froze onto the cold antenna, and the idea that rocket exhaust was observed directly. Briefly, the idea of synchronous pickup is unlikely because it requires a remarkable coincidence. The phase of the sky signal was essentially the same as the phase of the large optical signal from the calibrator. The frozen goo idea is unlikely because the measured time dependence of the temperature of the antenna was incompatible with the condensation of the amount of material required to give the required emissivity. Also, it is hard to understand why the goo would be warm enough emit at ~500 GHz. The idea of the direct observation of rocket exhaust is more difficult to discount completely. Contamination carried aloft by the rocket was observed, but it disappeared rapidly on ascent. Only a very long lived component of the rocket exhaust could produce the observed excess.

The Berkeley-Nagoya team was well aware of the need to check the submillimeter excess. A new rocket payload was constructed with a similar horn antenna and dichroic photometer, but the bolometric detectors were cooled with $^3$He. This extra sensitivity was useful for a careful search for systematic errors in channels 2 and 3, where the signal-to-noise ratio in the first experiment was smaller than for the other channels. The attitude-controlled payload was programmed for a scan pattern that would include regions of strong and weak emission from both galactic and zodiacal dust. This payload was flown in September 1989. A factor ten improvement in bolometer sensitivity was achieved compared with the previous flight. The data were contaminated, however, by electronic pickup that was synchronous with the chopper. Despite much effort devoted to modeling and removing this pickup, no conclusions regarding the submillimeter excess have been produced.

In January of 1990, the preliminary measurement of the spectrum of the CMB obtained by the FIRAS experiment on the COBE satellite was announced (Mather et al., 1990). It shows that the spectrum accurately fits a Planck curve from 30 to 600 GHz with a temperature of 2.735±0.06 K. This result is judged likely to be correct because the high quality team produced a sound design which takes advantage of the major benefits of the environment of space. They also had the resources to execute the design in exhaustive detail and to build in a wide variety of diagnostic tests. Finally, the result is consistent with conventional cosmology and is unlikely to have been produced by accident.

Spectacular as the data are, the FIRAS experiment has not ended worries about undetected errors. The detector noise is low enough to make a significantly more accurate test for deviations than was reported in the preliminary announcement. When this is done, it is likely that deviations from the Planck curve will be found at low signal-to-noise ratio. The difficult and uncertain process of searching for undetected systematic errors will once again become important.

The question arises of whether the COBE data can ever be checked. It seems unlikely that another FIRAS type experiment will be flown in the foreseeable future. Atmospheric windows at and below 90 GHz are sufficiently good from balloon altitude, however, that comparable accuracy could probably be obtained from a well designed experiment. The best submillimeter detectors are sensitive enough that a sounding rocket can be significantly more sensitive than FIRAS on a limited number of pixels. The ability of such rockets to measure the zodiacal dust at 180° solar elongation may prove especially helpful in exploration of the submillimeter window. Because of the limited array of diagnostic tests that can be carried on one sounding rocket, however, it seems clear that a well orchestrated series of flights will be required to achieve sufficient reliability.

A collection of recent measurements of the CMB are shown in Fig. 1. With the COBE result included, these measurements suggest that there might be a minimum in the CBR near 6 GHz. The obvious next step in improving our understanding of the spectrum of the CMB is to make more accurate low frequency measurements.

Finally, it seems worthwhile to point out that those groups measuring the anisotropy of the CMB will soon be observing features on a routine basis. They will then be faced with the questions "is it real" and "is it cosmological". Undetected systematic errors at the level of one part in $10^5$ can be expected to become very difficult.

## 5. REFERENCES

Bernstein, G.M. et al., 1990, Ap. J. (submitted).

Crane, P. et al., 1989, Ap. J. 346, 136.

Gush, H.P. 1981, Phys. Rev. Letters 47, 745.

Johnson, D.G. and Wilkinson, D.T. 1986, Ap. J. Letters 313, L1.

Kogut, A. et al., 1990, Ap. J. (to be published 5/90).

Lange, A.E. et al., 1987, Appl. Opt. 26, 401.

Lange, A.E. et al., 1990, Ap. J. (submitted 1990).

Mather, J.C. et al., 1990, Ap. J. Letters (to be published).

Matsumoto, T. et al., 1988, Ap. J. 329, 562.

Muehler, D. and Weiss, R., 1973, Phys. Rev. Letters 30, 757.

Peterson, J.B., Richards, P.L., and Timusk, T. 1985, Phys. Rev. Letters 55, 332.

Robson, E.I. et al., 1974, Nature 251, 591.

Woody, D.P. and Richards, P.L. 1981, Ap. J. 248, 18.

# Theoretical implications of the CMB spectral distortions

L. Danese, C. Burigana, L. Toffolatti
Dipartimento di Astronomia
Padova, Italy

G. De Zotti, A. Franceschini
Osservatorio Astronomico
Padova, Italy

The spectral distortions of the Cosmic Microwave Background (CMB) are quite informative on the processes that might (or had to) occur in the primordial plasma after the thermalization time, during an epoch which is crucial for the formation of the observed structures of the universe. Here we review the relevant processes generating CMB spectral distortions. We compare the presently available data on the spectrum with the predictions of a variety of theoretical scenarios. We also review the connections among the spectral shape of the CMB, the dipole anisotropy and the Sunyaev-Zeldovich effect.

## 1. INTRODUCTION

In the twenty five years since its discovery (Penzias and Wilson, 1965) and its immediate interpretation as a vestigial element of a very hot and dense early universe (Dicke $et$ $al.$, 1965), the cosmic background radiation (CMB) has served as the prime interface between cosmological theories and the "real" universe. The knowledge of the specific entropy $s = 4aT^3/3nk \simeq 3.6n_\gamma/n \simeq 5.1 \times 10^8 (T_0/2.7\,K)^3 \widehat{\Omega}_b^{-1}$ [$\widehat{\Omega} = (H_0/50)^2\Omega$; $\Omega_b$ is the baryon density in units of the critical density.] has allowed remarkably successful predictions of the abundances of light elements (see Boesgaard and Steigman, 1985). In turn, primeval nucleosynthesis entails important constraints on the very early universe (e.g. on shear and anisotropy of the expansion, Barrow, 1976), on the mean density of baryons (Kawano $et$ $al.$, 1988), as well as on quantities of interest to particle physics, such as the number of light, weakly interacting particles or the degeneracy of primordial neutrinos (Boesgaard and Steigman, 1985).

While the above results stem essentially from the determination of the photon to baryon ratio, much more can be learned from accurate measurements of the spectrum (for general reviews see Danese and De Zotti, 1977; Sunyaev and Zeldovich, 1980; De Zotti, 1986; Bond, 1988). In this paper, after a summary of the basic theoretical framework, we will discuss the implications of recent measure-

$N.$ $Mandolesi$ $and$ $N.$ $Vittorio$ $(eds.),$ $The$ $Cosmic$ $Microwave$ $Background:$ $25$ $Years$ $Later,$ 153–172.

ments of the spectrum with special emphasis on the sub–mm excess reported by the Berkeley–Nagoya experiment (Matsumoto et al., 1988b).

## 2. OVERVIEW OF EXPECTED SPECTRAL DISTORTIONS

### 2.1  Early distortions

As is well known, since the density of protons and electrons is so small in comparison to the density of photons, the frequency of $p$–$e$ interactions yielding bremsstrahlung radiation, is far smaller than that of $e$–$\gamma$ collisions (Compton scattering). Kinetic equilibrium between radiation and matter is achieved on a timescale

$$t_C = t_{\gamma e}\frac{mc^2}{kT_e} \simeq 4.5 \times 10^{28}\,(T_o/2.7)^{-1}\,(T_e/T_R)^{-1}\,\widehat{\Omega}_b^{-1}\,(1+z)^{-4}\,s \qquad (1)$$

corresponding to $y = \int (kT_e/m_ec^2)d\tau \approx 1$. Here $t_{\gamma e} = 1/(n_e\sigma_T c)$ is the photon–electron collision time and $\tau$ is the optical depth of the universe for Thomson scattering; $kT_e/mc^2$ is the mean fractional change of photon energy in a scattering of cool photons off hot electrons ($T_e \gg T_R$). So long as newly produced photons can be neglected, kinetic equilibrium corresponds to a Bose–Einstein (BE) spectrum characterized by a chemical potential $\mu$, which is a measure of the fractional amount of extra energy injected in the radiation field.

True termodynamic equilibrium requires a substantially longer time and is achieved through the combined effect of photon emitting processes (bremsstrahlung and radiative Compton scattering) which are very efficient in producing soft photons, and of Compton scattering, which moves them to higher energies. The minimum redshift $z_{therm}$ at which full equilibrium can be re–established depends on the baryon density and on the amount of energy released. Detailed numerical solutions of the equation governing the evolution of the chemical potential $\mu$ [eq. (130) of Danese and De Zotti (1977)] after an energy injection have been obtained by Burigana (1989), taking into account both bremsstrahlung and radiative Compton. The results are summarized in Fig. 1.

For $\widehat{\Omega}_b \leq 0.3$ radiative Compton scattering dominates over bremsstrahlung in the thermalization of large distortions ($\mu \gg 1$) and we have $z_{therm} \simeq 2.3 \times 10^6\widehat{\Omega}_b^{-0.35}$, in very good agreement with the analytical estimate by Danese and De Zotti (1982). For $\widehat{\Omega}_b = 1$ we have $z_{therm} \simeq 2.2 \times 10^6$. In the case $\mu \leq 1$ we find $z_{therm} \simeq 1.9 \times 10^6\widehat{\Omega}_b^{-0.38}\mu^{0.17}$ for $\widehat{\Omega}_b \lesssim 0.3$. (All the above numerical coefficients correspond to $T_0 = 2.74\,K$; they are roughly proportional to $T_0$.)

Relaxation to a Bose–Einstein spectrum is possible only at redshifts larger than some critical value $z_{BE}$, corresponding to $t_C \approx t_{exp}$, $t_{exp}$ being the expansion timescale. More precisely, assuming a primordial helium abundance of 25% by mass and three species of massless neutrinos, we have:

$$z_{BE} \simeq 4.9 \times 10^4 \left(\frac{T_0}{2.7\,K}\right)^{1/2}\left(\frac{T_e}{T_R}\right)^{-1/2}\widehat{\Omega}_b^{-1/2}y_1^{1/2}, \qquad (2)$$

where $y_1$ is a slowly varying function of the baryon density, which increases from $\simeq 2$ to $\simeq 3$ as $\widehat{\Omega}_b$ decreases from 1 to 0.1 (Illarionov and Sunyaev, 1974; Sunyaev and Zeldovich, 1980; Burigana, 1989).

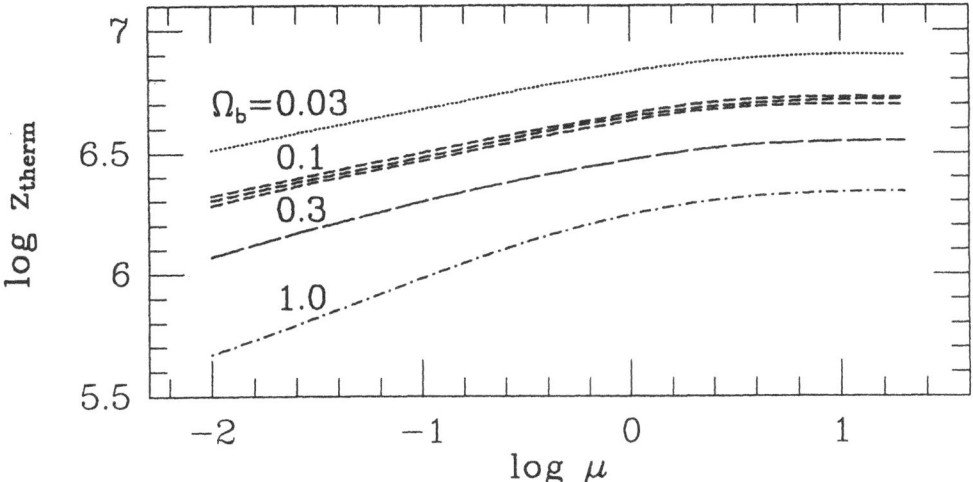

**Figure 1.** Minimum redshift for complete thermalization of a distortion characterized by a chemical potential $\mu$, for several values of $\Omega_b$ ($H_0 = 50$) and for $T_0 = 2.74$ K. The upper and lower dashed curves show the estimated uncertainty on $z_{therm}$ for $\Omega_b = 0.1$. Uncertainties for other values of $\Omega_b$ are of the same order.

For $z < z_{therm}$ photon emission is important only at low frequencies ($x = h\nu/kT_e \ll 1$). A blackbody spectrum is established below some critical frequency $x_C$ defined be the condition $t_{abs} \approx t_C$, where $t_{abs}$ is the timescale for photon absorption [actually, a more refined analysis, taking into account the energy gain of photons in each scattering, yields a value of $x_C$ a factor of $\sqrt{2}$ smaller than given by the above condition (Danese and De Zotti, 1980; Lightman, 1981)]. In the particularly interesting case of small distortions, the photon occupation number takes on the simple form (Sunyaev and Zeldovich, 1970a):

$$\eta_C = \{\exp[x + \mu(x)] - 1\}^{-1} ,$$
$$\mu(x) \simeq \mu_0 \exp(-x_C/x) \quad (\mu_0 \ll x_C). \tag{3}$$

The observed spectrum for $x \gg x_C$ is essentially the one formed at $z_{BE}$. As first pointed out by Illarionov and Sunyaev (1974), however, photon emission at $z < z_{BE}$ is able to blur out distortions up to a frequency significantly larger than $x_C$. An analytical approximation formula, taking this effect into account, reads (Danese and De Zotti, 1980):

$$\eta = \{1 - [1 - \eta_C (\exp(x) - 1)] \exp(-y_{abs})\} [\exp(x) - 1]^{-1} , \tag{4}$$

where $\eta_C$ is given by eq. (3) with $x_C$ computed at $z_{BE}$ [eq. (2)] and $y_{abs}$ is the optical depth of the universe for photon absorption at the effective redshift $z_*$. As illustrated by Figs. 2 and 3, a very good approximation to the results of direct numerical integrations of the kinetic equation (Burigana, 1989) is obtained with $z_* = 0.8 z_{abs} + 0.2 z_{BE}$, $z_{abs}$ being the redshift where $x_C$ coincides with the maximum frequency at which photon emission processes, taken alone, can re-establish a Planck spectrum.

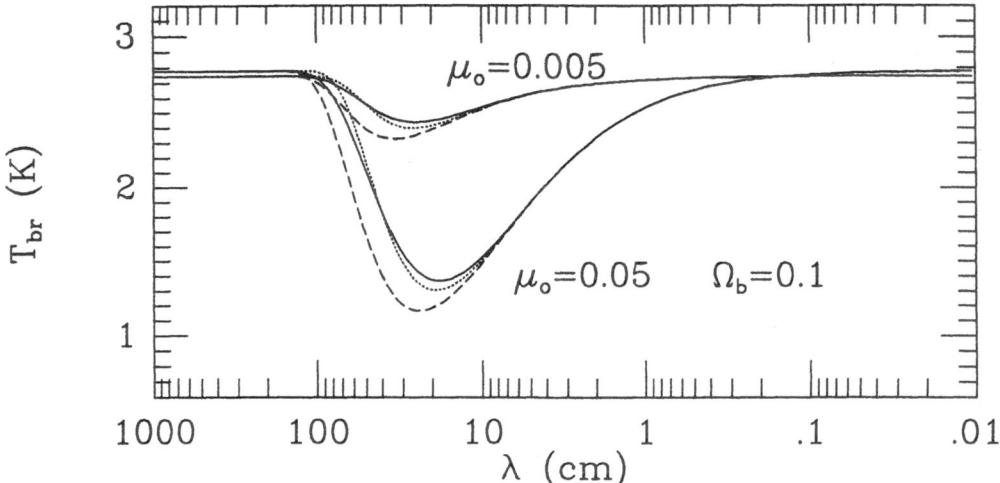

**Figure 2.** A comparison of analytical approximation formulae for BE–like distortions with the results of numerical integrations of the kinetic equation for $\widehat{\Omega}_b = 0.1$ and two values of the chemical potential $\mu_0$. The solid lines show the numerical results; the dashed lines correspond to eq. (4) with $z_* = z_{abs}$ (Danese and De Zotti, 1980); the dotted curve to eq. (4) with $z_* = 0.8 z_{abs} + 0.2 z_{BE}$.

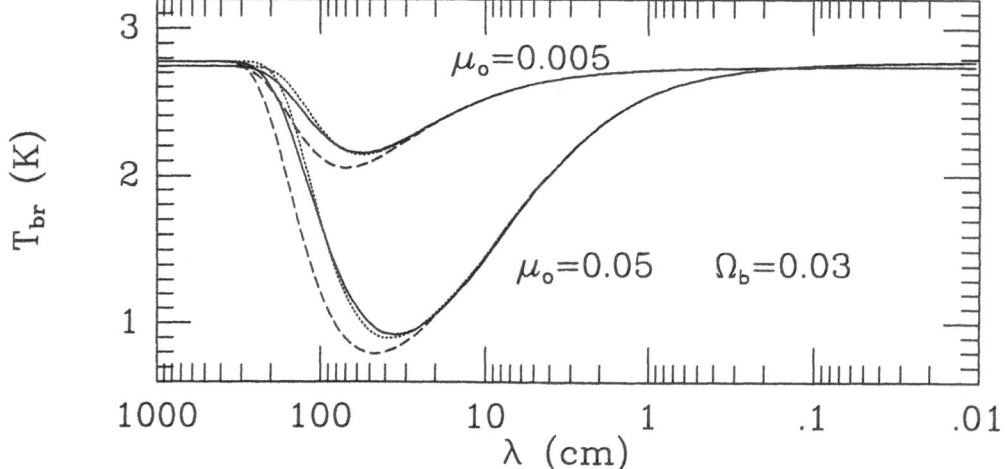

**Figure 3.** Same as in Fig. 2 but for $\widehat{\Omega}_b = 0.03$.

The distorted spectrum is characterized by a dip in the brightness temperature occuring at

$$\lambda_{max} \approx 4.5 \widehat{\Omega}_b^{-5/8} \ cm, \tag{5}$$

and whose amplitude is a measure of $\mu_0(z_{BE})$, which, in turn, is related to the amount of energy released. Note that $\lambda_{max}$ depends only on $\widehat{\Omega}_b$: a detection of a BE–like distortion would amount to a determination of the mean baryon density in the primeval plasma.

## 2.2 Distortions generated at $z < z_{BE}$

If the energy injection occurs after $z_{BE}$, kinetic equilibrium between radiation and matter cannot be achieved. The non–equilibrium occupation number of photons, $\eta(\nu, t)$, can be generally represented by means of a linear superposition of Planck spectra (Zeldovich et al., 1972), with a temperature distribution function specific to each particular situation. In the case of small distortions, however, the resulting spectrum is, to a good approximation, characterized by a single parameter

$$\widetilde{y} = \int \frac{k(T_e - T_R)}{m_e c^2} d\tau. \tag{6}$$

In fact, taking into account that for a Planck spectrum $\eta_P + \eta_P^2 = -\partial \eta_P / \partial x$ ($x = h\nu/kT_R$), the Kompaneets equation simplifies, at the first order on $\delta\eta$ and neglecting photon production processes, to the diffusion equation, whose exact solution has the form (Sunyaev and Zeldovich, 1980):

$$\eta(x, \widetilde{y}) = (4\pi\widetilde{y})^{-1/2} \int_0^\infty \eta_0(x') \exp\left[-\frac{(\ln(x/x') + 3\widetilde{y})^2}{4\widetilde{y}}\right] \frac{dx'}{x'}, \tag{7}$$

where $\eta_0(x)$ is the initial occupation number.

For large enough electron temperatures ($T_e \gg T_R$) direct and induced Compton scattering are negligible, so that the diffusion approximation holds even if the distortions are not very small. In this case, significant distortions may arise even at low redshifts. If, however, electron heating occurred at early times and the baryon density is not too low, free–free emission may be observed at radio frequencies (Zeldovich and Sunyaev, 1969; Stebbins and Silk, 1986), particularly around $\lambda \approx 40\,cm$ where the other backgrounds are low.

On the other hand, if electrons are very hot ($T_e \geq 3 \times 10^8\,K$) relativistic effects become significant (Wright, 1979; Sunyaev, 1980; Fabbri, 1981) and the distortion of the CMB spectrum must then be calculated using the exact frequency redistribution function of scattered photons.

Finally, as stressed by several groups (Bond et al., 1989, and references therein) radiation from dust at high redshift is expected to show up at sub–mm wavelengths. Hence measurements of the CMB spectrum would also be informative on the possibility of early generations of metal–forming stars.

## 2.3 Distortions due to the recombination.

The standard theory (Peebles, 1968; Zeldovich et al., 1968) predicts that the excess radiation due to recombination of the primeval plasma should contribute to the presently observed spectrum for $\lambda \leq 200\mu$m. Krolik (1989) has shown that previous calculations omitted important physical processes, which, however, nearly cancel each other for the commonly accepted values of the cosmological parameters. Unfortunately both the emission from interstellar dust, even in the direction of the galactic pole, and the integrated emission from unresolved galaxies exceed by at least four orders of magnitude the recombination radiation.

A number of lines, corresponding to excited states of the hydrogen and helium atoms are also expected in the Rayleigh–Jeans region (Dubrovich, 1975), but their intensities are extremely low in the standard model (Bernshtein *et al.*, 1977): $\Delta I/I \lesssim 10^{-6}$. However, if the spectrum of the CMB was distorted at the recombination epoch, transitions between excited levels occur with much higher frequencies and spectral features are correspondingly more prominent (Lyubarsky and Sunyaev, 1983): their fractional intensity can reach $\approx 10^{-3}$ in the case of BE–like spectra, and may be even larger in the case of comptonization distortions (Sect. 2.2). As emphasized by Lyubarsky and Sunyaev (1983), a detection of these features would provide interesting pieces of information on the epoch of energy dissipation, on the mean baryon density in the primeval plasma, on the primordial helium abundance and would confirm the relic nature of the CMB.

## 3. IMPLICATIONS OF MEASUREMENTS OF THE CMB SPECTRUM

### 3.1 The submillimeter excess

The most spectacular recent result is the detection (Matsumoto *et al.* 1988b), of a large excess in the Wien region, qualitatively similar to that reported earlier by Gush (1981), comprising $\approx 10\%$ of the total energy of the CMB. The upper limits set by Bernstein *et al.* (1989) on intermediate–to–large scale fluctuations of this excess sharply constrain the possibility of a local origin. On the assumption that it is real and non–local, various possible explanations have been proposed and discussed (Hayakawa *et al.*, 1987; Carr, 1988; Smoot *et al.*, 1988; Bond, 1988; Bond *et al.*, 1989; De Zotti and Toffolatti, 1989).

#### 3.1.1 Non–relativistic comptonization

A first possibility is comptonization by non–relativistic electrons. Indeed, a distortion of this kind was predicted by Field and Perrenod (1977) as associated to a hot intergalactic plasma yielding the $2$–$50\,keV$ X–ray background. The best fit to the currently available data as listed in Table 1 of De Zotti *et al.* (1989), corresponds to $T_0 \simeq 2.82\,K$ and to a comptonization parameter $y \simeq (2.02 \pm 0.16) \times 10^{-2}$. The remarkably close agreement with Field and Perrenod's prediction ($y \simeq 1.5 \times 10^{-2}$) is to be taken as fortuitous, however (see Sect. 3.1.2). Still, distortions of this kind may be expected in many different frameworks: dissipation of primeval large-scale turbulence (Chan and Jones, 1975a,b) or strong density fluctuations (Sunyaev and Zeldovich, 1970b; Daly, preprint); large–scale explosions (Yoshioka and Ikeuchi, 1987); superconducting cosmic strings (Ostriker and Thompson, 1987); decaying particles which reionize the medium and heat the electrons (Fukugita, 1988; Fukugita *et al.*, 1989; Field and Walker, 1989; Wang and Field, 1989); decaying vacuum (Freese *et al.*, 1987; Bartlett and Silk, 1989).

General energetic constraints have been investigated by Lacey and Field (1988), who conclude that the energy cannot be provided by stars formed with the same mass function as is observed in the solar neighborhood.

Radiative decays of massive particles must also comply with limitations from other electromagnetic backgrounds (Silk and Stebbins, 1983, and references therein). For example, current limits on the isotropic UV background set an upper bound to lifetimes of massive particles with number densities comparable to those of standard neutrinos, whose decay photons can inject enough energy to yield $y \simeq 2 \times 10^{-2}$ ($z_{decay} > 5700$; Field and Walker, 1989; see also Dar *et al.*, 1989). In the case of shorter lifetimes ($z_{decay} \approx 10^4$–$10^5$), decay photons would show up, today, at $\lambda \gtrsim 1\,\mu$m with a flux consistent with available limits (Fukugita *et al.*, 1989); with

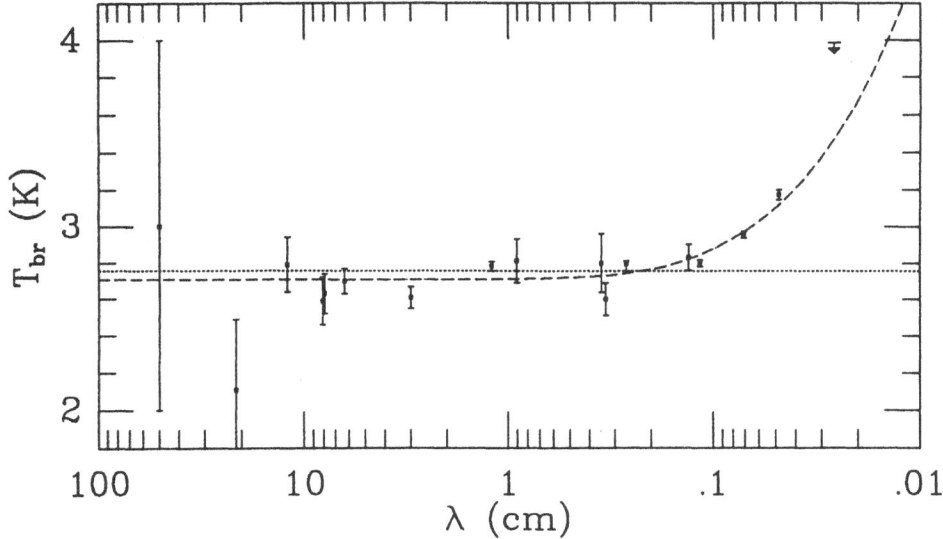

**Figure 4.** Best fit non–relativistic Compton distortion (dashed line): $y = 2.02 \times 10^{-2}$, $T_0 = 2.82 \, K$, $\chi^2_{min} = 31.4$ for 15 degrees of freedom.

a suitable choice of the branching ratio, the decay photons could also account for the infrared background intensity reported by Matsumoto *et al.* (1988a).

As discussed by Fukugita *et al.* (1989) and Field and Walker (1989), these decay models are consistent with constraints coming both from stellar evolution models (Raffelt *et al.*, 1989) and from the absence of $\gamma$–rays from SN 1987A at the time of the detection of the neutrinos by the Kamiokande II detector. On the other hand, they entail dramatic effects on the dynamics and on the thermal history of the universe (Dar *et al.*, 1989). For representative values of the parameters (comoving number density $n_{X_0} \approx 100 \, cm^{-3}$, i.e. approximately equal to that of conventional neutrinos, lifetime $\tau \approx 10^3$ yr, mass $m_X \approx 10 \, keV$), a matter dominated era sets in at $z_{RD} \simeq 4 \times 10^6 (n_{X_0}/100 \, cm^{-3})(m_X/10 \, keV)$. When particles decay at $z_{decay} \approx 10^4$, the universe turns into a second radiation dominated era which ends at $z_{MD} \approx 25(z_{decay}/10^4)/[(n_{X_0}/100 \, cm^{-3})(m_X/10 \, keV)]$. An obvious constraint on $n_{X_0} \cdot m_X$ follows from the condition that $z_{RD} \ll 10^9$ in order to avoid perturbing the standard big bang nucleosynthesis. Also, in order to get a CMB temperature of $\simeq 2.75 \, K$ and a present age of the universe $t_0 \geq 1.3 \times 10^{10}$ yr we must have $(m_X/10 \, keV)(\tau/10^3 \, yr)^{1/2} \leq 12(t_0/1.3 \times 10^{10} \, yr)^{1/2}/(n_{X_0}/100 \, cm^{-3})$. Finally, but perhaps more importantly, these models entail a very late beginning of the final matter dominated era. Correspondingly, the growth factor $f_g$ for linear perturbations inside the horizon [$f_g \approx z_{MD}$ (cf. Peebles, 1980, §12)] is bound to be relatively small, implying severe problems for galaxy and cluster formation, at least in the framework of the gravitational instability picture (remember that in a radiation dominated universe these perturbations essentially do not grow).

It may also be noted that the fit is not great ($\chi^2_{min} = 31.4$ for 15 degrees of freedom), mainly because there is an apparent conflict between the two sub–mm data points, favouring an even larger value for $y$, and the two most accurate results at $\lambda > 1\,\text{mm}$ (Johnson and Wilkinson, 1987; Crane *et al.*, 1989), which are consistent with $y = 0$. This clearly illustrates the importance of new accurate measurements in the Rayleigh–Jeans region as a test for this class of models. In addition, models assuming electron heating at high redshifts also predict a substantial bremsstrahlung emission at long wavelengths ($\lambda > 10\,\text{cm}$; cf. Bartlett and Silk, 1989); improving our knowledge of the CMB spectrum in this range would allow important constraints on the temperature history of the IGM (Zeldovich and Sunyaev, 1969; Stebbins ans Silk, 1986).

If the submillimeter excess is attributed to something different from comptonization by non–relativistic electrons, the data yield: $y = (8.3 \pm 4.0) \times 10^{-3}$, or a $2\sigma$ upper limit $y \leq 1.63 \times 10^{-2}$.

### 3.1.2  *Hot IGM models for the X–ray background*

As first pointed out by Wright (1979) the original Field and Perrenod's calculation was not entirely self–consistent, since non–relativistic formulae were used even though most of the X–ray emission came from mildly relativistic electrons.

Using the formulae derived by Fabbri (1981), we have computed the comptonized microwave background spectra corresponding to the detailed models for the X–ray background (XRB) worked out by Guilbert and Fabian (1986), Barcons (1987) and Barcons and Fabian (1988). The initial radiation temperature has been taken as a free parameter, to be determined from a fit to the measurements of the CMB spectrum. Some of our results are presented in Figs. 5 to 7. A few points are worth noticing. First, comptonization by a hot intergalactic medium (IGM) yielding the X–ray background does not produce the correct shape for the spectral distortion, as first stressed by Hayakawa et al. (1987). In general, the predicted CMB brightness temperatures at $709\,\mu\text{m}$ and $481\,\mu\text{m}$ are well below the observational data, while the upper limit at $262\,\mu\text{m}$ tends to be exceeded.

We may, however, assume that the sub–mm excess is due to a different process, so that the corresponding measurements only provide upper limits to a comptonization distortion. Then, the best fit initial CMB temperature decreases, and the fit to the points at $\lambda \geq 1\text{mm}$ is much better (see Fig. 8).

Consistency with the upper limit at $262\,\mu\text{m}$ is much more difficult to achieve. A simple physical argument indicates that, to this end, the effective electron temperatures must be kept as low as possible. In fact, relativistic effects increase the average frequency shift of microwave photons in a single scattering (Rybicki and Lightman, 1979): when electron are non–relativistic $\langle \Delta\nu/\nu \approx 4kT/m_e c^2 \ll 1 \rangle$, but $\langle \Delta\nu/\nu \rangle \approx \left(4kT/m_e c^2\right)^2 \gg 1$ when they are ultrarelativistic. In the two–phase model for the IGM discussed by Barcons and Fabian (1988), the scattering by the high temperature component is actually so efficient to yield a $100\,\mu\text{m}$ flux exceeding the upper limit set by Boulanger and Pérault (1988) on the isotropic background at this wavelength. (We note, in passing, that, for the same reason, Compton cooling of the IGM by CMB photons cannot be neglected in the case of the hot phase, even for $z < 3.5$; correspondingly, the required amount of energy and the amplitude of the predicted sub–mm excess are much bigger than estimated by Barcons and Fabian).

The above expectation is fully born out by detailed calculations: the upper limit at $262\,\mu\text{m}$ is exceeded by a factor increasing with increasing $z_H$, i.e. with increasing effective electron temperature. In the case of Guilbert and Fabian's (1986) models the ratio of predicted to observed flux at $262\,\mu\text{m}$, $(F_{pred}/F_{obs})_{262\,\mu\text{m}}$, is $\geq 2.2$ for model (a) ($z_H = 3.6$), $\geq 2.6$ for model (f) ($z_H = 4.6$), $\geq 3.6$ for model (g) ($z_H = 5.5$) and is $\geq 3.9$ for model (b) ($z_H = 6.1$). Barcons' (1987) models

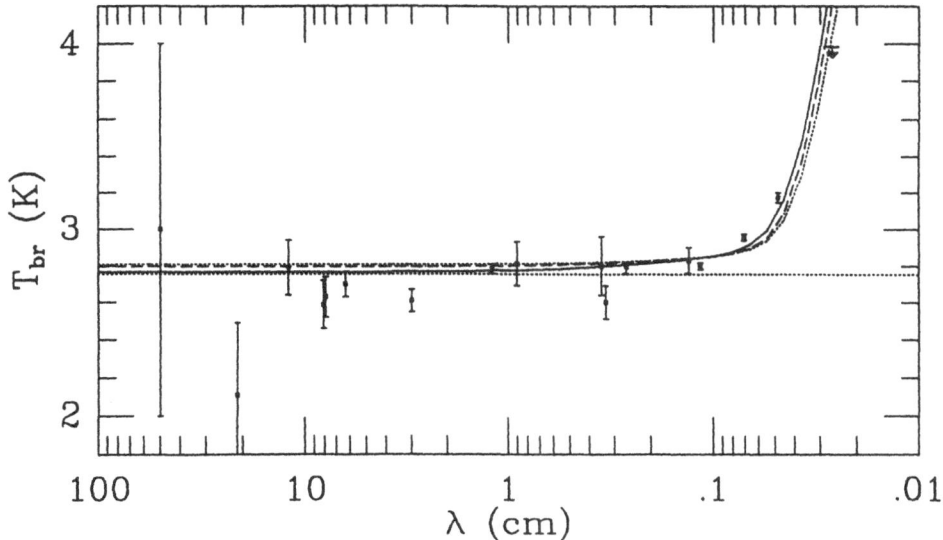

**Figure 5.** CMB distortions predicted by Guilbert and Fabian's (1986) models (a) (dashed curve; $\chi^2_{min} = 87.8$ for 16 degrees of freedom), (b) (solid curve; $\chi^2_{min} = 63.5$) (c) (dotted curve; $\chi^2_{min} = 102.8$). These models assume instantaneous heating at $z_H = 3.6$, 6.1 and 3.6, respectively. Models (a) and (b) take into account cooling by both adiabatic expansion and Compton scattering; model (c) ignores Compton cooling.

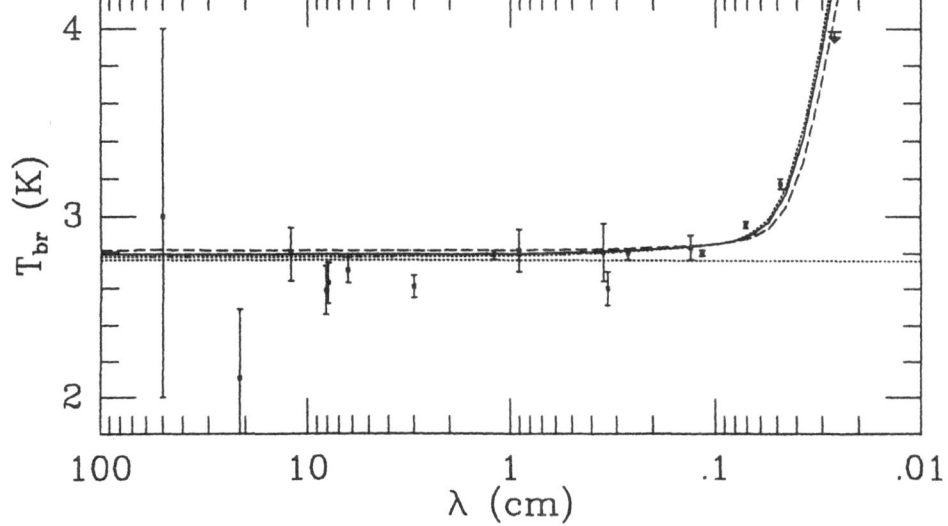

**Figure 6.** CMB distortions predicted by Guilbert and Fabian's (1986) models (e) (dashed curve; $\chi^2_{min} = 104$ for 16 d.o.f.), (f) (solid curve; $\chi^2_{min} = 72.3$) and (g) (dotted curve; $\chi^2_{min} = 63.3$). These models assume a Field–Perrenod temperature profile, with $z_H = 3$, 4.6 and 5.5, respectively. They correspond to a more extended heating era, in comparison to models (a), (b) and (c) (Fig. 5); also the cooling at $z \geq 3.5$ is slower than in the case of models taking into account Compton losses.

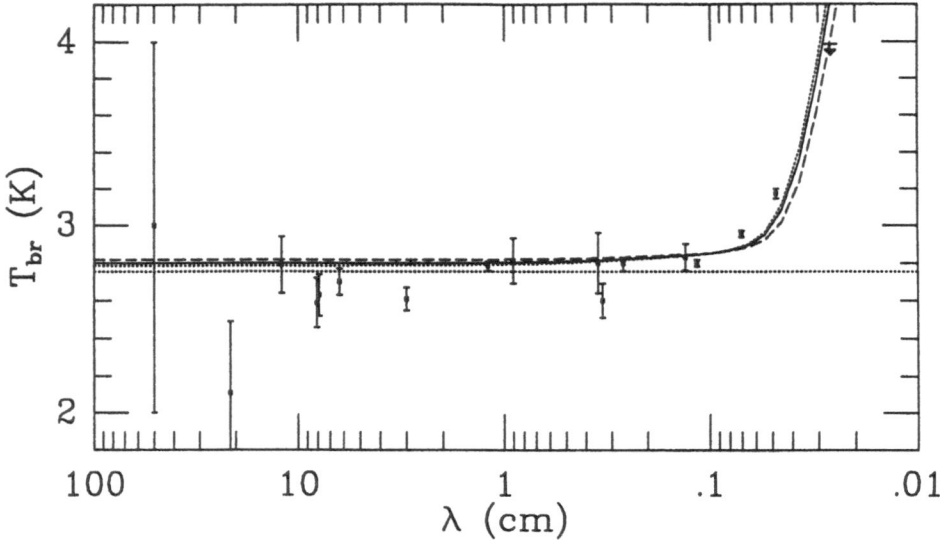

**Figure 7.** CMB distortions predicted by Barcons's (1987) models with $z_H = 2$ (dashed curve; $\chi^2_{min} = 124$ for 16 d.o.f.), $z_H = 3$ (solid curve; $\chi^2_{min} = 89.9$) and $z_H = 4$ (dotted curve; $\chi^2_{min} = 80.2$).

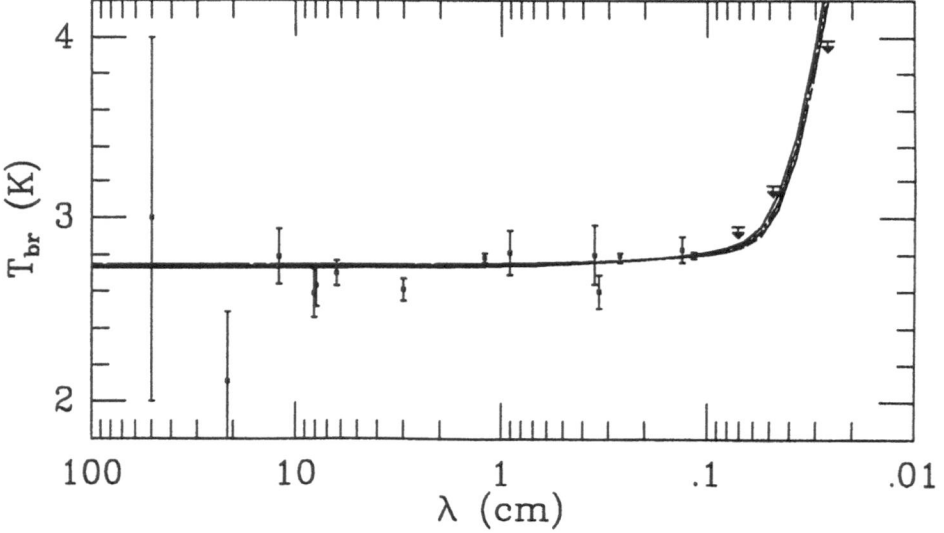

**Figure 8.** Best fit CMB distortions associated to some hot IGM models for the X-ray background, computed treating the sub–mm points by Matsumoto et al. as upper limits. Plotted are predictions of Guilbert and Fabian's (1986) models (a) ($\chi^2_{min} = 22.4$ for 16 d.o.f.), (b) ($\chi^2_{min} = 32.35$), (f) ($\chi^2_{min} = 24.14$) and (g) ($\chi^2_{min} = 29.95$), and of Barcons's models with $z_H = 3$ ($\chi^2_{min} = 22.1$) and $z_H = 4$ ($\chi^2_{min} = 27.5$). Differences between the values models are too small to be discernable in the figure; what can be seen is that in all cases the upper limit at 262 $\mu$m is exceeded (see text).

yield $(F_{pred}/F_{obs})_{262\,\mu m} \geq 2.1$ for $z_H = 3$ and $(F_{pred}/F_{obs})_{262\,\mu m} \geq 3.2$ for $z_H = 4$. The increasing disagreement with the data at $262\,\mu m$ translates in an increase of $\chi^2_{min}$ with increasing $z_H$ (see caption to Fig. 8). The quality of the fit depends also, to some extent, on the details of the models; in general, models assuming a more extended heating era (for example, those adopting a temperature history of the kind proposed by Field and Perrenod) score better.

On the other hand, the energy spectral index of the bremsstrahlung emission from the IGM below $\simeq 20\,\mathrm{keV}$ (in the observers frame) must be flat enough to match the *residual* XRB spectrum (foreground emission from clusters of galaxies and AGN's subtracted; see Leiter and Boldt, 1982). To achieve this, relativistic effects are essential. If, as is commonly accepted, the foreground contribution in the 2–10 keV is $\geq 30\%$ and has an effective energy spectral index $\simeq 0.7$, the effective temperature of the IGM, $T_*$, must be $\geq 200\,\mathrm{keV}$ (Boldt, 1987; Fabian, 1988) or the effective redshift $z_* \gtrsim T_*/T_{XRB} - 1 \geq 5.6$.

We conclude that a hot intergalactic gas cannot account for both the XRB and the submillimeter excess reported by Matsumoto *et al.* (see also Rogers and Field, 1989; Taylor and Wright, 1989). Hot IGM models for the X–ray background can be made consistent with the measurements by Matsumoto *et al.* (1988b) only if relativistic effects are very small; in this case, however, it is very difficult to account for the residual XRB spectrum unless the foreground contribution is lower than currently believed. Anyway, the forthcoming measurements of the microwave background spectrum at sub–mm wavelenths are expected to provide a definitive test on hot IGM models for the XRB.

### 3.1.3  *High redshift dust*

Bond *et al.* (1986), Negroponte (1986) and McDowell (1986) pointed that dust at high redshifts would emit in the far–IR. The peak wavelength, redshifted to the present, turned out to be only very weakly sensitive to the parameter values; for a reasonable choice of the latter, it was predicted to be $\lambda_{pk} \approx 700\,\mu m$, in close agreement with the peak wavelength of the excess of Matsumoto *et al.*. Hayakawa *et al.* (1987) showed that a very good fit can indeed be obtained with a single temperature dust emission spectrum superimposed on a Planck spectrum at $T = 2.74\,K$. De Bernardis *et al.* (1985, 1989) emphasized that the redshifted light of stars heating the dust could explain the near–IR background intensity reported by Matsumoto *et al.* (1988a).

A test for a dust origin of the submillimeter excess could be provided by polarization measurements: in the presence of a large scale magnetic field, the emission from non–spherical grains would be appreciably polarized (Rudak and Panek, 1989; de Bernardis *et al.*, 1989)

On the other hand, stars formed with the same mass function as in the solar neighborhood cannot supply the required energy (Lacey and Field, 1988); other sources such as Very Massive Objects, accreting black holes or decaying particles are needed. Also, Draine and Shapiro (1989) have shown, independently of specific assumptions on the optical properties of grains (but with the exception of conducting grains which are either highly elongated or flattened or of extremely low internal density), that very large amount of dust must be present at $z \gtrsim 10$. Now, it is extremely difficult to generate enough dust at sufficiently high redshifts both in the framework of scale invariant biased Cold Dark Matter models and of pancake models; hierarchical models with more power at small scales could, however, do the job (Bond *et al.*, 1989).

As already mentioned, the lower limit on the redshift at which dust must be present (and the corresponding constraint on the amount of energy required to account for the sub–mm excess) can be relaxed if the grains are long conducting needles (whiskers) as proposed by Hawkins and Wright (1988). However, no convincing observational evidence has been found so far that such needles really exist. The good agreement between predictions of the standard grain model and observations of the dust emission spectrum of our Galaxy [see Cox and Mezger (1989) for a recent review] suggests that whiskers are not an important component of galactic dust.

Furthermore, very special conditions must be met in order to account for the detailed shape of the observed excess in terms of emission from normal grains. For example, under the usual assumption of a dust absorption efficiency approximately constant at the effective frequencies of the heating radiation, and proportional to $\lambda^{-\alpha}$, , with $\alpha \approx 2$, at far–IR wavelengths, the temperature of a grain of size $r_g$ heated by a radiation field of intensity $I$, is roughly proportional to $(I/r_g)^{1/6}(1 + z)^{-1/3}$. Even if the intensity of the radiation field is strictly uniform, the standard grain size distribution, ranging from 50Å to 2500Å, translates into a temperature distribution having a width of at least $(2500/50)^{1/6} \simeq 2$. A further substantial smearing out is expected, due to inhomogeneities in the radiation field intensity. In fact, as illustrated by De Zotti and Toffolatti (1989), the dust emission spectra observed in a variety of astrophysical settings are much broader than the spectrum of the Matsumoto et al.'s excess (see also Fig. 9).

The additional broadening due to the redshift distribution of the emitting dust may be a minor effect if the relevant redshift range is not too large. Thus it is not a surprise that models assuming a single dust temperature at each redshift, but allowing for realistic dust temperature histories (Bond et al., 1989; Adams et al., 1989) yield a good fit.

A third severe constraint on dust emission models comes from the upper limit on fluctuations of the CMB set by Kreysa and Chini (1989) at $\lambda \simeq 1.3$ mm: $\delta T/T \leq 2.6 \times 10^{-4}$ (95% confidence) for a scale of 30''. As shown by Bond et al. (1989), this limit implies that if the dust is clumped, it must be housed in structures with comoving density $n \geq 4.6(H_0/50)^3$ Mpc$^{-3}$, much larger than the density of normal galaxies. If these structures are clustered with $\xi(r) = (r_0/r)^{1.8}$, their comoving clustering scale must be $r_0 \lesssim 0.12(50/H_0)$ Mpc, i.e. much smaller than that of normal galaxies.

### 3.1.4  *Radiation from decaying particles*

As already discussed in Sect. 3.1.1, decaying particles have been proposed as providing the energy needed to distort the CMB spectrum via comptonization. An alternative possibility is that the sub–mm excess is made up by direct decay radiation. Kawasaki and Sato (1987) considered decays occurring before recombination, but after the time at which the produced photons can be driven to thermal or kinetic equilibrium. Taking into account the effect of Compton scattering, bremsstrahlung and radiative Compton, they concluded that with an appropriate choice of the mass ($\approx 5$ eV) of the lifetime $\approx 10^{13}$ s and of the number density ($\approx 1/10$–$1/20$ of that of photons) it is possible to account for the observed CMB spectrum. Although their analysis is not entirely self consistent [their formula for computing the electron temperature in the perturbed radiation field is not accurate (De Zotti and Toffolatti, 1989) and their expression for the spectrum of decay photons is oversimplified (see Silk and Stebbins, 1983)], their conclusion essentially stands, only with somewhat

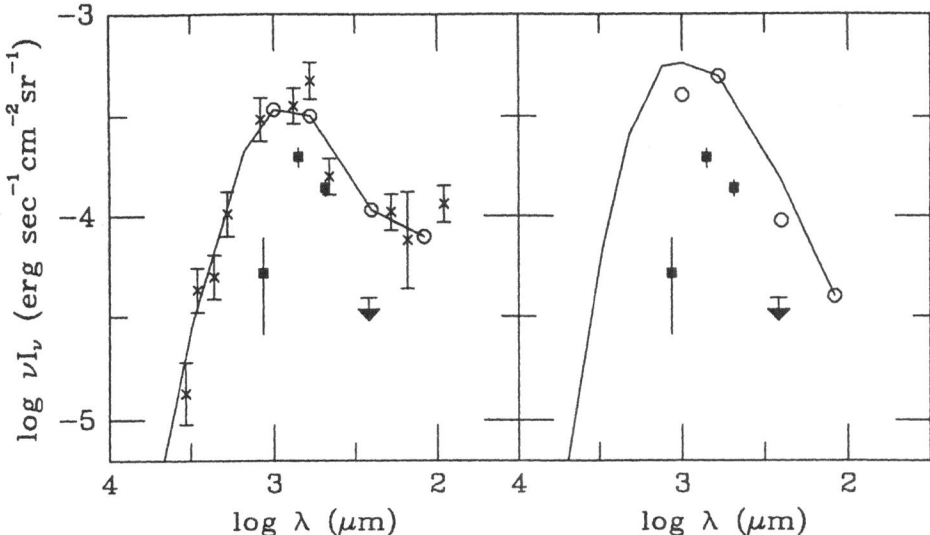

**Figure 9.** Comparison of the Matsumoto et al.'s (1988b) excess (filled squares and arrow) with observed shapes of dust emission spectra. The panel on the left hand side shows the average observed far–IR spectrum of non–Seyfert Markarian galaxies (○) and of our own Galaxy (taken from Cox and Mezger, 1989; ×) together with the best–fit model of Xu and De Zotti (1989) for Markarians (solid line). The other panel displays the mean far–IR spectra of compact HII regions (Chini et al. 1987; solid line) and of ultraluminous IRAS galaxies (○). Dust emission spectra have been arbitrarily redshifted and scaled in flux to ease the comparison with the data by Matsumoto et al. (1988b). The dotted line is shows the best fit to those data in terms of single temperature dust emission.

different values of the parameters (Burigana, 1989).

However, Raffelt *et al.* (1989) showed that the full range of masses and lifetimes over which models of this kind may work, is ruled out because it would imply a decrease of the helium burning lifetime of low mass stars beyond observational limits. The basic idea is that the same electromagnetic coupling which is responsible for the radiative decay of a particle $X$ ($X \to X' + \gamma$), may generate, in dense stellar cores, a pair $X'\overline{X}$ or $X\overline{X'}$ by plasmon decay. These weakly interacting particles may eventually freely escape causing an anomalous cooling which strongly affects the evolution on the red giant branch. This argument is very powerful, because it relies only on consideration of the energy losses from stars and are independent of the appearance of decay photons.

Further important constraints on radiative decays (see Dar *et al.*, 1989 and references therein) come from laboratory bounds on transition moments and from upper limits on the $\gamma$–ray flux from SN 1987A following its neutrino burst. All these are inconsistent with models for the sub–mm excess of the kind discussed above.

Yet another class of models involves very light ($m_X \approx$ few $\times 10^{-3}$ eV) unstable particles decaying close to the present time ($\tau \approx 10^{10}$ yr). As discussed by Dar *et al.* (1989), however, astrophysical and particle physics arguments indicate that the required transition rates are highly implausible. Furthermore, detailed evaluations of photon decay spectra show that those values of the parameters yielding a good fit

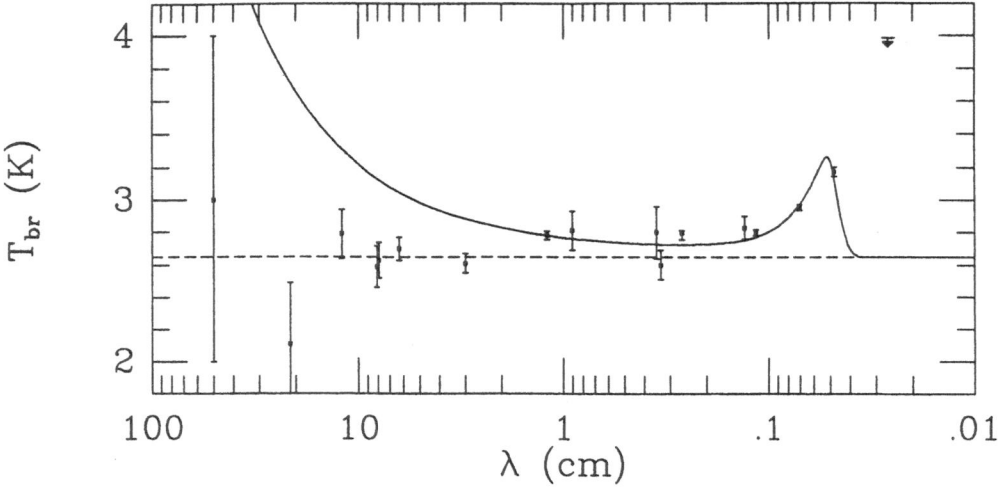

**Figure 10.** An example of CMB spectral distortions due to a late decay of a very light weakly interacting particle: $m_X = 5 \times 10^{-3}\,eV$, $\tau = 0.9 \times 10^{10}\,yr$, $n_{X_0} = 100\,cm^{-3}$, $T_0 = 2.65\,K$, $H_0 = 50$, $\Omega = 1$.

to the sub–mm excess, also imply an unacceptably large flux in the Rayleigh–Jeans region (see Fig. 10), due to the relatively rare decays ($\approx 10\%$ of the total) which produce photons in that spectral region, where the CMB has few photons and little energy.

In conclusion, all particle decay models envisaged to fit the sub–mm excess have serious drawbacks.

## 4. DIFFERENTIAL MEASUREMENTS OF THE CMB AS TOOLS TO SEARCH FOR SPECTRAL DISTORTIONS.

Two methods have been proposed which allow to gain information on the shape of a background radiation spectrum exploiting differential measurements. Although indirect, these methods have two important advantages: they bypass the difficulties associated to absolute calibrations and automatically distinguish between local and extragalactic contributions to the observed flux.

If the dipole anisotropy of the CMB is due to the Doppler effect associated to the peculiar motion of the observer with respect to the reference frame where the CMB is isotropic, the flux received at a fixed *observed* frequency from different directions, comes from slightly different *intrinsic* frequencies. Thus the amplitude of the dipole $(\Delta T)_d$ carries information on the slope of the spectrum around the observed frequency. Since $(\Delta T)_d$ is independent of frequency in the case of a Planck spectrum, any observed frequency dependence would be indicative of a spectral distortion (Danese and De Zotti, 1981; Lubin, 1982). In terms of the photon occupation number $\eta$, we have, to first order in $V/c$

$$\frac{\Delta\eta}{\eta} \simeq \frac{d\ln\eta}{d\ln\nu}\frac{V}{c}. \tag{8}$$

The anisotropy of the brightness temperature is

$$(\Delta T)_d = -\frac{h\nu}{k(1+\eta)} \ln^{-2}\left(1 + \frac{1}{\eta}\right) \frac{d\ln\eta}{d\ln\nu} \frac{V}{c}. \tag{9}$$

It is easy to see that in the case of a Planck spectrum eq. (9) simplifies to $(\Delta T/T)_d \simeq V/c$.

As shown by Fig. 11, different models for the sub–mm excess imply significantly different spectra of the dipole anisotropy, as expected since the latter is proportional to the first derivative of the radiation brightness and is therefore sensitive to the detailed shape of the distortion.

A general implication of the Berkeley–Nagoya excess is a *decrease* of $(\Delta T/T)_d$ at sub–mm wavelengths. This expectation is not borne out by the measurements of Halpern *et al.* (1988), although the disagreement is only at the $2\sigma$ level. Also the direction of the dipole found by these authors is about $4\sigma$ away from that determined by the most accurate experiment at longer wavelengths; it may be that emission from the Galaxy or some bright sub–mm source has somewhat contaminated the data.

The Sunyaev–Zeldovich (1972) effect in the direction of rich clusters of galaxies provides another way to test the spectral shape of the CMB by means of differential measurements (Gould and Rephaeli, 1978; Fabbri and Melchiorri, 1979; Rephaeli, 1980; Wright, 1983). The change in the CMB brightness temperature due to scattering of photons by hot $(kT_e \gg h\nu)$ electrons is essentially a second order Doppler effect, connected with the twofold transition from the radiation frame to the rest frame of the scattering electron and back (see Danese and De Zotti, 1977, §2.2.2.1). The amplitude of the effect is thus proportional to the second derivative of the intensity at the frequency of observations:

$$\frac{\partial I}{\partial y_C} = \frac{\partial^2 I}{\partial(\ln x)^2} - 3\frac{\partial I}{\partial \ln x}, \tag{10}$$

where $y_C$ is the comptonization parameter of the cluster and $x = h\nu/kT_0$. If we assume $I \propto x^\alpha$, then $\partial I/\partial y_C \propto \alpha(\alpha - 3)$. In the Rayleigh–Jeans region, $\alpha = 2$; it then decreases with increasing frequency and becomes negative in the Wien region. Correspondingly, the Sunyaev–Zeldovich effect changes it sign around the peak of the CMB. Again, the spectrum of this effect is sensitive to the detailed shape of the CMB spectrum (Fig. 12).

Obviously any local contribution to the observed intensity does not have a dipole pattern and is not affected by the Sunyaev–Zeldovich effect. Hence, detailed measurements of this kind can single out truly extragalactic contributions to the observed flux.

## 5. THE SPECTRUM AT $\lambda > 1\,\mathrm{mm}$

Even ignoring the results of Matsumoto *et al.* (1988b) the fit to an unperturbed blackbody spectrum is not fully satisfactory. We find $\chi^2_{min} = 18.5$ for 11 degrees of freedom, for a best–fit blackbody temperature $T_{BB} = 2.755 \pm 0.017\,K$. Although the significance of such large $\chi^2_{min}$ should not be over–emphasized in view of the possibility of unrecognized systematic errors, it may be noted that there is a tendency for the brightness temperatures to be slightly lower at *cm* than at *mm*

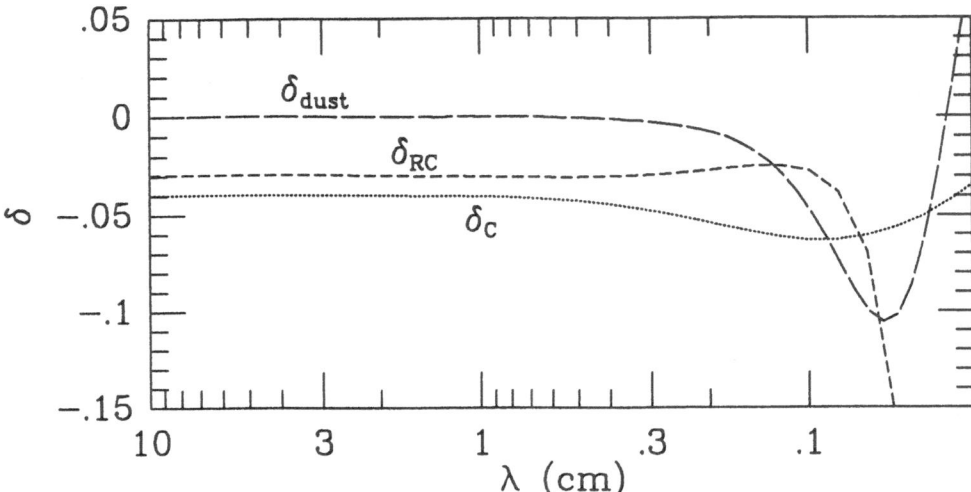

**Figure 11.** Predicted fractional variations of the dipole amplitude $\{\delta = [(\Delta T/T)_d - V/c]/(V/c)\}$ with wavelength corresponding to different best fit models for the sub-mm excess: single temperature dust emission (long dashes), relativistic comptonization (short dashes), non-relativistic comptonization (dotted line).

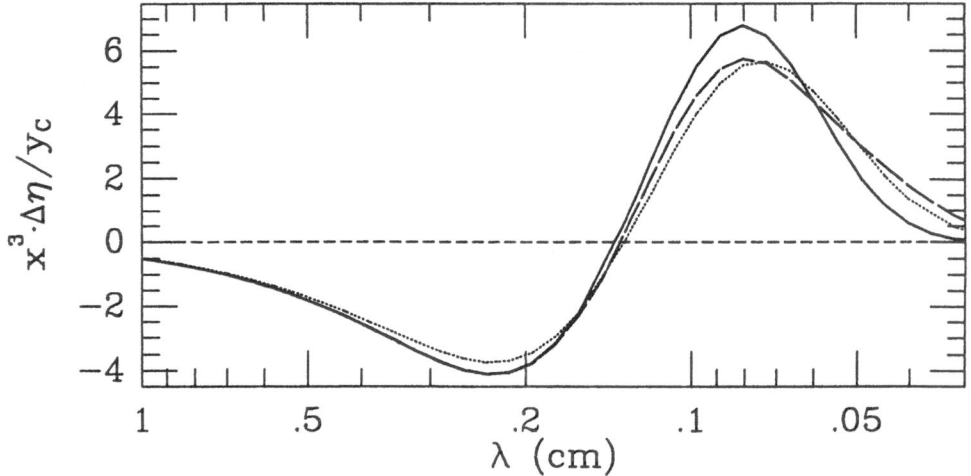

**Figure 12.** Predicted Sunyaev–Zeldovich effect corresponding to a Planck spectrum (solid line) and two models for the sub-mm excess: single temperature dust emission (long dashes) and non-relativistic comptonization (dotted line). Here $x = h\nu/kT_0$, $x^3 \Delta\eta$ is a quantity proportional to the change of the CMB intensity, $y_C$ is the comptonization parameter for the intergalactic medium in the cluster.

wavelengths. The weighted mean of data at $\lambda \geq 3\,\mathrm{cm}$ is $T_l = 2.632 \pm 0.037\,K$, while for data at $3 < \lambda \leq 0.1\,\mathrm{cm}$ we find $T_s = 2.793 \pm 0.01\,K$.

It is of interest to see whether this is indicative of a Bose–Einstein distortion. Leaving aside the data by Matsumoto *et al.* (1988b), the best fit values of the parameters for $\widehat{\Omega}_b = 0.1$ are: $\mu_0 = (4.1 \pm 1.7) \times 10^{-3}$, $T_0 = 2.793 \pm 0.019\,K$;

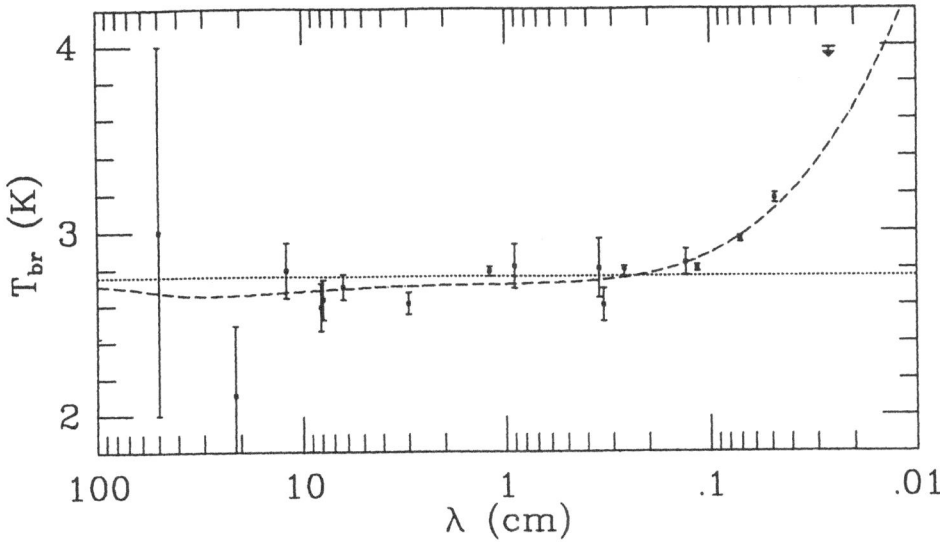

**Figure 13.** Non–relativistic comptonization of a Bose–Einstein like spectrum for $\widehat{\Omega} = 1$, $\widehat{\Omega}_b = 0.1$. The best fit values of the parameters are $T_0 = 2.82\,K$, $y = 0.02$, $\mu_0 = 0.8 \times 10^{-3}$; $\chi^2_{min} = 31.1$ for 14 degrees of freedom.

$\chi^2_{min} = 11.8$ for 10 degrees of freedom. The minimum $\chi^2$ is slightly larger both for $\widehat{\Omega}_b = 1$ [$\chi^2_{min} = 12.4$, $\mu_0 = (1.08 \pm 0.44) \times 10^{-2}$] and for $\widehat{\Omega}_b = 0.01$ [$\chi^2_{min} = 12.4$, $\mu_0 = (3.4 \pm 1.5) \times 10^{-3}$] (the results are insensitive to the value of $\widehat{\Omega}$, as a consequence of the fact that Bose–Einstein distortions arise in the radiation–dominated era). A non–zero chemical potential is suggested at the $\simeq 2.4\sigma$ level.

If the latter indication, however weak, is taken seriously, comptonization models for the sub–mm excess must take into account that the initial spectrum was not planckian. An example of the resulting best fit spectrum for the case of non–relativistic comptonization is shown in Fig. 13. The overall quality of the fit is not significantly improved in comparison with the case of comptonization of a Planck spectrum: we have one additional parameter and $\chi^2_{min}$ is decreased by only 0.3.

## 6. CONCLUSIONS

The thermalization redshift of arbitrarily large distortions is now well determined. For the most commonly accepted values of the baryon density ($\Omega_b \leq 0.3$) it is given by a very simple formula: $z_{therm} \simeq 2.3 \times 10^6 [(H_0/50)^2 \Omega_b]^{-0.35}$. An analytic formula accurately describing early distortions has been obtained.

So far, no fully satisfactory explanation of the sub–mm excess has been found. Models invoking radiatively decaying particles either face serious difficulties with astrophysical and particle physics constraints or involve embarassingly drastic changes of the dynamical and thermal history of the universe. The mildly relativistic comptonization implied by hot IGM models for the X–ray background (XRB) cannot account for the excess; the upper limit at 262 $\mu$m entails so severe constraints on the temperature history of the IGM, that it is doubtful whether models of this kind are actually consistent with the residual spectrum of the XRB (after subtraction of the contribution of known discrete sources). The two most accurate measurements at $\lambda > 1\text{mm}$ disagree with non–relativistic comptonization models; also they require somewhat exotic energy sources. Dust reradiation models are difficult to reconcile with Kreysa and Chini (1989) upper limit on small scale fluctuations at 1.3 mm. Also very special conditions must be met in order to account for the detailed shape of the excess.

Further accurate data in the Rayleigh–Jeans region would be very important to test some of the proposed explanations for the sub-mm excess and to search for Bose–Einstein or bremsstrahlung distortions.

Measurements of dipole anisotropy and of the Sunyaev–Zeldovich effect at many frequencies would help to understand the shape of the spectrum and would allow to discriminate among the possible models for the distortions.

Work supported in part by MPI, CNR (through GNA and PSN), and ASI.

## 7. REFERENCES

Adams, F.C., Freese, K., Levin, J., and McDowell, J.C., 1989. *Ap. J.*, **344**, 24.

Barcons, X., 1987. *Ap. J.*, **313**, 547.

Barcons, X., and Fabian, A.C., 1988. *M.N.R.A.S.*, **230**, 139.

Barrow, J.D., 1976. *M.N.R.A.S.*, **175**, 359.

Bartlett, J.G., and Silk, J., 1989. In *Proc. Workshop on Particle Astrophysics: Forefront Experimental Issues*, ed. E.B. Norman, World Scientific, Singapore, p. 132.

Bernshtein, I.N., Bernshtein, D.N., and Dubrovich, V.K., 1977. *Astr. Zh.*, **54**, 727 [*Soviet Astr.*, **21**, 409].

Bernstein, G.M., Fischer, M.L., Richards, P.L., Peterson, J.B., and Timusk, T. 1989. *Ap. J. Letters*, **337**, L1.

Boesgaard, A.M., and Steigman, G., 1985. *Ann. Rev. Astr. Ap.*, **23**, 319.

Boldt, E., 1987. *Phys. Rept.*, **146**, 215.

Bond, J.R., 1988. In *The Early Universe*, eds. W.G. Unruh and G.W. Semenoff (Dordrecht: Reidel), p. 283.

Bond, J.R., Carr, B.J., and Hogan, C.J., 1986. *Ap. J.*, **306**, 428.

Bond, J.R., Carr, B.J., and Hogan, C.J., 1989. *Ap. J.*, submitted.

Boulanger, F., and Pérault, M., 1988. *Ap. J.*, **330**, 964.

Burigana, C., 1989. Thesis, University of Padua.

Carr, B.J., 1988. In *Comets to Cosmology, Lecture Notes in Physics, Vol. 297*, ed. A. Lawrence (Berlin: Springer–Verlag), p. 265.

Chan, K.L., and Jones, B.J.T., 1975a. *Ap. J.*, **200**, 454.

Chan, K.L., and Jones, B.J.T., 1975b. *Ap. J.*, **200**, 461.

Chini, R., Krügel, E., and Wargau, W., 1987, *Astr. Ap.*, **181**, 378.

Cox, P., and Mezger, P.G., 1989. *Astr. Ap. Rev.*, **1**, 49.

Crane, P., Kutner, M.L., Hegyi, D.J., Blades, J.C., Palazzi, E., and Mandolesi N. 1989. In *Proc. Moriond Astrophysics Meeting*, March 1989.

Danese, L., and De Zotti, G., 1977. *Rivista Nuovo Cimento*, **7**, 277.

Danese, L., and De Zotti, G., 1980. *Astr. Ap.*, **84**, 364.

Danese, L., and De Zotti, G., 1981. *Astr. Ap.*, **94**, L33.

Danese, L., and De Zotti, G. 1982. *Astr. Ap.*, **107**, 39.

Dar, A., Loeb, A., and Nussinov, S., 1989. *Ap. J. Letters*, **338**, L41.

Draine, B.T., and Shapiro, P.R., 1989. *Ap. J. Letters*, **344**, L45.
de Bernardis, P., Masi, S., Malagoli, A., and Melchiorri, F., 1985. *Ap. J.*, **288**, 29.
de Bernardis, P., Masi, S., Melchiorri, F., and Moreno, G., 1989. *Ap. J.Letters*, **340**, L45.
De Zotti, G. 1986. *Prog. Part. Nucl. Phys., Vol. 17, The Early Universe and its Evolution*, ed. A. Faessler (Oxford: Pergamon Press), p. 117.
De Zotti, G., Danese, L., Toffolatti, L. and Franceschini, A., 1989. In *Proc. IAU Symp. 139, "Galactic and Extragalactic Background Radiation"*, in press.
De Zotti, G., and Toffolatti, L. 1989. In *Highlights of Astronomy*, Vol. 8, D. McNally ed., p. 681.
Dicke, R.H., Peebles, P.J.E., Roll, P.G., and Wilkinson, D.T., 1965. *Ap. J.*, **142**, 414.
Dubrovich, V.K., 1975. *Pis'ma Astr. Zh.*, **1**, 3 [*Soviet Astr. Letters*, **1**, 196 (1976)].
Fabbri, R. 1981. *Ap. Space Sci.*, **77**, 529.
Fabbri, R., and Melchiorri, F., 1979. *Astr. Ap.*, **78**, 376.
Fabian, A.C., 1988. In *The Post-Recombination Universe*, ed. N. Kaiser and A.N. Lasenby (Dordrecht: Kluwer), p. 51.
Field, G.B., and Perrenod, S., 1977. *Ap. J.*, **215**, 717.
Field, G.B., and Walker, T.P., 1989. *Phys. Rev. Letters*, **63**, 117.
Freese, K., Adams, F.C., Frieman, J.A., and Mottola, E., 1987. *Nuclear Phys. B*, **287**, 797.
Fukugita, M., 1988. *Phys. Rev. Letters*, **61**, 1046.
Fukugita, M., Kawasaki, M., and Yanagida, T., 1989. *Ap. J. Letters*, bf 342, L1.
Gould, R.J, and Rephaeli, Y., 1978. *Ap. J.*, **219**, 12.
Guilbert, P.W., and Fabian, A.C., 1986. *M.N.R.A.S.*, **220**, 439.
Gush, H.P. 1981. *Phys. Rev. Letters*, **47**, 745.
Halpern, M., Benford, R., Meyer, S., Muehlner, D., and Weiss, R., 1988. *Ap. J.*, **332**, 596.
Hawkins, I., and Wright, E.L., 1988. *Ap. J.*, **324**, 46.
Hayakawa, S., Matsumoto, T., Matsuo, H., Murakami, H., Sato, S., Lange, A.E., and Richards, P.L., 1987. *Publ. Astron. Soc. Japan*, **39**, 941.
Illarionov, A.F., and Sunyaev, R.A., 1974. *Astron. Zh.*, **51**, 1162 [*Sov. Astr.*, **18**, 691 (1975)].
Johnson, D.G., and Wilkinson, D.T., 1987. *Ap. J. Letters*, **313**, L1.
Kawano, L., Schramm, D., and Steigman, G. 1988. *Ap. J.*, **327**, 750.
Kawasaki, M., and Sato, K., 1987. *Publ. Astr. Soc. Japan*, **39**, 837.
Kreysa, E., and Chini, A., 1989. In *Proc. 3rd ESO/CERN Symp. on Astronomy, Cosmology and Fundamental Particles*, M. Caffo et al. (eds.), Kluwer, Dordrecht, p. 433.
Krolik, J.H., 1989. *Ap. J.*, **338**, 594.
Lacey, C.G., and Field, G.B., 1988. *Ap. J. Letters*, **330**, L1.
Leiter, D., and Boldt, E.A., 1982. *Ap. J.*, **260**, 1.
Lightman, A.P. 1981. *Ap. J.*, **244**, 392.
Lubin, P., 1982. In *Proc. International School of Physics, Course on Gamow Cosmology*, Varenna.
Lyubarsky, Yu.E., and Sunyaev, R.A., 1983. *Astr. Ap.*, **123**, 171.
Matsumoto, T., Akiba, M., and Murakami, H., 1988a. *Ap. J.*, **332**, 575.
Matsumoto, T., Hayakawa, S., Matsuo, H., Murakami, H., Sato, S., Lange, A.E., and Richards, P.L. 1988b. *Ap. J.*, **329**, 567.
McDowell, J.C., 1986. *M.N.R.A.S.*, **223**, 763.

172

Negroponte, J., 1986. *M.N.R.A.S.*, **222**, 19.
Ostriker, J.P., and Thompson, C. 1987. *Ap. J. (Letters)*, **323**, L97.
Peebles, P.J.E., 1968. *Ap. J.*, **153**, 1.
Peebles, P.J.E. 1980. *The Large-Scale Structure of the Universe* (Princeton: Princeton University Press).
Penzias, A.A., and Wilson, R.W. 1965. *Ap. J.*, **142**, 419.
Raffelt, G., Dearborn, D., and Silk, J. 1989. *Ap. J.*, **336**, 61.
Rephaeli, Y., 1980. *Ap. J.*, **241**, 858.
Rogers, R., and Field, G.B., 1989. Preprint.
Rudak, B., and Panek, M., 1989. Preprint.
Rybicki, G.B., and Lightman, A.P., 1979. *Radiative Processes in Astrophysics*, Wiley, New York.
Silk, J., and Stebbins, A., 1983. *Ap. J.*, **269**, 1.
Smoot, G.F., Levin, S.M., Witebsky, C., De Amici, G., and Rephaeli, Y. 1988. *Ap. J.*, **331**, 653.
Stebbins, A., and Silk, J. 1986. *Ap. J.*, **300**, 1.
Sunyaev, R.A., 1980. *Pis'ma Astr. Zh.*, **6**, 387 [*Sov. Astr. Letters*, **6**, 213].
Sunyaev, R.A., and Zeldovich, Ya.B., 1970a. *Ap. Space Sci.*, **7**, 20.
Sunyaev, R.A., and Zeldovich, Ya.B., 1970b. *Ap. Space Sci.*, **9**, 368.
Sunyaev, R.A., and Zeldovich, Ya.B., 1972. *Comm. Ap. Space Phys.*, **4**, 173.
Sunyaev, R.A., and Zeldovich, Ya.B., 1980. *Ann. Rev. Astr. Ap.*, **18**, 537.
Taylor, G.B., and Wright, E.L., 1989. *Ap. J.*, **339**, 619.
Wang, B., and Field, G.B., 1989. *Ap. J. Letters*, **345**, L9.
Wright, E.L., 1979. *Ap. J.*, **232**, 348.
Wright, E.L., 1983. In *Early Evolution of the Universe and Its Present Structure*, G.O. Abell and G. Chincarini (eds.), Reidel, Dordrecht, p. 113.
Xu, C., and De Zotti, G., 1989. *Astr. Ap.*, in press.
Yoshioka, S., and Ikeuchi, S., 1987. *Ap. J. Letters*, **323**, L7.
Zeldovich, Ya. B., Illarionov, A.F., and Sunyaev, R.A., 1972. *Zh. Eksp. Teor. Fiz.*, **62**, 1216 [*Soviet Phys. JETP*, **35**, 643].
Zeldovich, Ya. B., Kurt, V.G., and Sunyaev, R.A., 1968. *Zh. Eksp. Teor. Fiz.*, **55**, 278 [*Soviet Phys. JETP*], **28**, 146].
Zeldovich, Ya. B., and Sunyaev, R.A., 1969. *Ap. Space Sci.*, **4**, 301.

# COSMIC INSTABILITY FROM RADIATION PRESSURE

Craig J. Hogan
Steward Observatory
University of Arizona
Tucson, AZ 85721 USA

The Cosmic Background Explorer has recently confirmed the blackbody character of the microwave background to high accuracy (Mather et al. 1990), and will have the capability to detect other cosmic backgrounds throughout the infrared. A detection of cosmic background radiation dating from the pregalactic era would have important consequences for theories of cosmic structure. During the creation of such a background the pressure of the radiation itself causes an instability which leads inevitably to the growth of large-scale structure in the matter distribution. In contrast to conventional gravitational-instability models, the statistical properties of this structure are determined primarily by the self-organizing dynamics of the instability rather than details of cosmological initial conditions. The behavior of the instability is described here.

## 1. INTRODUCTION

A fundamental process in astrophysics is the production of short wavelength radiation which is quickly absorbed by matter (for example, in the Lyman continuum) and eventually degraded (often with a small amount of

*N. Mandolesi and N. Vittorio (eds.), The Cosmic Microwave Background: 25 Years Later*, 173–185.
© 1990 *Kluwer Academic Publishers.*

cosmic dust) to longer wavelengths which interact more weakly with the matter. In such a situation the radiation pressure leads to a dynamical instability in the absorbing medium (Spitzer 1941, Field 1971, 1976). It is interesting to speculate about the possible role of this instability in the origin of the observed large-scale galaxy distribution (Hogan and White 1986, Hogan 1989a,b). Because of the instability, the production of some structure is practically inevitable as a byproduct of the production of any radiation. Moreover, the character of this structure is determined primarily not by initial conditions but by the self-organizing properties of the instability itself, and displays quantitative and qualitative similarities to the observed galaxy distribution. I outline here a model in which the nonlinear astrophysical dynamics of this instability determines the cosmic distribution of galaxies on the largest scales.

This view represents a major departure from the conventional gravitational instability models (Peebles 1980), in which primordial fluctuations in binding energy are laid down on supergalactic scales during the first picoseconds of the big bang, and eventually evolve into the observed structures. If the ideas developed here are correct, such primordial fluctuations might play no direct role in the formation of cosmic large-scale structure.

This paper aims to characterize the behavior of the instability and to sketch its potential role in shaping the observed galaxy distribution. Analysis of the instability has previously been confined to either the optically-thin limit, or to linear perturbation theory of small amplitude perturbations, or to artificially symmetric initial conditions (i.e. "bubbles"). Although these treatments of the instability were by and large technically correct, their approximations obscured some key features. It is shown here that if the linear growth rate of the instability is very rapid compared to the expansion rate then the dynamics of the instability guarantee that the system will in general evolve to a state where none of these approximations is adequate. The typical structures on large scales are most like the bubble solutions, but of course without the exact spherical symmetry. Most of the volume of system fills up with voids or cavities, which are separated by geometrically thin, but optically thick walls. Using the bubble models as a guide to the behavior, it is conjectured that the scale and other general statistical properties of the structure evolve to a state which is very insensitive to the specific initial conditions (although the *specific* final state is of course extremely sensitive to these conditions). In this "saturated" state, the instability sweeps material into the largest possible structures: the radiation field becomes very anisotropic, and the matter acquires a systematic organized motion with momentum density comparable to the entire radiation momentum density. Such behavior occurs because the matter distribution modifies the radiation field in such a way that the radiation pressure amplifies the structures in the matter. This paper is devoted to formulating and motivating this "saturation conjecture" and its application to the formation of large-scale structure.

## 2. Models of the instability

The system considered here is highly idealized. A universe with scale factor $a \propto (1 + z)^{-1} \propto t^{2/3}$ expanding at the Hubble rate $\dot{a}/a \equiv H$ is filled with a uniformly distributed radiation source with luminosity density $\ell(t)$. A non-uniform absorbing medium of "particles" of mass $m$ with mean density $\rho = \bar{n}m$ ($\bar{n}$ is the mean number density) absorbs the primary radiation from this source with mean frequency-averaged absorption cross section per mass $\sigma/m$. The absorbing medium is assumed to cool efficiently, re-radiating the absorbed energy at wavelengths where the secondary photon interactions with matter are much weaker and can be neglected. The absorbing medium has negligible coupling to the radiating medium. For the purposes of most of this discussion of the large-scale absorber configuration the gravitational interactions of both components may be neglected, and both may be assumed to have negligible pressure compared to the radiation. (The role played by these forces on small scales is discussed below).

An example of a real physical luminosity source close to that envisioned would b e a population of unclustered decaying nonrelativistic particles; this would produce $\ell(t) \propto e^{-t/\tau}a^{-3}$, where $\tau$ is the particle lifetime. An example of a real absorbing medium behaving like the idealized one would be dissipative ionized gas tightly dynamically coupled to absorbing dust grains, or atomic gas with enough electrons in bound states to provide significant continuum opacity. In both of these examples the primary photons should be ultraviolet or soft X-rays for efficient absorption. In general, scattering opacity is not sufficient for the instability to occur.

For definiteness, it is convenient to adopt a particular form for $\ell(t)$; although the specific form makes no difference in the general results, this will allow us to quote exact analytical solutions. Suppose that $\ell$ has a power-law behavior

$$\ell/\rho = (\ell_1/\rho_1)(t/t_1)^\gamma \tag{1}$$

where $\gamma = 0$ corresponds to constant comoving luminosity density, and $t_1$ is a fiducial epoch. The total amount of energy which has been radiated up to the time $t_*$ is then

$$\rho_\gamma = \int_0^{t_*} dt\ell(t)a^4 = \ell_* t_*/(\gamma + \frac{5}{3}). \tag{2}$$

For numerical examples we will assume that $\ell = 0$ after some such epoch $t_*$, corresponding to $1 + z_* \simeq 100$, and that the integrated background today has an integrated energy density $\rho_\gamma(t_*)(1 + z_*)^{-4} = \rho_{irb} = 10^{-13}$erg cm$^{-3}$; if it is much smaller than this, the instability does not have much hope of generating large scale structure. A background this intense will not be able to evade the scrutiny of COBE.

There are three characteristic length scales in this system. (If gas pressure and gravity are included, there are in addition the Jeans and the Mock Jeans scales, which are discussed briefly below. We assume that these are small enough not to enter into large-scale dynamics).

The first is associated with the inhomogeneity of the absorbers. Let $x_\rho$ denote the scale where the rms absorber density fluctuations are of order unity: $< (\delta\rho/\rho)^2 >_{x_\rho} = < (\delta\rho/\rho) >_{x_\rho}^2$, where the averages are taken over randomly placed volumes of size $x_\rho^3$. Averaged on scales larger than this, the density displays statistical uniformity; on smaller scales, the density displays large amplitude excursions above and below the mean.

The second scale characterizes the absorption of radiation. We denote by $x_\gamma$ the mean photon interaction length. If $x_\rho < x_\gamma$, then $x_\gamma = x_\tau \equiv (\bar{n}\sigma)^{-1}$— a typical photon travels a distance where the mean optical depth $\tau$ is unity. On the other hand if $x_\rho > x_\tau$ then in a highly clustered medium typically $x_\gamma \simeq x_\rho$, because photons can travel across very underdense regions without being absorbed. The mean primary radiation pressure (not including the reradiated or secondary component, which does not interact) is the amount radiated over a photon interaction time, $p_\gamma = \ell x_\gamma/c$.

Finally, we define a momentum equipartition scale $x_p \simeq \ell/\rho c H^2$. The Hubble momentum density $\simeq \rho H x_p$ of the absorbing material on this scale is about equal to the radiation momentum density integrated over a Hubble time,$\simeq \ell/H$. This is the maximum scale over which structure *could* be generated by the radiation pressure if it were anisotropic; we will argue that something approaching this scale is actually achieved.

Three useful analytical approximations describing this idealized system in different regimes defined by these scales show that there is an instability and display different features of the true underlying instability. In what follows we will use these approximate descriptions to argue that for a wide range of initial conditions these three scales converge under the action of the instability, a conjectured behaviour we call "saturation;"

$$x_\rho \to x_\gamma \to x_p. \tag{3}$$

*Mock Gravity.* In the limit of an optically thin system of absorbers with a negligible internal radiation energy density and isotropic incident radiation, the equations of motion for the absorbers are identical to Newtonian gravity, with an effective gravitational constant $G_{Mock} = I_\nu(\sigma/m)^2$ where the incident radiation intensity is $I_\nu$. This approximation was used by Spitzer (1941) in his pioneering studies of the effects of radiation pressure on dust in the galactic interstellar medium. This approximation was also used by Hogan and White (1986) for their discussion of the nonlinear dynamics of the instability. In the cosmological setting this approximation is reasonably accurate for optically thin systems, even nonlinear ones, which have a size $<< x_\gamma$; the mean incident intensity is given by $I_\nu = \ell x_\gamma/4\pi$. As we shall see however, this treatment is not adequate to understand the most interesting later stages of the instability, because the dominant structures are optically thick, and the radiation is highly anisotropic almost everywhere.

*Linear perturbation theory.* For small-amplitude plane wave perturbations of a nearly uniform absorbing medium, the instability can be studied at arbitrary optical depth $\tau$. This model can be used on all scales $>> x_\rho$, where the unperturbed medium i s statistically uniform. We will find however that unlike the gravitational equivalent, this approximation sheds little light on the growth of structures which are actually nonlinear. Consider a plane-wave perturbation in gas density, $\delta n/\bar{n} = \delta_k e^{i\underline{k}\cdot\underline{x}}$. Define an optical depth parameter for the scale $k$ by $\tau_k \equiv \bar{n}\sigma/k$. Then the equation of motion for the growth of these perturbations is (Hogan and White 1986, Hogan 1989b)

$$\ddot{\delta}_k + (2\dot{a}/a)\dot{\delta}_k - (\ell\sigma/mc)[1 - \tau_k \arctan(\tau_k^{-1})]\delta_k - 4\pi G(\bar{n}m)\delta_k = 0 \tag{4}$$

The second term is Hubble slowing; the third term represents the action of radiation pressure; and the fourth represents the ordinary gravitational instability. Were this term acting on its own we would recover the gravitational growth $\delta_k \propto t^{2/3}$. The radiation-driven term leads to a growth rate

$$\omega_k = (\ell\sigma/mc)^{1/2} \left[1 - \tau_k \arctan(\tau_k^{-1})\right]^{1/2} \equiv \omega_\infty^{1/2} F(\tau_k) \tag{5}$$

which is strongly suppressed for optically-thick perturbations. The linear radiation instability on its own looks mathematically identical to the linear gravitational instability, except that the growth rate depends on $\tau$ (for large $\tau$) and can be much larger than $H$. In the limit where $H$ is negligible, as in the original Jeans analysis of gravitational instability in a static medium, the mode amplitudes grow as $\delta_k \propto e^{\omega_k t}$. For the power-law adopted above with $\gamma = 0$ (with constant $\omega_k/H$), there is an exact analytical solution:

$$\delta_k \propto t^\beta, \beta \equiv [(2\omega_k/3H)^2 + \frac{1}{36}]^{1/2} - \frac{1}{6}. \tag{6}$$

Notice the contrast with the gravitational instability; here the growth in a single octave of expansion can be enormous. As an numerical example suppose that $\sigma/m$ is that of neutral hydrogen at the Lyman edge, in which case

$$(2\omega_\infty/3H)_* = \omega_\infty t_* = \sqrt{\frac{\gamma + \frac{5}{3}}{4\pi} \frac{\rho_\gamma}{\rho c^2} \frac{c\sigma H}{Gm}} = 605(\Omega_g h^2/.01)^{-1/2} h^{1/2} \left(\frac{1+z_*}{100}\right)^{5/4}, \tag{7}$$

using a Hubble constant of $H_0 = 50h$km sec$^{-1}$ Mpc$^{-1}$. We have adopted here the usual notation, with $\Omega_g$ denoting the comoving gas or absorber density in units of the present-day cosmic critical density, $\rho_c(1+z)^3 = (1+z)^3(3H_0^2/8\pi G)$.

We are ignoring a number of effects which have been more thoroughly treated in previous work (Field 1971, Wang and Field 1989) using linear theory: the drag effect of the cosmic microwave background, the effect of scattering, the possibility of another fluid viscously coupled to the absorbing medium (as in gas and poorly coupled dust grains), and relativistic effects (the tendency of the radiation anisotropy to disappear in the rest frame of the moving particle). It will become clear however that the actual behavior of the instability in the rapid-growth limit is dominated by nonlinear dynamics, so no attempt will be made here to incorporate these refinements of the linear theory.

In the linear approximation, the state of the system at any time is explicitly and fundamentally dependent on the initial conditions as well as the parameters $\ell$, $\rho$, $H$, and $\sigma/m$; the instability amplifies the initial perturbation by a certain amplitude-independent factor (which can be very large.) If an initial perturbation amplitude in a region of size $\simeq k^{-1}$ exceeds $\exp[-\omega_k/H]$, it will grow to of order unity in less than a Hubble time $H^{-1}$. If $\omega_\infty/H$ is much larger than unity, then a wide range of initial conditions would produce nonlinearity on the scale $x_\gamma$ in a time much less than $H^{-1}$: $x_\rho \to x_\gamma$, thereby insuring that the Mock Gravity approximation cannot thereafter be used to describe the large-scale development of the instability. As the following model shows, the linear theory itself also begins to produce misleading results at this point.

*Bubble model.* The third approximation (Hogan 1989a,b, Wang 1990) allows both the scale and the amplitude of the perturbation to be arbitrarily large, but with the restriction of spherical symmetry and with considerably less flexibility than the linear approach allows for adding new physical effects. The important point illuminated by this model however is the tendency for the scale of structure to grow to saturation—$x_\rho \to x_p$—which does not appear at all in the other models. Even though the source of the radiation remains uniform throughout, the radiation redistributes the absorbers, and the absorbers in turn redistribute the radiation, in such a way as to create a self-reinforcing, runaway instability.

The absorbing material is assumed to be dissipative gas. The model describes the expansion of a spherical bubble completely evacuated of absorbing material, with all the evacuated gas piled up in a thin, cold, optically-thick wall. The interior of the bubble is however still filled with the luminous material, and all of this radiation is absorbed by the bubble wall. The expansion of the bubble is slowed by redshifting and by plowing up fresh gas at the mean background density $\rho$. Still ignoring the gravity of the gas, the equation of motion for the comoving radius $x$ of the bubble is

$$\ddot{x} + 2(\dot{a}/a)\dot{x} + 3\dot{x}^2/x = \beta\ell/\rho ca. \tag{8}$$

The parameter $\beta$ denotes the ratio of the outward-directed component (normal to the bubble wall) of the radiation momentum flux to the total energy density of the absorbed radiation. For a point source in the center (Wang 1990), all the radiation is radial, and $\beta = 1$; but for the case of primary interest here, a bubble uniformly filled with the radiation source, we can show that $\beta = 3/4$. The momentum flux per solid angle incident on any point of the wall, coming from an angle $\phi$ from normal, is

$$f(\phi) = x\ell \cos^2 \phi / 2\pi c. \tag{9}$$

The total energy absorbed is then $2\pi c \int_0^{\pi/2} f(\phi) \sin \phi d\phi = x\ell/3$, and the normal component of the momentum flux is $2\pi c \int_0^{\pi/2} f(\phi) \cos \phi \sin \phi d\phi = x\ell/4$.

For our power law $\ell(t)$, the comoving bubble radius then has a self-similar power-law solution (Hogan 1989a,b)

$$a_1 x_b(t) = \frac{\beta}{4\alpha^2 + \alpha/3} \left( \frac{\ell_1 t_1^2}{\rho_1 c} \right) \left( \frac{t}{t_1} \right)^\alpha = \frac{\alpha + 1/3}{4\alpha^2 + \alpha/3} \frac{\beta \rho_\gamma(t_1)}{\rho(t_1)} \left( \frac{t}{t_1} \right)^\alpha ct_1 \tag{10}$$

where $\alpha = \gamma + \frac{4}{3}$. The mass in the bubble wall at time $t$ is $4\pi \rho_1 a_1^3 x_b^3 / 3$.

The crucial point is that this solution is an attractor[13]—smaller bubbles accelerate to approach it with a response rate $\simeq (\ell/\rho c a x)^{1/2}$. It hardly matters at all how small the bubbles are when they begin; they always end up at about the same size. The reason is clear from equation (8): the rate of acceleration relative to the bubble size, $\sqrt{\ddot{x}/x} \simeq H\sqrt{x_b/x}$, increases for smaller bubbles, and the braking terms become relatively less important. All solutions converge to the "attractor" with a response rate $\simeq (\ell/\rho c a x)^{1/2}$. Since $(\ell/\rho c a x)^{1/2} \gtrsim H$ if $x < x_b$, the approximate bubble radius is uniquely determined by $\ell(t)$, and is insensitive to the initial bubble size. (Such an attractor exists also in the general case where $\ell(t)$ is not a power-law, but it may not be described by an analytical expression.) As a bubble approaches the attractor and its growth rate approaches $H$, the rate of runaway growth becomes universally regulated. At this point it is said to become "saturated"—a term we will apply also more generally to the entire radiation-unstable cosmic system as it develops into an analogous state.

The interesting point for large scale structure is the scale of the saturated bubbles. The observable energy density of the infrared background is given by the integral of $\ell a^4$ over cosmic time, so the radius of the attractor can be computed with only one significant parameter—the characteristic cosmic scale factor $a_*$ at which $\ell a^4$ is maximized, that is the epoch at which the submillimeter excess is mainly generated. For the above model with a power law $\ell$ terminating at $z_*$, the attractor has a comoving radius at $z_*$, in present-day Hubble velocity units,

$$
\begin{aligned}
H_0 a_0 x_{b*} &= \frac{2}{3} \frac{\alpha + 1/3}{4\alpha^2 + \alpha/3} \left( \frac{\beta \rho_{irb}}{\rho} \right) (1 + z_*)^{1/2} c \\
&= 1100 \text{ km sec}^{-1} \beta \left( \frac{\Omega_g h^2}{.01} \right)^{-1} \left( \frac{1 + z_*}{100} \right)^{1/2}
\end{aligned}
\tag{11}
$$

where the final numerical value is evaluated for the $\gamma = 0$ dust model of $\ell(t)$. This radius is close to the typical size of observed voids in the galaxy distribution, if we use the value of $\Omega_g \simeq .005 h^{-1}$ observed to be in stars in galaxies.

This estimate does not take into account any growth after $t_*$; the attractor bubble described above is expanding so fast it eventually grows in radius by another factor of $3\gamma + 5$ just by coasting (without any additional acceleration), if no more material is swept up. Depending on the details of the model, a background up to five times smaller than we have assumed could generate the observed structure.

Thus any bubble, regardless of its initial size, grows to the attractor radius $x_b$, that is to say, to the saturation scale $x_p$. The momentum of the matter becomes comparable to the integrated radiation momentum $\rho_\gamma$, rather than the instantaneous primary photon pressure; the reason is that once a bubble forms the light generated in it is mostly moving outwards, and the absorption maintains this situation over many light-crossing times. In the bubble model, we can compute the approximate radius as a function of time, once we adopt a model for $\ell(t)$, without knowing the initial radius. This is in contrast to the linear model, where the amount of growth is independent of the perturbation and the state of affairs depends entirely on the initial conditions. The amount of amplification here is a nonlinear function of the initial perturbation, and the final state almost independent of it.

Another important conceptual difference from the linear model is that the bubble displays considerable growth even though it is an optically-thick system. The nonlinearity of the void enables radiation to propagate and amplify structure over distances where the linear growth rate would be negligible. This shows that we should no longer trust the linear theory once $x_\rho \simeq x_\gamma$; the bubble is a better characterization of the actual behavior.

## 3. Saturation of the instability

The bubble model describes an isolated structure, not a statistically uniform medium, and it cannot be rigorously generalized to a statistical description. But we can use these models to motivate the conjecture that the absorbing medium under the action of the instability organizes itself into bubble-like structures, and formulate a quantitative picture of how the structure develops.

*Saturation Conjecture. Provided that $\ell$ and $\sigma/m$ are high enough that $\omega_\infty >> H$, then the instability will cause any generic initial condition to develop into a similar "saturated" state, in which the absorbing material becomes concentrated in thin walls around empty regions of radius $\simeq ax_b(t)$, so that $x_\rho$ and $x_\gamma$ approach $x_p$: $x_\rho \to x_\gamma \to x_p$.*

A "generic initial condition" here is one in which there is at least a certain minimum fluctuation amplitude—specifically, in each volume of size $x_p^3$ there must be at least one region of size $x_r^3$ with a fractional perturbation in absorber or luminosity density greater than one part in $\exp[\omega_\infty/H]$. As shown above from linear theory, this is the minimal requirement for the linear instability to form at least one seed bubble (an empty region $x_\gamma$ across) per $x_p^3$. This is much smaller than the perturbation amplitudes usually contemplated in cosmology, and may be produced by a wide variety of small-scale processes—the details of which do not matter in the end.

The idea is that once fluctuations become nonlinear on the scale $x_\gamma$, the subsequent behavior is like the bubble solution, and $x_\rho$ grows to approach an attractor defined by the scale $x_p$, in the same way that the bubbles do. Whereas the linear theory would predict that the growth should stop at this time, the very fact that there are nonlinearities allows growth to continue, by extending the photon trajectories to the whatever the current scale of structure is. This allows a sort of "bootstrapping"...the bigger the bubbles get, the more easily they can grow bigger. The final structure organizes itself around the largest bubbles, which have a similar properties independent of any particular distribution of seed perturbations.

· The saturation conjecture, if it is correct, implies some remarkable and attractive features of the instability for forming cosmic structure. For example, the initial distribution of sources and absorbers does not affect the final outcome. This feature is practically unique in current cosmological scenarios. It suggests that many important features of the structure may be realistically calculable with conceptually simple models of the astrophysical instability. The very early universe may have no direct role at all to play in the process.

The validity of the conjecture can be demonstrated by appealing to the above models. To start with the linear theory tells us that any perturbation $\lesssim x_\gamma$ in size grows by a very large factor $\simeq \exp[\omega_\infty t]$ in a time $t$. Thus a generic negative seed perturbation as just defined becomes nonlinear in a time $< H^{-1}$. Once this occurs it acts like a bubble model until it encounters other bubbles. We must then show that the bubble-like behavior persists for the typical or dominant structures even after the bubbles from different seed perturbations meet each other. If this occurs the scale of structure grows to the bubble attractor size, $x_\rho \rightarrow x_b \simeq x_p$, as conjectured.

To see that this must occur, consider first a collision of two bubbles. If the walls of two bubbles of different size meet, the small one is taken over by the big one. Although the growth rate $\dot{x}/x \propto x^{-1/2}$ is bigger for the small bubble, the velocity is bigger for the big one ($\dot{x} \propto x^{1/2}$); it also has a larger column density so it always "wins" in a collision. Moreover the radiation pressure is bigger on the side of the wall towards the big bubble ($p_\gamma = x\beta\ell/3c$). The smaller bubble is therefore incorporated into the expanding wall of the larger one, and is squeezed out of existence by the pressure from the big bubble on one side and the ram pressure of the absorbing material on the other. Note that the internal radiation pressure of the smaller bubble, and hence its ability to resist compression, decrease as it gets smaller. So, a large bubble propagating into a medium of smaller bubbles still behaves in pretty much the same way as it would propagating into a smooth medium.

A wall between two bubbles always experiences a net acceleration away from the larger one. Thus, a small bubble surrounded by somewhat larger ones shrinks, whatever the scale. A distribution of bubbles of different sizes therefore gradually becomes dominated by the largest ones. This trend continues until the universe is filled with bubbles approaching the attractor size; in the competition for volume, small bubbles continue to get squeezed out. The whole system develops in a way similar to a bubble bath, which starts as a small-scale foam. The small bubbles gradually consolidate into big ones, so that big ones eventually dominate. (Of course the physics here is entirely different, since the evolution is not driven by surface tension, but by the feedback of the optically-thick bubble walls on the radiation field which accelerates them.) Eventually, the largest bubbles take over, expanding in volume in approximately the way they would in an isolated spherically-symmetric solution. This means that the scale of structure $x_\rho$ approaches the bubble attractor $x_b$: the instability saturates.

This behavior is an interesting contrast to the conventional gravitational instability. For the gravitational instability, linear theory has provided a good intuitive guide to the behavior of the system and an accurate description of th e growth of perturbations on scales large enough that the matter is smooth. Small-scale nonlinearities do not seem to dominate or even influence the behavior on large scales. Although such "gravitational bootstrapping" behavior was considered a real possibility at one time (Press and Schechter 1974), subsequent work has shown that it does not occur (Peebles 1980). The reason is simple enough: gravity is a conservative force, and the gravitational instability of linear theory is just a reflection of energy conservation. The growth rate of the amplitude of a linear plane wave perturbation can be derived from the condition that the gravitational binding energy per comoving volume does not change with time, and small-scale nonlinearities cannot change this fundamental conservation principle. Thus the gravitational instability is not really an instability; any bound system has always had the same binding energy. In the radiation instability however, the energy of the absorbing medium is not conserved; the free energy of the radiation field is being diverted to the motion of matter. In this situation, bootstrapping can and does occur: the actual growth is far greater than that predicted on the basis of linear theory, because of nonlinear effects.

As in nonlinear gravitational dynamics the detailed statistical properties of the medium are most usefully studied by direct simulation. Simulations of a one-dimensional system have been used to verify that the above description of saturation is indeed qualitatively correct (Hogan and Woods 1990), and that the scale of structure indeed appears to follow the prediction of the bubble attractor solution. These experiments have also verified of the result that the final distribution depends hardly at all on initial conditions, but ends up statistically almost the same over a wide range of small initial displacements of the particles.

## 4. Predictions for COBE

The ideas sketched above form only the bare bones of a real theory, and are intended only to isolate some attractive features of radiation pressure a s an agent for generating structure. No real radiative transfer is as simple as we have assumed. Nevertheless, there is one clear prediction which can falsify or confirm the whole scheme, namely, the presence of an intense pregalactic infrared background (Bond, Carr and Hogan 19 86). It is physically implausible to arrange for the degraded radiation to be lost in the blackbody radiation without leaving residual distortions exceeding COBE limits, so the radiation must still exist in a distinct extragalactic background with an energy density $\rho_{irb} \geq 2 \times 10^{-14}$, if the proposed instability is to generate large scale galaxy clustering. Such an intense background is detectable with COBE at any wavelength longwards of $1\mu$, and its anisotropy in this scenario should also be detectable (Bond et al. 1990).

*Acknowledgements.* This work was supported by NASA grant NAGW-1703 and by the Alfred P. Sloan Foundation.

5. REFERENCES

Bond, J. R., Carr, B. J., and Hogan, C. J., 1986. *Astrophys. J.* **306** 428–450.
Bond, J. R., Carr, B. J. and Hogan, C. J., 1990. *Astrophys. J.,* in press.
Hogan, C., 1989a. *Nature* **338** 123–126.
Hogan, C. J., 1989b. *Astrophys. J.* **340** 1–10.
Hogan, C. J. and White, S. D. M., 1986. *Nature* **321** 575–578.
Hogan, C. and Woods, J., 1990. In preparation.
Field, G. B., 1971. *Astrophys. J.* **165** 29–40.
Field, G. B., 1976. In *Stars and Stellar Systems*, **Vol. 9**, *Galaxies and the Universe* (eds. Sandage, A., Sandage, M. & Kristian, J.) 359–40 7 (University of Chicago Press, Chicago, 1976).
Mather, J. et al., 1990. COBE preprint No. 90–01.
Peebles, P. J. E., 1980. *The Large Scale Structure of the Universe* (Princeton University Press, Princeton).
Press, W. H. & Schechter, P., 1974. *Astrophys., J.* **187** 425–438.
Spitzer, L., 1941. *Astrophys. J.* **94** 232–244.
Wang, B., 1990. Preprint, Center for Astrophysics.
Wang, B. & Field, G. B., 1989. *Astrophys. J.* **346** 3–11.

# Infrared Cosmic Background Radiation

Toshio Matsumoto

Department of Astrophysics, Nagoya University

Chikusa-ku, Nagoya, Japan 464-01

**Abstract**

Observation of the cosmic background radiation in the infrared region is reviewed.

The redshifted light from stars of the first generation forms diffuse cosmic background radiation in the near-infrared region. Measurement of the sky fluctuation at $2.2\mu m$ gives a very low upper limit. The rocket observation of the near-infrared diffuse emission reveals isotropic emission which is possibly ascribed to an extragalactic origin. The observed brightness and fluctuation are not consistent with the standard scenario of the primeval galaxies.

In the far-infrared region, integrated light of dust emission of the distant galaxies forms another cosmic background radiation. IRAS and the Nagoya-Berkeley rocket experiment found a clear correlation between HI column density and far-infrared sky brightness, however, there remains an uncorrelated isotropic emission component. If we ascribe this emission to be extragalactic origin, a fairly big evolution effect is required.

In the submillimeter region, the Nagoya-Berkeley rocket experiment has shown that the submillimeter cosmic background is much brighter than expected from the $2.74K$ blackbody spectrum. The excess energy corresponds to about 10% of the $2.74K$ blackbody, which requires the vast energy generation in the early universe.

# 1 Introduction

Cosmic background radiation (CBR) usually means extragalactic diffuse emission. CBR exists in all electromagnetic energy range, although 3K background radiation is most famous. In the X-ray and ultra-violet region, the background radiation have been studied as a probe of the hot intergalactic matter. In the visible region, the light of the night sky, that is, an integrate light of distant stars and galaxies, has been one of the important astronomical objects, concerning the Olbers's paradox. This CBR is sometimes called as an extragalactic background radiation (EBL). Modern cosmology threw a new light on this issue to investigate the formation and evolution of the galaxies (Partridge and Peebles 1967). Since then, many

*N. Mandolesi and N. Vittorio (eds.), The Cosmic Microwave Background: 25 Years Later, 187–201.*
© 1990 *Kluwer Academic Publishers.*

people have tried to detect the EBL, however, observations were limited in the optical region. The recent development of space experiments has made it possible to observe the infrared CBR and has revealed its importance to cosmology.

The infrared observation could provide an additional and a different kind of information for cosmology. Compared with the optical region, near-infrared observation is advantageous because of less bright foreground emission. Redshifted UV and optical radiation from the primeval galaxies could be observed in the near-infrared region ($1 - 5\mu m$), if the galaxy formation occurred at high redshift era. In the mid-infrared region ($10 - 50\mu m$) thermal emission of the interplanetary dust is too strong to observe CBR, but foreground emission becomes very dark at $\lambda > 100\mu m$. The far-infrared and submillimeter regions provide an opportunity to observe a different kind of CBR, which was originated in the dust emission of the distant galaxies. The far-infrared CBR has a complementary nature to the near-infrared CBR, and both will render observational evidences on the galaxy formation. At the wavelength region close to $1mm$ Wien part of the $3K$ CBR is observable. It has been believed that the distortion of the spectrum in this wavelength region could reveal the activities in the early universe.

In this paper we review the present status of the observation of the infrared CBR. In section 1, the foreground emission is summarized. Section 2 and 3 present the observational results on the near-infrared and the far-infrared CBR, respectively. Future Japanese space mission to observe the infrared CBR are briefly introduced in section 4.

# 2  Foreground emission

The CBR is supposed to be isotropic over the whole sky. In order to obtain the CBR component from the observed sky brightness, foreground diffuse emission must be carefully subtracted. In infrared region terrestrial emission causes serious problems even at the atmospheric windows. The airglow emission of atmospheric molecules ($OH, CO, CO_2$ etc.) is prominent in the near-infrared region, while thermal emission of the atmosphere is conspicuous in the far-infrared region. At the balloon altitude atmospheric transmission is much better, however, residual airglow and atmospheric emission are still prominent compared with the expected CBR brightness. The instrument itself also radiates huge amount of thermal emission, producing additional background emission. These emission components not only mask the CBR but also reduce the detector sensitivity due to the photon fluctuation noise. On the other hand space observation using the rocket and satellite is free from terrestrial emission. Observation in space with cooled optics is ideal to observe the extended diffuse object such as the CBR. In this section review of the foreground emission is limited on the extraterrestrial emission assuming the space observation.

Fig.1 indicates the estimated sky brightness of several extraterrestrial emission components for the darkest region of the sky.

Interplanetary dust causes two kinds of diffuse emission. One is a zodiacal light, that is, a scattered sunlight. ZL-scatter in Fig.1 is obtained by extrapolating the optical data at the ecliptic pole assuming the solar color (Hofmann et al. 1970, Hayakawa et al. 1970). Based on the recent near-infrared rocket observation, Akiba et al. (1989) reported that the infrared color of the zodiacal light is bluer than the solar color at the ecliptic pole region, as is shown by dashed line in Fig.1. The near-infrared feature and its spatial distribution of the zodiacal light is not so well understood yet and more extensive studies are required.

Another interplanetary emission component is thermal emission of the interplanetary dust (ZL-thermal in Fig.1). IRAS extensively observed the ZL-thermal and the result at the ecliptic pole is shown in Fig.1, adopting dust temperature of 184K and emissivity index of 1.0 (Hauser et al. 1984). Since IRAS observed only the sky at the solar elongation angle of 90 degree, the spatial distribution is not well established yet. Nevertheless, it should be mentioned that ZL-scatter is so bright œthat CBR at the wavelength region of $10 - 100\mu m$ is heavily masked. This makes it extremely difficult to detect the far-infrared CBR.

Integrated light of the faint stars , star light (SL), forms diffuse extended emission. Fig.1 shows the estimated brightness of the SL at the galactic pole region. The spectrum of the SL is taken from the color of $S_{bc}$ type galaxies (Yoshii and Takahara 1988) and the absolute value observed by Matsumoto et al. (1988b). Brightness of the SL strongly depends on the magnitude of the faintest star which can be identified as a point source. Two cases corresponding to the limiting magnitude $m_K = 3.0$ mag and 10.0 mag, are indicated in Fig.1. In order to reduce the SL component, a large aperture with a small beam is desirable. At high galactic latitude, spatial distribution of the SL is essentially represented by $cosec(b)$ law (Matsumoto et al. 1988b).

Interstellar dust (ISD) heated by the interstellar radiation field also forms far-infrared and submillimeter diffuse emission. IRAS and Nagoya-Berkeley rocket experiment have revealed that the ISD emission is fairly well correlated with the column density of the neutral hydrogen, N(HI). In Fig.1 ISD emission is estimated assuming N(HI) of $1.5 \times 10^{20} cm^{-2}$ and the dust temperature of 19K (Lange et al. 1989).

At the right end of the Fig.1 the spectrum of the $2.74K$ blackbody is also shown for comparison. Dashed line over $2.74K$ blackbody spectrum indicates the result of the Nagoya-Berkeley rocket experiment (Matsumoto et al. 1988a).

Fig.1 clearly shows that there are two windows through the Galaxy to observe CBR. One is the near-infrared and the other is the submillimeter region.

# 3   Near-infrared observations

At the wavelength shorter than the K band, observations of the CBR from the ground-based telescope have been carried out. In this region atmospheric emis-

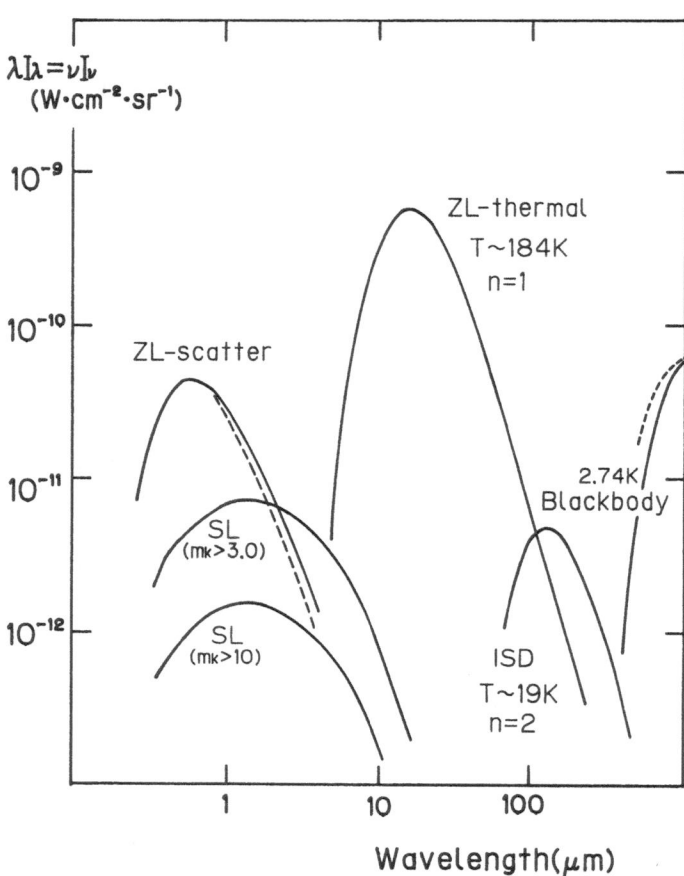

Figure 1: The estimated brightness of the extraterrestrial foreground emission. The SL (star light) means the integrated light of the faint stars in the Galaxy. The ZL-scatter and the ZL-thermal indicate the scattered sunlight by the interplanetary dust (IPD) and the thermal emission of the IPD, respectively. The ISD implies the thermal emission of the interstellar dust which correlates with column density of the neutral hydrogen. The $2.74K$ blackbody spectrum and the recently observed submillimeter excess (dashed line) are also indicated.

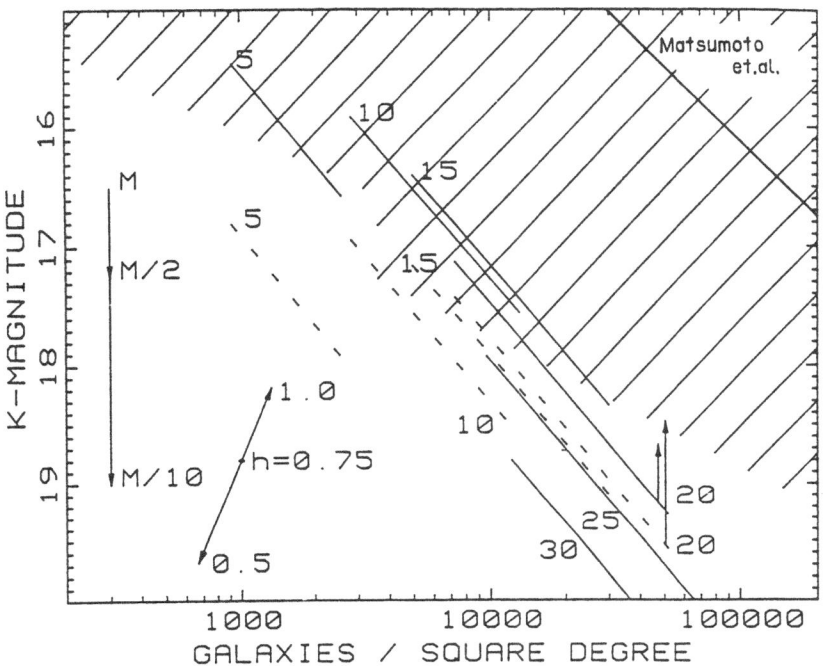

Figure 2: Models of primeval galaxies excluded by the measurement of the smoothness of the sky at $2.2\mu m$ (shaded area). The solid line labeled as Matsumoto et al. corresponds to the K band brightness observed by them (Matsumoto et al. 1988b).

sion is rather weak but airglow is still prominent. Boughn and Kuhn (1986) have tried to detect the CBR using the eclipse œby dark clouds. They found that the dark cloud was brighter than the nearby reference sky and its level was $4 \times 10^{-12} W cm^{-2} sr^{-1}$. This is due to the reflected galactic light and the contribution of the CBR is not certain. Boughn et al. (1986) observed the smoothness of the sky at $2.2\mu m$ and obtained the upper limit of $10^{-12} W cm^{-2} sr^{-1}$ for 10"-30" beam and $4 \times 10^{-13} W cm^{-2} sr^{-1}$ for 60"-300" beam. Assuming a model for primeval galaxies they excluded the shaded area of Fig.2 in which the K-magnitude and the number density of primeval galaxies are plotted. Obtained upper limits constrain the model but are still marginally consistent with the model.

As described before, space observation with cooled optics is very advantageous even in observing the near-infrared CBR. Harwit et al. (1966) first tried the rocket observation, however, detector sensitivity was not good enough, resulting in a rather high upper limit for the CBR. Hofmann and Lemke (1978) analyzed $cosec(b)$ dependence for the data of the balloon-born telescope and obtained an upper limit of $6 \times 10^{-10} W cm^{-2} sr^{-1}$ at $2.4\mu m$. This implies that balloon altitude is not high enough to make an absolute measurement of the near-infrared CBR.

Matsumoto et al. (1988b) have recently made a rocket experiment to search for the near-infrared CBR with liquid nitrogen cooled telescope. Fig.3 shows the observed sky brightness at the galactic north pole region. Open circles represent the results of the standard astronomical filter bands, H, J ,K, L and M whose data were taken during whole the time of the flight. Crosses represent the data of the filter bands on the rotating wheel, which were changed intermittently during the flight. This caused slightly larger error bars for the crosses than those for the open circles. As is shown in Fig.3, the J band data can be explained well as the sum of the SL and the ZL-scatter. The spatial variation of the observed sky brightness indicates that the SL essentially obeys a $cosec(b)$ law, which is consistent with the model of the Galaxy. At the K band, the observed brightness is considerably brighter than that extrapolated from the J band. The observed spatial variation indicates that the SL is consistent with the model and the ZL-scatter has a little bluer color than the solar color. As a result there remains an isotropic emission component in the K band. Similar isotropic emission components are found in the wavelength region of $3 - 4\mu m$.

Fig.4 indicates the isotropic emission thus obtained. There are several significant data points, however, a careful study is required to conclude the extragalactic origin for the observed isotropic emission. One serious issue is the environmental emission surrounding the rocket body. Before apogee, some emission lines were observed which had a decay time constant of 41 sec. The environmental emission was prominent at the wavelength longer than $3\mu m$, especially $3.5\mu m$ band due to CH-bond of the organic materials was clearly seen. The contribution of the environmental emission was estimated assuming exponential decay, which resulted in the negligible effect after apogee. Nevertheless, the environmental emission could not be excluded completely as a origin for the isotropic emission at the wavelength longer than $3\mu m$. On the other hand, the isotropic emission around the K band is almost free from the environmental emission and there is no bright atomic and molecular line in this band. As for the K band, a small possibility is left that the ZL-scatter has an isotropic offset brightness only in the K band, although this is not consistent with previous measurements of the interplanetary dust and cometary dust. Taking these ambiguities in mind, Matsumoto et al. (1988b) concluded that the isotropic emission at the K band can be attributed to be extragalactic origin, but other isotropic emission could be a marginal detection of the CBR.

In the case that the isotropic emission observed by Matsumoto et al. (1988b) is extragalactic in origin, its effect on cosmology is enormous. In Fig.4 models of the CBR for two extreme cases by Partridge and Peebles (1967) are shown. Model 4 assumes that all helium were synthesized during the galaxy formation era, while model 1 assumes no evolution for galaxies. The isotropic emission in Fig.4 is close to the model 4, which indicates the vast energy generation in the early universe. There seems to be a line feature at the K band showing an anomalous spectral feature. Matsumoto et al. (1988b) suggested its origin to be a redshifted Lyman

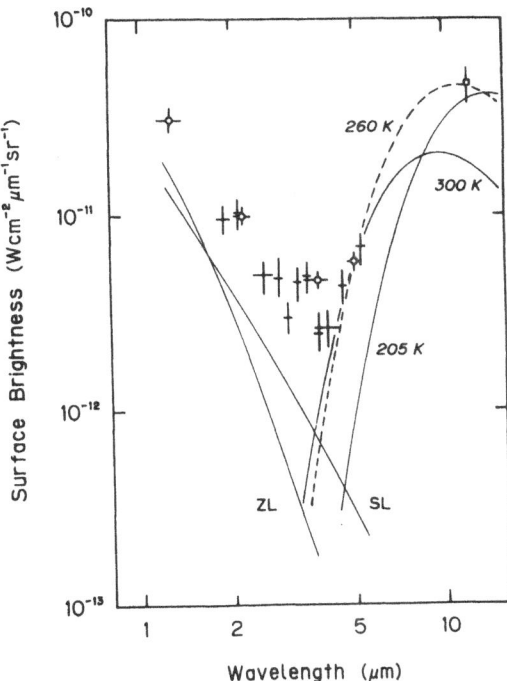

Figure 3: Near-infrared sky brightness at the galactic north pole region (Matsumoto et al. 1988b). Open circles indicate the data for the standard astronomical filter bands ; J, K, L, and M, while crosses represent data for narrow filter bands on the rotating filter wheel. The square shows the IRAS data for ZL-thermal at $12\mu m$. The emission component around $5\mu m$ is fitted by three types of blackbody emission.

$\alpha$ or Lyman-limit. In this case redshift corresponds to $\sim 17$ and 24, respectively. At this high redshift light sources must be pregalactic, and Population III objects proposed by Carr et al. (1984) may be responsible.

The isotropic emission at the K band should be compared with the smoothness of the sky. In Fig.2, a solid line labeled as Matsumoto et al. shows the allowable position which gives the brightness for the observed isotropic emission. The observed brightness is so bright that it lies in the excluded region (shaded area) by the model. Further, the observed isotropic emission should be consistent with the fluctuation of the $2.2\mu m$ sky brightness, which requires the number density of the primeval galaxies to be $\geq 100$ galaxies in a 10" beam. In this case, the apparent K magnitude of the primeval galaxies should be larger than 21.3 mag. The feature of the primeval galaxies thus obtained is far from that in Fig.2 and could hardly be explained by the standard scenario on the formation and evolution of the galaxies.

194

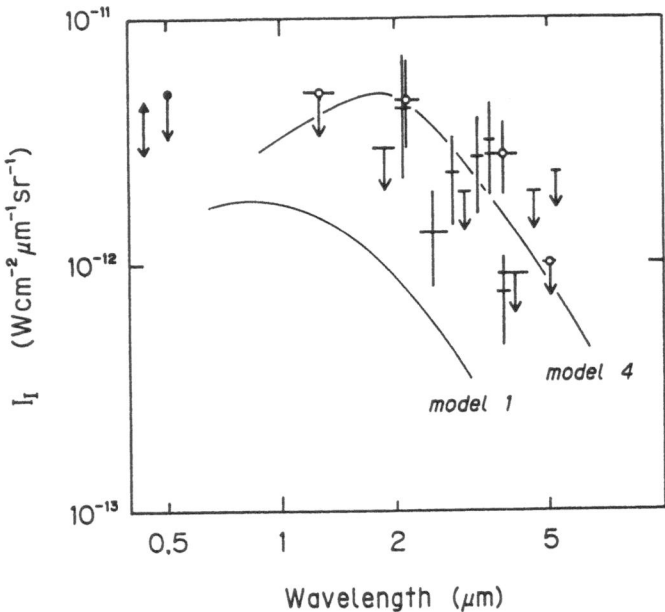

Figure 4: The isotropic emission observed by Matsumoto et al. (1988b). Upper limits in the optical region by Dube et al. (1977):filled cycle; and Toller (1983):triangle are also shown. The CBR calculated by Partridge and Peebles (1967) for two extremely cases is indicated by solid lines.

## 4    Far-infrared and submillimeter observations

At the wavelength longer than $100\mu m$ a possibility to detect the CBR appears again. IRAS has attained whole sky survey and found a clear correlation between the diffuse infrared emission and the column density of the neutral hydrogen, N(HI), in all wavelength bands. In these correlation diagrams, linear fits do not cross the origin and there remains considerable amount of the isotropic emission which does not correlates with N(HI). In the three short wavelength bands, this isotropic emission could be ascribed to the ZL-thermal, however, isotropic emission at the $100\mu m$ can hardly be explained by known astronomical emission components (Boulanger and Perault 1988).

Although IRAS was not designed well to observe the diffuse isotropic emission, a diffuse sky brightness in the far-infrared and submillimeter region was more extensively observed by the Nagoya-Berkeley rocket experiment (Matsumoto et al. 1988a, Lange et al. 1989). In this experiment, a circle on the sky centered at $l = 203.2 \pm 1.5$ degree, $b = 34.9 \pm 1.5$ degree of a radius of $15.7 \pm 1$ degree was

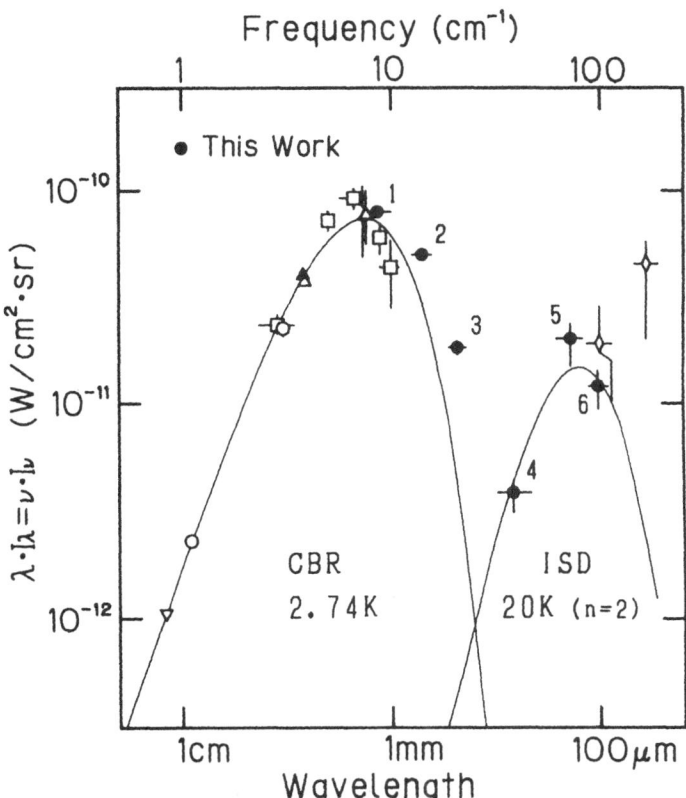

Figure 5: Average sky brightness observed by the Nagoya-Berkeley rocket experiment ((Matsumoto et al. 1988a) is shown by filled circles. Previous data, $2.74K$ blackbody spectrum and $20K$ blackbody spectrum with spectral index 2 are shown for comparison.

scanned with 7.6 degree beam. Whole optical system and detectors were cooled down to $1K$ and the sky brightness between $100\mu m$ and $1mm$ was measured with 6 channel wide band photometer.

Fig.5 shows the sky brightness observed by the Nagoya-Berkeley rocket experiment. Filled circles indicate average sky brightness, and $2.74K$ blackbody spectrum and $20K$ blackbody spectrum of spectral index 2 are shown for comparison. The observed brightness of the three short wavelength channels indicates a typical feature of thermal emission of the interstellar dust. Errors in two short wavelength channels are due to the systematic uncertainty of the absolute calibration and observed signal to noise ratios were fairly high. This enabled us to make the correlation study with other galactic objects. Fig.6 shows the correlation diagram between

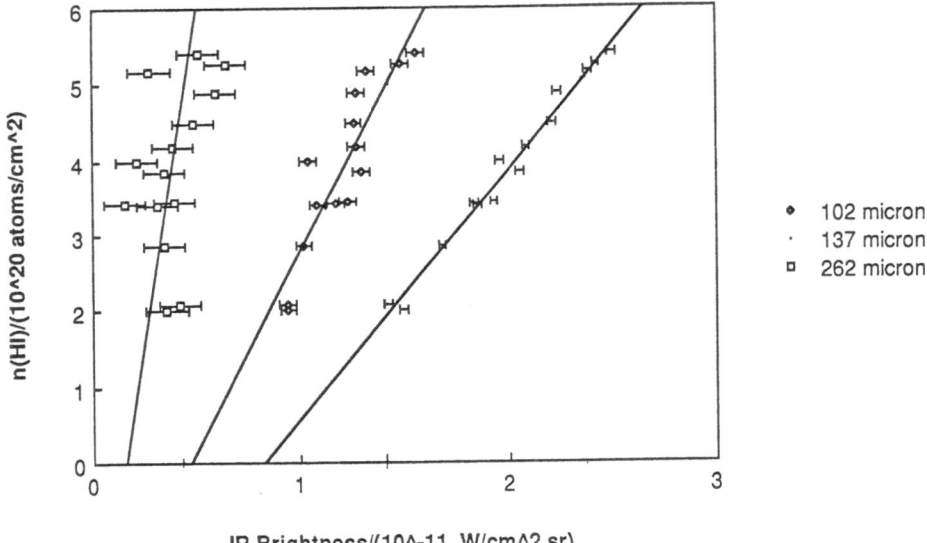

Figure 6: Correlation between the infrared sky brightness and the column density of the neutral hydrogen, N(HI) (Lange et al 1989).

N(HI) and the infrared brightness for three short wavelength channels, in which error bars represent statistical errors only. Fig.6 indicates a tight correlation same as observed by IRAS. The gradients of the fitted straight lines provide ratios of the infrared brightness to N(HI). The result at $102\mu m$ is consistent with that by Terebey and Fich (1986) but a little lower than that by Boulanger and Perault (1988) (Lange et al. 1989). If we assume that the infrared brightness correlated with N(HI) is due to the thermal emission of the interstellar dust, the observed spectrum can be fit with temperature of $19 \pm 3K$ and spectral index of 2.

Fig.6 indicates there remains an offset brightness even if N(HI)=0. Concerning to this residual isotropic emission, scatter of the data points in Fig.6 provides valuable information. At $102\mu m$, scatter from the fitted straight line is larger than that at $137\mu m$, but correlated each other. This suggests that the isotropic emission at $102\mu m$ is due to the thermal emission of the interplanetary dust, that is, ZL-thermal. Since the observed sky was anti-solar region, direct comparison with IRAS is not possible, however, it seems plausible to attribute whole of the residual emission at $102\mu m$ to the interplanetary origin. At $137\mu m$, the situation is not so simple, since even the graybody spectrum for the ZL-thermal can not explain the residual isotropic emission. There remains unknown isotropic emission component at $137\mu m$. At $262\mu m$, separation to individual emission components is not easy because of a rather low signal to noise ratio, and only upper limit was obtained for

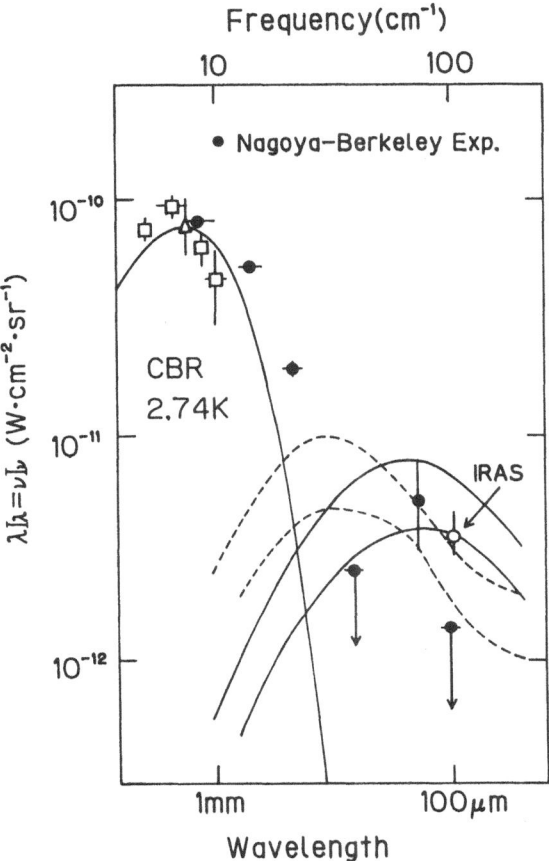

Figure 7: Isotropic emission observed by the Nagoya-Berkeley experiment (filled circles). IRAS background by Boulanger and Perault (1988) is shown by open circle. Models which include a large evolution effect for galaxies are indicated by the solid and dashed lines (Rowan-Robinson and Carr 1988).

the isotropic emission.

Fig.7 shows the residual isotropic emission obtained by the Nagoya-Berkeley experiment, in which the data of the three long wavelength channels are also included. It should be mentioned that the isotropic emission in Fig.7 is for the lowest case. For example, it can not be excluded to attribute the main part of the residual emission at $102\mu m$ to this unknown component. In this case the isotropic emission thus obtained is consistent with the IRAS isotropic emission, but the ZL-thermal at the anti-solar region must be much darker than that expected from the model.

It should be mentioned that isotropic emission at $100-300\mu m$ has fairly narrow spectral width. Rowan-Robinson and Carr (1988) proposed a model to explain the IRAS background by the integrated light from distant galaxies, including a large evolution effect. The solid lines and dashed lines in Fig.6 represent the contribution of the star burst galaxies and cirrus emission of the normal galaxies, respectively. As is seen in Fig.7, the integrated light of the galaxies with continuous redshift distribution inevitably results in a very broad spectral width. In the case the $100 - 300 \mu m$ background is cosmological, a very narrow spectrum feature at the fixed redshift is required.

The far-infrared background may be of galactic origin. One possibility is to assume a dust component which distributes uniformly in the Galaxy and has no correlation with neutral hydrogen. In order to search for the CBR, more detailed observations on the spectrum and the spatial variation are necessary. Especially, observations of the sky around $300 \mu m$ seem very important, since the sky is very dark and the galactic emission has small contribution.

The most impressive result of the Nagoya-Berkeley rocket experiment is the excess brightness of the CBR over $2.74 K$ blackbody spectrum in the three long wavelength channels. No significant spatial fluctuation was found in these channels, suggesting the cosmological origin for the excess brightness. Fig.8 shows equivalent blackbody temperatures for these channels, which are compared with previous measurements. The solid and dashed lines represent the models assuming the Compton scattering and dust emission in the early universe, respectively. One important point to construct the model is very low brightness at $262 \mu m$ in Fig.7. For example, the Compton scattering model with relativistic electrons could be excluded by this data point (Wright 1981, Hayakawa et al. 1987). Even the dust emission model is fairly restricted by the low sky brightness at $262 \mu m$.

More serious issue for the model may be how to explain the excess energy which is estimated to be an order of 10% of the energy density of $2.74 K$ blackbody radiation. Since energy density of the radiation field is proportional to $(1+z)^4$, vast amount of energy release is required if the excess were originated at high redshift. Details of the models and theories are described in this proceedings.

# 5 Future space mission in Japan

As described in the previous section, space observation is essential to study the infrared CBR. In the rocket experiment advantage is that cycle time is relatively short and new technology can be applied rather easily. On the other hand observing time is too short to measure wide sky area and to integrate data for a long time. Further, residual environmental emission sometimes causes serious contamination. The satellite experiment is absolutely necessary which have a complementary role to the rocket experiment. COBE is planned so as to observe the infrared CBR thoroughly and will provide valuable data on the cosmology.

Japan is now developing a small liquid helium cooled infrared telescope (IRTS) onboard the small space platform (SFU : Space Flyer Unit). Although the telescope has only $15 cm$ diameter, IRTS is specially designed to observe the diffuse extended

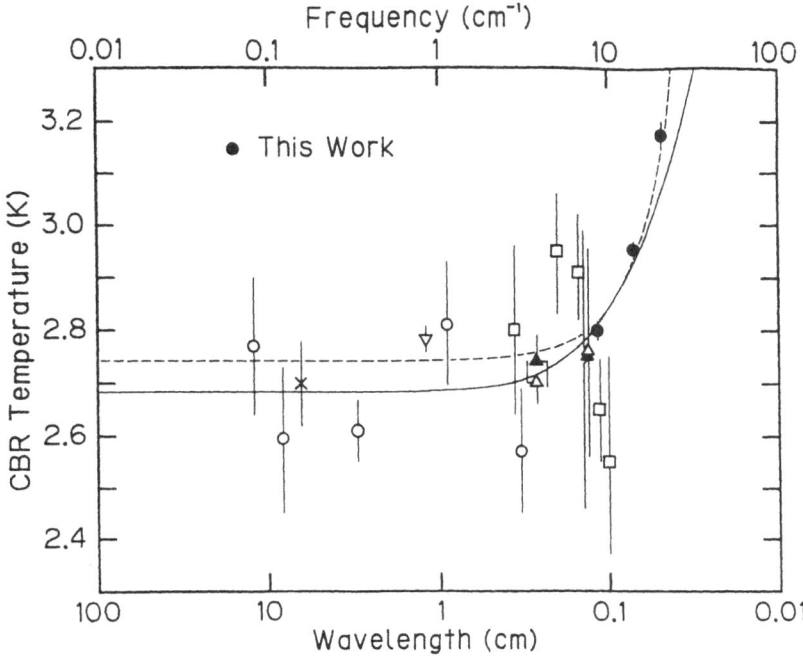

Figure 8: The equivalent blackbody temperatures for three long wavelength channels (filled circles). Models of the Compton scattering and dust emission are shown by solid and dashed lines, respectively.

emission from the near-infrared to the submillimeter range. 4 focal plane instruments are being prepared in collaboration with U.S.A. Table 1 shows the details of instruments. Near-infrared spectrometer (NIRS) and Far-infrared photometer (FIRP) are especially powerful instruments for the cosmological observation. NIRS is designed to observe the near-infrared CBR with very high sensitivity but with coarse spectral resolution. Nature of the isotropic emission observed in the rocket experiment will be delineated by NIRS. FIRP covers wide wavelength range from $100\mu$ to $1mm$ with very sensitive He3 cooled bolometer. Spatial fluctuation of the submillimeter CBR with half degree scale and Sunyaev-Zeldovich effect for a fairly large cluster of galaxies will be main targets of the FIRP.

The IRTS is scheduled to be launched on Jan. 1994. Although only three weeks are allocated, IRTS will play an important role as an advanced mission of the COBE.

| INSTRUMENT | NIRS | MIRS | FILM | FIRP |
|---|---|---|---|---|
| OPTICAL SYSTEM | GRATING SPECTROMETER | GRATING SPECTROMETER | GRATING SPECTROMETER | MULTO-BAND PHOTOMETER |
| WAVELENGTH RANGE ($\mu$m) | 1.2 - 4.1 | 5 - 13.5 | 63.2 (O I) 157.7 (C II) | 60 - 1000 |
| SPECTRAL RESOLUTION | 0.15$\mu$m | 0.26$\mu$m | $\lambda/\Delta\lambda$ =400 350 | $\lambda/\Delta\lambda$ =0.4 |
| BEAM SIZE | 0.14° ×0.14° | 0.14° ×0.14° | 0.15° ×0.3° | 0.5° |
| DETECTOR | InSb ARRAYS | SiGa ARRAYS | Ge:Ga × 4 | He BOLOMETER ×7 |
| PRE-AMP | DRO/CIA | DRO/CIA | TIA | TIA |
| DETECTION LIMIT W/CM$^2\mu$m STR | $2.8 \times 10^{-13}$ (1 SEC, 1$\sigma$) | $3 \times 10^{-13}$ (1 SEC, 1$\sigma$) | $5 \times 10^{-14}$ (1 SEC, 1$\sigma$) | $5 \times 10^{-14}$ (1 SEC, 1$\sigma$) |
| MAIN OBJECTIVES | Cosmic Bkg Galaxy, Dust | Gases, Dust Zod, Light | Interstellar Gases | Cosmic Bkg, IR-Cirrus |

Table 1: Focal plane instruments of IRTS

# References

Akiba,M., Matsumoto.T., and Murakami,H. 1989, submitted to *Astr.Ap.*

Boughn,S.P., Saulson,P.T., and Uson,J.M. 1986, *Ap.J*, **301**, 17.

Boughn,S.P. and Kuhn,J.R., *Ap.J.*, **309**, 33.

Boulanger,F and Perault,M. 1988, *Ap.J.*, **330**, 964.

Carr,B.J., Bond,J.R., and Arnett,W.D. 1984, *Ap.J.*,**277**, 445.

Dube,R.R., Wickes,W.C., and Wilkinson,D.T. 1977, *Ap.J.(Letters)*, **215**, L51.

Harwit,M., McNutt,D.P., Shivanandan,K., and Zajak,B.J. 1966, *A.J.*, **71**, 1026.

Hauser,M.G. et al. 1984, *Ap.J.(Letters)*, **278**, L15.

Hayakawa.S., Matsumoto,T., and Nishimura,T. 1970, *Space Res.*, **10**, 248.

Hayakawa,S., Matsumoto,T., Matsuo,H., Murakami,H., Sato,S., Lange,A.E.,
   and Richards,P.L. 1987, *Pub.Astr.Soc.Japan*, **39**, 941.

Hofmann,W and Lemke,D. 1978, *Astr.Ap.*, **68**, 389.

Hofmann,W., Lemke,D., Thum,C., and Fahrbach,U. 1973, *Nature Phys. Sci.*,
   **243**,140.

Lange,A.E., Richards,P.L., Hayakawa,S., Matsumoto,T., Matsuo,H., Murakami,H.,
   and Sato,S. 1989, submitted to *Ap.J.*

Matsumoto,T., Hayakawa,S., Matsuo,H., Murakami,H., Sato.S., Lange,A.E., and
   Richards,P.L. 1988a, *Ap.J.*, **329**, 567.

Matsumoto,T,, Akiba,M., and Murakami,H. 1988b, *Ap.J.*, **332**, 575.

Partridge,R.B. and Peebles 1967, *Ap.J.*, **148**, 377.

Rowan-Robinson,M and Carr,B 1988, *Post-Recombination Universe*,
   ed. N.Kaiser and A.N.Lasenby, p.125.

Terebey,S and Fich,M 1986, *Ap.J.Letters*, **309**, L73.

Toller,G.N. 1983, *Ap.J.(Letters)*, **266**, L79.

Yoshii,Y. and Takahara,F 1988, *Ap.J.*, **326**, 1.

Wright,E.L. 1981, *Ap.J.*, **250**, 1.

# THE ORIGIN OF THE EXTRAGALACTIC X-RAY BACKGROUND

G. SETTI
*European Southern Observatory*
*Karl-Schwarzschild-Str. 2*
*D-8046 Garching bei München*
*Federal Republic of Germany*

ABSTRACT. Although the shape of the spectrum in the 3–100 keV interval is suggestive of an optically thin bremsstrahlung at $\sim$ 40 keV, it is well known that the interpretation in terms of a hot intergalactic gas (IGG) requires a rather extreme energy supply and a gas density conflicting with the baryon density upper limit derived from primordial nucleosynthesis calculations. A summary discussion of the estimated contributions from the integrated X-ray emission of known classes of extragalactic discrete sources at a reference energy of 2 keV is given. Although these estimates are still uncertain, the subtraction of a "minimum" contribution drastically modifies the 40 keV thermal shape, prima facie evidence of a hot IGG. AGNs are the main contributors. Low luminosity AGNs at redshift $z = 1$–2 may in fact saturate the 2 keV XRB, but their observed hard X-ray spectra are on the average unlike (much too steep) that of the XRB. This has led a number of authors to postulate new classes of sources and some exotic models which are briefly summarized. However, if a recently proposed unified scheme of AGNs holds, then the bulk of the XRB intensity can be explained independently of the observed spectral differences and with a mild cosmological evolution.

## 1. Introduction

The cosmic X-ray background (XRB) has been the subject of many reviews (e.g. Boldt, 1987). It is well known that in the energy interval 3–100 keV the data are well fitted by an optically thin thermal bremsstrahlung spectrum at a temperature $kT \simeq 40$ keV, while at higher energies up to several hundred keV the energy spectrum can be represented by a power law of index $\alpha \simeq -1.8$ (Gruber *et al.*, 1984). At lower energies the spectrum of the XRB is essentially unknown because the galactic emission begins to be substantial and difficult to separate out. Below $\sim 20$ keV the energy spectrum is consistent with a power law of index $\alpha \simeq -0.4$, much flatter than the typical spectra of known classes of extragalactic X-ray sources, a fact that, as will be seen later, plays a key role in the overall discussion on the origin of the XRB.

The intensity integrated over the spectrum corresponds to a present energy density in the radiation field of $\simeq 5 \times 10^{-5}$ eV cm$^{-3}$ of which about 2/3 is below 40 keV. Although

*N. Mandolesi and N. Vittorio (eds.), The Cosmic Microwave Background: 25 Years Later*, 203–213.
© 1990 *Kluwer Academic Publishers.*

this value is not large in absolute terms compared to what is found in other regions of the electromagnetic spectrum, it remains unexplained why so much energy has been channeled in high energy photons. In fact, from the beginning it was clear that the bulk of the XRB must have been produced at higher redshifts either as a result of diffuse processes taking place in the intergalactic space and/or as the superposition of the emission from discrete sources.

The large scale distribution of the intensity of the XRB appears to be highly isotropic when a (small) galactic component is removed. A dipole component with a fractional amplitude of $5 \times 10^{-3}$ has been suggested but rms fluctuation encompassing large solid angles are $\gtrsim 3 \times 10^{-3}$ (Boldt, 1988). If real and interpreted in terms of the Compton-Getting effect, it would correspond to an observer's velocity vector (Shafer, 1983; Shafer and Fabian, 1983) consistent within the error with that inferred from the cosmic microwave background (CMB) dipole anisotropy (Smoot et al., 1977; Cheng et al., 1979).

Of particular interest to us here are the measurements of the small-scale intensity fluctuations which may be directly related to the statistical fluctuations in the source populations contributing to the XRB and their clustering properties and/or to the clumpiness of the medium if diffuse emission is important.

At energies $\gtrsim 3$ keV the surface brightness fluctuations are $\approx 0.02$ on angular scales of several degrees (Shafer, 1983). These fluctuations are consistent with an extrapolation of the number counts ($N$) vs. flux ($f$) relationship with an Euclidean power law of the form $N(> f) \propto f^{-1.5}$ down to a flux about one order of magnitude fainter than the survey limit of the bright sources found in the HEAO-1 complete sample. It is also found that any population of sources which would contribute most of the XRB should have a surface density $\gtrsim 50$ sources deg$^{-2}$.

In the 1–3 keV energy band, arcminute scale fluctuations of the XRB have been analyzed by Hamilton and Helfand (1987) using deep survey fields obtained by the Einstein Observatory. The measured granularity requires that the X-ray source counts extend well below the High Sensitivity Survey limit of the Einstein Observatory. However, if a population of discrete sources had to account for most of the XRB in this energy band, then the source counts should flatten well below the Euclidean slope and the derived surface density of such sources should be no less than $\sim 5 \times 10^3$ deg$^{-2}$.

The fact that the observed spectrum of the XRB between 3 and 100 keV can be so well approximated by an optically thin bremsstrahlung has renewed an interest in the proposal of Cowsik and Kobetich (1972) that the XRB is due to a hot diffuse intergalactic gas (IGG) with a (present) temperature of $\sim 5 \times 10^8$ K (Field and Perrenod, 1977). Because of the relative inefficiency of the bremsstrahlung mechanism, the cooling time is much larger than the Hubble time for any reasonable value of the baryon density parameter and, as a result, the energy deposited in the gas is comparable to that in the CMB. To cope with the huge amount of energy to be supplied one has relegated the reheating of the IGG to earlier epochs for which at least there is observational evidence that non-thermal explosive phenomena in galaxies were much more frequent and powerful as shown by the dramatic cosmological evolution associated with quasars and radio galaxies. Subsequent work has simply assumed that the IGG is suddenly heated up at a redshift $z_m$, and then cools down because of the expansion and Compton losses with the photons of the CMB. Since Compton cooling starts to progressively dominate at redshift $z > 3.5$ it becomes energetically unfavourable to assume that the IGG has been reheated at the required temperature at even larger redshifts, while at the same time the low energy spectrum from a Compton cooled optically thin gas would be much steeper than the observed XRB (Guilbert and Fabian, 1986). Therefore, in order to fit the shape of the XRB spectrum, the gas should have been reheated at a temperature $T \approx (1 + z_m)40$ keV, with $z_m \approx 4$, and relativistic corrections to the bremsstrahlung emission and to the gas thermodynamics must be taken into account. But even with the enhanced emission due to the relativistic corrections, it turns out that, in order to meet the XRB intensity, the IGG density should correspond to a density parameter $\Omega_{IGG} > 0.2$ (Barcons and Lapiedra, 1985; Guilbert and Fabian, 1986; Barcons, 1987). For instance, a recent discussion (Taylor and Wright, 1989) indicates $\Omega_{IGG} \simeq 0.27$ for a model in which the IGG was heated at $z_m = 5$ so that its energy content reached $\sim 50\%$ of that in the CMB and then cooled down to a present temperature of $\sim 10$ keV. So in addition to the energy problem there is a further difficulty with the hot IGG interpretation of the XRB in the framework of the standard hot big-bang model of the universe, because the requirement that the primordial nucleosynthesis produces the observed abundances of light elements, in particular deuterium, places an upper bound of $\Omega_B < 0.20$ to the total density of baryonic matter (Boesgaard and Steigman, 1985; Reeves et al., 1989).

Since the bremsstrahlung emission is proportional to the square of the gas density, a way out of the above difficulty would be to assume that the IGG is clumped, thereby reducing the mean density required to fit the XRB intensity (Field and Perrenod, 1977; Guilbert and Fabian, 1986). Then, one would have to assume a two-phase model of the IGG in which hot, relatively dense clumps of gas responsible for the observed XRB between 3 and 100 keV are in pressure equilibrium with a surrounding thinner but hotter gas. Obviously, the observed surface brightness fluctuations in the XRB can be used to constrain the physical parameters of this two-phase model of the IGG, but much more stringent constraints are imposed from the analysis of the temperature fluctuations of the CMB induced by Compton scattering the photons out of the Rayleigh-Jeans portion of the CMB spectrum (Guilbert and Fabian, 1986). In a recent work Barcons and Fabian (1988) come to the conclusion that the size of the clumps should be less than a few tens of Kpc, not to exceed the upper limits obtained with the VLA (Fomalont et al., 1984; Knoke et al., 1984) on the fluctuations of the CMB temperature on sub-arc minute angular scales. If this result holds, it is certainly not easy to imagine how this fine-structured two-phase plasma could have originated and maintained, and the whole idea of explaining the XRB by an optically thin bremsstrahlung seems to meet very serious difficulties.

In the hot IGG hypothesis one has implicitly assumed that the integrated contribution of known classes of extragalactic X-ray emitting objects to the XRB in the 3–100 keV energy interval is small so as not to substantially deform the optically thin thermal bremsstrahlung shape which is the prima facie evidence in support of the existence of a hot IGG from which the bulk of the XRB originates. How large this contribution is, will be the subject of the next section.

## 2. Discrete Source Contribution to the XRB

Recent, detailed discussions on the contribution of known classes of extragalactic X-ray sources to the XRB can be found in Schmidt and Green (1986), Giacconi and Zamorani

(1987) and Setti (1987). Since the various classes of objects are characterized by different spectral types, and therefore their percentage contributions to the XRB in general depend on the photon energy, it is useful for the sake of discussion to choose a reference photon energy for normalization. Since most data have been collected by the Einstein Observatory, which was sensitive in the energy range 0.3–3.5 keV, and since the XRB spectrum and the bright source sample spectra were observed at energies $> 2$ keV by instruments such as HEAO-1 and EXOSAT, it has become convenient to refer the estimates to an (observed) energy of 2 keV where the intensity of the "reference" XRB is obtained by smoothly extrapolating the bremsstrahlung spectrum applicable to energies $> 3$ keV downward with a power law of index $-0.4$.

The estimates of the integrated contributions to the 2 keV XRB thus defined from known classes of sources are summarized in Table 1 following the discussion of Setti (1987). It should be noted that the contribution from the resolved sources down to the Einstein Observatory High Sensitivity Survey (HSS) limit is $(26 \pm 11)\%$ (Giacconi et al., 1979) on the assumption that the energy spectra of the sources in the observed band are represented by a power law of index $-0.5$. If, as will be discussed later, the spectral index $\alpha \lesssim -1.0$ then the HSS limit should be reduced by $\sim 40\%$.

The dominant contribution apparently comes from AGNs (quasars and Sy 1 nuclei), but the estimates are still affected by uncertainties.

TABLE 1: The contributions of known classes of discrete extragalactic sources to the nominal 2 keV XRB

| Objects | % | Remarks and References |
|---|---|---|
| Quasars | $8^a$–13 | $M_B < -23^b$. Fitting total # found in MSS survey. [1] |
| | $\sim 20$ | $M_B < -23.8$. From the optical counts. [2] |
| | $7^a$–11 | $M_B < -23.8$. Same as above but fitting the MSS survey. [2] |
| Seyfert 1 | 29 | $M_B > -23$. From HEAO-1 survey. No evolution. [1,3] |
| | $> 34$ | $M_B > -23.8$. Some evolution. [2,4] |
| Galaxy clusters | $\lesssim 10$ | From HEAO-1 survey. No evolution. [1,4] |
| Normal galaxies | $4^a$–9 | Einstein Obs. data. [1,4] |
| | 13 | Einstein Obs. data, but opt. counts from Tyson (1984). [5] |
| Other sources | $\lesssim 7$ | BL Lacs, radio galaxies, etc. [4] |

[a] Percentage values obtained from Einstein Obs. data by adopting a steep spectral index ($\alpha = -1.2$).

[b] Quoted $M_B$ values assume $H_0 = 50$ km s$^{-1}$ Mpc$^{-1}$.

Ref. : 1) Schmidt & Green (1986); 2) Setti (1987); 3) Piccinotti et al. (1982); 4) Setti & Woltjer (1982); 5) Giacconi & Zamorani (1987)

*Quasars.* The highest percentage contribution reported in Table 1 ($\simeq 20\%$) has been obtained by integrating over the X-ray source counts derived from the optical counts via the distribution of X-ray to optical flux ratios provided by the observations of the Bright Quasar Survey (BQS) sample with the Einstein Observatory (Tananbaum *et al.*, 1986). While the slope of the X-ray source counts thus derived agrees with that of the Einstein Observatory Medium Sensitivity Survey (MSS; Gioia *et al.*, 1984) in the region of overlap, their number exceeds by at least a factor 1.6 that of the MSS [see Setti (1987) for a detailed discussion]. The reason(s) for this discrepancy is unknown. If the MSS counts are enforced then one obtains a lower percentage value which is in very good agreement with that found by Schmidt and Green (1986). This is reassuring since the two estimates have been made by following completely different methodologies.

The spectra of quasars in the Einstein Observatory band (0.2–3.5 keV) have been the subject of recent studies (Wilkes and Elvis, 1987; Canizares and White, 1989). According to these the radio loud (RL) quasars with flat (F) radio spectra present X-ray spectra much flatter ($<\alpha> \simeq -0.4$) than those of the radio quiet (RQ) quasars ($<\alpha> \simeq -1.0$), while the RL quasars with steep (S) radio spectra provide some sort of intermediate case. It should be noted that the RL(F) quasars have an average spectral index which matches that of the XRB in the 3–10 keV interval. By splitting their sample in redshift bins Canizares and White (1989) do not find any indication of a redshift dependence of the average spectral indices, although the statistics are really good enough only for the RL(F) quasar sample showing that these objects maintain a flat X-ray spectrum up to an energy of $\sim 10$ keV.

From the standpoint of the contribution to the 2 keV XRB it is important to note that the Einstein Observatory observations have usually been reduced by assuming a spectral slope of $-0.5$, while if the actual slope is $\lesssim -1.0$ then the corresponding 2 keV monochromatic fluxes should be reduced by $\simeq 40\%$ (Tananbaum *et al.*, 1986). Since the great majority of the quasar population is composed of RQ objects ($\gtrsim 90\%$), then the lowest estimates reported in Table 1 are found. Keeping in mind that RL quasars are on average stronger X-ray sources than the RQ quasars ($\sim$ a factor 3 after Zamorani *et al.*, 1981) one can conclude that the contribution of quasars to the 2 keV XRB has an upper bound of 14% and could be as low as 8% if the MSS counts are matched.

The hard X-ray spectra of quasars are poorly known. Only very few had been measured until recently, notably 3C 273 with $\alpha \simeq -0.5$ in the 2–30 keV interval (Worral *et al.*, 1979), but the sample is now growing as a result of the data which are being obtained by the instruments on board the Japanese X-ray satellite GINGA in the 2–30 keV energy interval. The fitted power law spectra are consistent with an average slope and a dispersion typical of Sy type 1 nuclei to be discussed next (Inoue, 1989).

*Seyfert 1 Nuclei (or Low Luminosity AGNs).* A minimal contribution of 29.0% ± 6.5% is obtained from the HEAO-1 A2 complete sample of bright sources and uniform space distribution, but this figure may be substantially increased by the presence of cosmological evolution for which there is some indication at the bright end of the Sy 1 luminosity function (Setti, 1984). If the integral optical counts of AGNs were to continue with the slope $\simeq 1$, applicable to the B magnitude interval 21–23, down to B $\simeq 27$–28, then most of the 2 keV XRB would originate in low luminosity AGNs at redshifts $z \sim 1$–2 (Setti and Woltjer, 1982; Setti, 1987; Giacconi and Zamorani, 1987; Anderson and Margon, 1987; Hamilton and Helfand, 1987). This would correspond to extending the X-ray source counts down to a flux ~ 100 times fainter than the HSS limit with a correspondingly flat slope consistent with the arcminute scale fluctuation analysis of Hamilton and Helfand (1987).

As is well known, however, the main problem resides in the discrepancy between the average spectrum of this type of sources and that of the XRB. A sizable sample of bright nearby hard X-ray selected Sy type 1 nuclei has been measured with the detectors on board EXOSAT in the energy interval 0.1–10 keV (Turner and Pounds, 1989). In the 2–10 keV interval the spectra are well described by single power laws with a mean spectral index $<\alpha> \simeq -0.7$ and a dispersion $\sigma \simeq 0.17$ around the mean, in good agreement with previous findings based on HEAO-1 A2 data (Mushotzky, 1984). It is interesting to note that the soft X-ray spectra of this hard X-ray selected sample are complex but include several which show a turn up of the spectrum at ~ 1 keV in agreement with the average steep slope found in the Einstein Observatory band for corresponding low luminosity AGNs (Wilkes and Elvis, 1987).

Thus, if the AGNs were to supply the bulk of the XRB then the relevant portions of their spectra should be much flatter ($< \alpha > \gtrsim -0.4$) than measured in the nearby sample, which implies the presence of a very significant cosmological evolution (e.g., Morisawa and Takahara (1989) for recent modelling). In fact it has been shown (Danese et al., 1986) that any population with hard X-ray spectra typical of the nearby AGN sample cannot contribute more than ~ 30% to the total XRB energy flux in the 2–10 keV interval otherwise it would produce spectral wiggles incompatible with the general smoothness of the XRB spectrum (De Zotti et al., 1982). A statistical analysis of a fainter sample of X-ray selected AGNs from the Einstein Observatory Extended MSS yields an average spectral index $<\alpha> \simeq -1.0$ (Maccacaro et al., 1988), like for the bright sample of optically selected AGNs of Wilkes and Elvis (1987), and therefore no indication of spectral evolution is found.

However, GINGA observations of several Seyfert galaxies show a flattening of the spectra at about 10 keV raising the interesting possibility that luminosity evolution of Sy 1 nuclei with the redshift, together with a possible enhancement of the flattening of the spectra with increasing luminosities, may account for the shape and intensity of the XRB (Hayakawa, these proceedings).

The X-ray properties of a unified scheme of AGNs and their implications for the XRB are discussed by Setti and Woltjer (1989). According to this scheme (Barthel, 1989) the RL quasars and the strong radio galaxies are members of the same population, the RL quasar phenomenon showing up whenever the associated relativistic beams are aimed toward us within a certain angle from the line of sight. Similarly, it is proposed that the radio-quiet quasars and luminous infrared galaxies can be unified, the apparent morphological differences being attributable only to the geometrical orientation of the sources with respect to the observer. In this framework one has to assume that quasars and Sy type 1 nuclei remain hidden by optically thick tori of surrounding absorbing material whenever the line of sight is not favourably placed. Setti and Woltjer (1989) show that with very reasonable assumptions also the X-rays emitted by the central sources can be effectively absorbed up to 20 keV, or more, which can explain why quasars and Sy 1 nuclei are observed to be much stronger X-ray sources than radio galaxies and IR galaxies in the Einstein Observatory energy band. At energies $\gtrsim 10$ keV, depending on the precise value of the absorbing column of gas along the line of sight, the radio galaxies and IR galaxies would have to show the same hard X-ray spectra as quasars and Sy 1 nuclei if Barthel's unified picture holds. It is inferred that by assuming an appropriate distribution of the absorption cut-offs combined with a "mild" cosmological evolution of the global AGNs population up to a redshift $z \simeq 2.5$–3, one can account for the overall XRB even if the average hard X-ray spectrum of the sources is as steep as that observed in the bright sample of Sy 1 nuclei ($\alpha \simeq -0.7$). Incorporating the flattening of the spectra now found by GINGA should lead to a much improved model.

*Clusters of Galaxies.* Their emission is due to optically thin thermal bremsstrahlung of hot intracluster gas with typical temperature $\sim 6$ keV, much below that required for the XRB. The estimate of the contribution is based on the HEAO-1 A2 survey, without including possible cosmological evolution scenarios which may enhance the expected contribution (Blanchard, these proceedings).

*Normal Galaxies.* Very little is known about their hard X-ray spectra (Makishima and Ohashi, 1989). Their shapes will depend in general on the particular admixture of galactic sources making up their global emission. The resulting spectra could also be rather flat. Bookbinder *et al.* (1980) have discussed the possible contribution of massive X-ray binaries whose spectra are known to be flat, much like that of the XRB, and to extend up to $\sim 20$ keV. This possibility has been revived by Griffiths (1989) who argues that the integrated emission from an early population of low metallicity massive X-ray binaries in star-forming galaxies may contribute up to $\sim 50\%$ of the XRB in the 3–20 keV energy interval if most of the contributed flux originates in galaxies with redshifts confined to $z < 1$.

In any case normal galaxies and clusters together are bound to make a non-negligible contribution to the low energy part of the XRB spectrum.

## 3. Conclusions

Assuming that the spectra of the AGNs are on average as steep as those measured in the bright sample, then the discrete source contributions to the XRB above 3 keV can be subtracted out and one is left with the problem of explaining the "residual" background which still incorporates most of the energy flux (Leiter and Boldt, 1982; Setti, 1985). The low energy side of the residual XRB becomes very flat up to 20–30 keV (spectral index $\alpha \gtrsim -0.2$ below $\sim 10$ keV), the precise shape depending on the spectral distribution and percentage contributions of the source populations to be subtracted. Giacconi and Zamorani (1987) have modelled the residual XRB by making different assumptions on the relevant quantities, including "minimal" assumptions. The main conclusion one can draw is that the residual XRB can no longer be fitted by an optically thin thermal bremsstrahlung with a temperature of $\sim 40$ keV. Even the "minimal" contribution of the discrete sources destroys its extremely good representation of the observed spectrum in the 3–100 keV range, so that no physical meaning can be attached to it. The extreme flatness of the residual XRB spectrum excludes the possibility that it is due to an optically thin bremsstrahlung from a hot diffuse gas. The amount of comptonization required to flatten the spectrum leads in fact to rather stringent upper limits on the size of the thermally emitting regions and one falls back into the concept of compact sources.

So it appears unlikely that a hot IGG can make a major contribution to the XRB.

This in itself does not mean that there cannot be a hot diffuse IGG which might somewhat contribute to the XRB, but it will be difficult to detect its presence by X-ray observations alone.

The low luminosity AGNs still form the best candidate population amongst known classes of objects capable of supplying the bulk of the XRB. As has been pointed out, there are now reasons to believe that the main obstacle to this interpretation, that is the discrepancy between the average slope of the source spectra and that of the XRB, can be removed. For this interpretation to be viable a relatively modest cosmological evolution is required.

Many other models have been proposed in the attempt to explain the XRB in the course of almost three decades after its discovery (Giacconi $et$ $al.$, 1962). It is obviously impossible to summarize here all the work which has been done. We shall limit ourselves to briefly outlining some other proposals put forward in this decade to give a flavour of the ample possibilities which may remain open. These ideas are generally based on assumptions concerning the early evolution of the physical properties of normal galaxies or AGNs or on cosmological events of some form all of which are in general difficult to verify observationally.

Bookbinder $et$ $al.$ (1980) have made the interesting conjecture that the XRB can be essentially supplied by the thermal bremsstrahlung emission associated with hot galactic winds powered by an increased rate of supernovae in young galaxies. This requires that on average each galaxy produces $\gtrsim 10^{10}$ supernovae over a typical time interval of $10^7$ years and that the epoch of galaxy formation is rather recent, $z \simeq 2$, otherwise the temperature of the wind would not be high enough to cope with the redshift effect. There are other difficulties with this model: Giacconi and Zamorani (1987) noted that the associated optical emission would be such that the surface density of young galaxies at faint optical magnitudes would be much larger than what has actually been found, while it is also likely that the $\gamma$-ray emission associated with these energetic events would produce a $\gamma$-ray background flux far in excess of the observed one (Setti and Woltjer, 1982).

Leiter and Boldt (1982) have proposed that the sources of the XRB could be found in precursor AGNs, where gas accretion onto a central supermassive black hole first generates a very compact, hot, slightly comptonized thermal source prevalently emitting hard X-rays (luminosity $L \sim 1/10 L_E$, with $L_E$ the Eddington luminosity), which then evolves to a less compact Sy type nucleus ($L \sim 1/100 L_E$) in a characteristic time interval corresponding to the Hubble time at $z \sim 3$ (Boldt and Leiter, 1987). These authors have assumed a total contribution from known extragalactic X-ray objects ($\sim 50\%$ at 2 keV) such that the residual XRB would have a spectral slope $\alpha \simeq -0.2$ at the low energy end. This model provides in a way a somewhat extreme example of the spectral evolution for AGNs required to explain the XRB.

Daly (1987) has proposed a unified picture in which the XRB, the formation of galaxies and voids are all parts of the same scenario. The XRB would be produced by thermal bremsstrahlung emission associated with large ($10^{15-16} M_\odot$) gravitationally bound condensates of matter whose formation and successive evolution is governed by the properties of an (unspecified) unstable dark matter particle candidate. The main difficulty here, at least as far as the origin of the XRB is concerned, is that the surface density of the condensates is much smaller than the lower bound imposed by the XRB fluctuations, unless, as the author argues, the condensates are constrained to form following a very regular pattern.

That the XRB could be linked to the formation of structures in the universe has also been proposed by Ostriker *et al.* (1986) in a different scenario, where superconducting cosmic strings are assumed to release enormous amounts of energy to the surrounding intergalactic medium, thereby producing huge expanding shells of hot gas on the edge of which galaxies could form after cooling. This picture predicts the existence of a very hot intergalactic gas with a bubble-like structure and temperatures of $10^8$–$10^9$ K which may contribute to the X-ray background.

Fabian *et al.* (1988) propose that the hard X-ray shape of the XRB spectrum can be explained by a non-thermal process in which the acceleration mechanism of the electrons is "loaded" by the electron-positron pairs created in photon-photon collisions. This feedback mechanism would "naturally" produce a sharp break in the emitted photon spectrum at the electron rest mass energy, much larger however than the observed $\sim 40$ keV. As a result these hypothetical sources, identified as young active galaxies, should be placed at redshifts $z \simeq 10$–30, emitting on average $\sim 10^{46}$ erg s$^{-1}$ each.

If on the one hand the broad astrophysical and cosmological implications involved in the extragalactic XRB justify the continuing interest in its study, on the other hand the variety of models which have been (and are being) proposed is in itself a demonstration of the need for new and deeper observations in order to constrain the various possibilities. As a result of our discussion it is clear that the basic problem still remains the determination of the contribution of known classes of extragalactic sources, in particular AGNs and related objects, over the energy range of interest for the XRB.

# References

Anderson, S.F., and Margon, B. 1987, *Astrophys. J.*, **314**, 111.
Barcons, X., and Lapiedra, R. 1985, *Astrophys. J.*, **289**, 33.
Barcons, X. 1987, *Astrophys. J.*, **313**, 547.
Barcons, X., and Fabian, A.C. 1988, *Mon. Not. R. astr. Soc.*, **230**, 189.
Barthel, P.D. 1989, *Astrophys. J.*, **336**, 606.
Boesgaard, A.M., and Steigman, G. 1985, *Ann. Rev. Astr. Astrophys.*, **23**, 319.
Boldt, E. 1987, *Phy. Reports*, **146**, No. 4, 215.
Boldt, E. 1988, NASA Report #88-037.
Boldt, E., and Leiter, D. 1987, *Astrophys. J.*, **322**, L1.
Bookbinder, J., Cowie, L.L., Krolik, J.H., Ostriker, J.P., and Rees, M.J. 1980, *Astrophys. J.*, **237**, 647.
Canizares, C.R., and White, J.L. 1989, *Astrophys. J.*, **339**, 27.
Cheng, E.S., Saulson, P.R., Wilkinson, D.T., and Corey, B.E. 1979, *Astrophys. J.*, **232**, L139.
Cowsik, R., and Kobetich, E.J. 1972, *Astrophys. J.*, **177**, 585.
Daly, R.A. 1987, *Astrophys. J.*, **322**, 20.
Danese, L., De Zotti, G., Fasano, G., and Franceschini, A. 1986, *Astrophys. J.*, **161**, 1.
De Zotti, G., Boldt, E.A., Cavaliere, A., Danese, L., Franceschini, A., Marshall, F.E., Swank, J.H., and Szymkowiak, A.E. 1982, *Astrophys. J.*, **253**, 47.
Fabian, A.C., Done, C., and Ghisellini, G. 1988, *Mon. Not. R. astr. Soc.*, **232**, 21P.
Field, G.B., and Perrenod, S.C. 1977, *Astrophys. J.*, **215**, 717.
Fomalont, E.B., Kellermann, K.I., and Wall, J.V. 1984, *Astrophys. J.*, **277**, L23.

Giacconi, R., Gursky, H., Paolini, F., and Rossi, B. 1962, *Phys. Rev. Letters*, **9**, 439.

Giacconi, R., *et al.* 1979, *Astrophys. J.*, **234**, L1.

Giacconi, R., and Zamorani, G. 1987, *Astrophys. J.*, **313**, 20.

Gioia, I.M., Maccacaro, T., Schild, R.E., Stocke, J.T., Liebert, J.W., Danziger, I.J., Kunth, D., and Lub, J. 1984, *Astrophys. J.*, **283**, 495.

Griffiths, R.E. 1989, NATO–ASI on *The Epoch of Galaxy Formation*, eds. C.S. Frenk *et al.* (Kluwer, Dordrecht), p. 235.

Gruber, D.E., Rothschild, R.E., Matteson, J.L., and Kinzer, R.L. 1984, in *X-Ray and UV Emission from Active Galactic Nuclei*, eds. W. Brinkmann and J. Trumper, MPE Report **184**, p. 129.

Guilbert, P.W., and Fabian, A.C. 1986, *Mon. Not. R. astr. Soc.*, **220**, 439.

Hamilton, T.T., and Helfand, D.J. 1987, *Astrophys. J.*, **318**, 93.

Inoue, H. 1989, in *Big Bang, Active Galactic Nuclei and Supernovae*, (Universal Ac. Press, Tokyo), p. 301.

Knoke, J.E., Partridge, R.B., Ratner, M.I., and Shapiro, I.I. 1984, *Astrophys. J.*, **284**, 479.

Leiter, D., and Boldt, E. 1982, *Astrophys. J.*, **260**, 1.

Maccacaro, T., Gioia, I.M., Wolter, A., Zamorani, G., and Stocke, J.T. 1988, *Astrophys. J.*, **326**, 680.

Makishima, K., and Ohashi, T. 1989, in *Big Bang, Active Galactic Nuclei and Supernovae*, (Universal Ac. Press, Tokyo), p. 371.

Morisawa, K., and Takahara, F. 1989, *P.A.S.J.*, to be published.

Mushotzky, R.F. 1984, in COSPAR/IAU Symp., *High-Energy Astrophysics and Cosmology*, eds. G.F. Bignami and R.A. Sunyaev, *Adv. Space Res.*, **3**, p. 157.

Ostriker, J.P., Thompson, C., and Witten, E. 1986, *Phys. Letters*, **B180**, 231.

Piccinotti *et al.* 1982, *Astrophys. J.*, **253**, 485.

Reeves, H., Richer, J., Sato, K., and Terasawa, N. 1989, preprint.

Schmidt, M., and Green, R.F. 1986, *Astrophys. J.*, **305**, 68.

Setti, G., and Woltjer, L. 1982, in *Astrophysical Cosmology*, eds. H.A. Bruck, G.V. Coyne and M.S. Longair (Pontificia Academia Scientiarum, Vatican City), p. 315.

Setti, G. 1984, in *X-Ray and UV Emission from Active Galactic Nuclei*, eds. W. Brinkmann and J. Trumper, MPE Report **184**, p. 243.

Setti, G. 1985, in *Non-thermal and Very High Temperature Phenomena in X-ray Astronomy*, eds. G.C. Perola and M. Salvati (Istituto Astronomico, Universita "La Sapienza", Rome), p. 159.

Setti, G. 1987, in IAU Symp. No. 124, *Observational Cosmology*, eds. A. Hewitt, G. Burbidge and L.Z. Fang (Reidel, Dordrecht), p. 579.

Setti, G., and Woltjer, L. 1989, *Astron. Astrophys.*, **224**, L21.

Shafer, R.A. 1983, NASA Tech. Mem. 85029.

Shafer, R.A., and Fabian, A.C. 1983, in IAU Symp. No. 104, *Early Evolution of the Universe and its Present Structure*, eds. G.O. Abell and G. Chincarini (Reidel, Dordrecht), p. 333.

Smoot, G.F., Gorenstein, M.V., and Muller, R.A. 1977, *Phys. Rev. Letters*, **39**, 898.

Tananbaum, H., Avni, Y., Green, R.F., Schmidt, M., and Zamorani, G. 1986, *Astrophys. J.*, **305**, 57.

Taylor, G.B., and Wright, E.L. 1989, *Astrophys. J.*, **339**, 619.

Turner, T.J., and Pounds, K.A. 1989, *Mon. Not. R. astr. Soc.*, submitted.

Tyson, J.A. 1984, in IAU Colloq. 78, *Astronomy with Schmidt Type Telescopes*, ed. M. Capaccioli (Reidel, Dordrecht), p. 489.

Wilkes, B.J., and Elvis, M. 1987, *Astrophys. J.*, **323**, 243.

Worral, D.M., Mushotzky, R.F., Boldt, E.A., Holt, S.S., and Serlemitsos, P.J. 1979, *Astrophys. J.*, **232**, 683.

Zamorani, G., *et al.* 1981, *Astrophys. J.*, **245**, 357.

# COSMIC BACKGROUND RADIATION: COMPONENTS IN OTHER WAVELENGTH RANGES AND THEIR RELEVANCE TO MICROWAVE BACKGROUND

Satio Hayakawa
Nagoya University
Furo-cho, Chikusa-ku, Nagoya, Japan 464-01

## SUMMARY

The background radiation is usually contaminated by local components of galactic and solar system origins. The local components can be separated out of the cosmological component by utilizing their anisotropies in different wavelength ranges, and their contributions to spatial fluctuations are comparable to the upper limits observed to date. The cosmological components in different wavelength ranges are genetically related to each other. Models for explaining the submillimeter excess predict the background radiation in other ranges, such as in the infrared and X-ray ranges. The X-ray background is discussed by reference to recent results obtained by Ginga.

## 1. INTRODUCTION

The background radiation of astronomical origin cannot easily be separated from the contributions of instrumental and environmental components, as often debated at this conference. Concerning measurements of the spectrum in the microwave and submillimeter ranges, Richards (1989) discussed

215

*N. Mandolesi and N. Vittorio (eds.), The Cosmic Microwave Background: 25 Years Later*, 215–228.
© 1990 *Kluwer Academic Publishers*.

problems of obtaining errors associated with the flux value and demonstrated that many of the errors should be regarded as systematic (see, also Hayakawa 1989).

Once the astronomical component is obtained with sufficient reliability, we meet the next problem of distinguishing the extragalactic component from the solar system and galactic components. Matsumoto (1989) showed how the contribution of local components due to interstellar dust (ISD) emission and interplanetary dust (IPD) emission were subtracted from the far infrared flux observed by his rocket experiment. An apparently finite flux in the 140mm range left after subtraction remains to be investigated.

If the local components are subtracted, we are left with components of cosmological interest. These components may be divided into two groups, one being a superposition of discrete sources astronomically well identified, such as galaxies, clusters of galaxies and possibly superclusters, and the other arising in the pregalactic era; radiation in thermal equilibrium with matter before decoupling is now observed as the cosmic microwave background (CMB) with a Planckian spectrum, and photons of various origins and reprocessed are considered to distort the spectrum. These components of radiation have been found highly isotropic, and only the dipole component has been positively observed, though a smaller degree of anisotropy has been reported in a few observation.

The spectral distortion and the spatial fluctuations on different angular scales have been discussed since the discovery of CMB, as they reveal the history of the universe. Various theories thus far proposed are relevant also to observed phenomena in other wavelength ranges. For example, the Sunyaev-Zeldovich (S-Z) effect is based on the X-ray emission from a hot intergalactic plasma which can distort the spectrum of such radiation that penetrates the plasma. For this reason the present conference includes sessions for the background radiation in other wavelength ranges than the microwave range.

In this discussion paper I will first summarize problems concerning the local components referring to several important points made by participants in the open discussion. Then a number of models that attempts to explain the submillimeter excess will be discussed in connection with their consequences to be observed in other wavelength ranges. Among them the X-ray background will be separately discussed by reference to recent results obtained by Ginga.

## 2. LOCAL COMPONENTS

The background radiation of cosmological and extragalactic origin is inevitably contaminated by local components of galactic and solar system origins. Each local component is of astrophysical interest in itself, and a superposition of local sources may contribute to the background radiation and particularly to its fluctuations.

In the low and high frequency sides of the CMB spectrum, the galactic radio waves and the far infrared emission from ISD are considered to give some contributions, respectively. Both are supposed to depend approximately on the galactic latitude $b$ and cosec$b$. One might suspect if the submillimeter excess could be accounted for in terms of the emission from cold ISD of temperatures below 10 °K. However, a dip in the spectrum observed in the 260mm band by Matsumoto $et$ $al.$, (1988a) does not favor the cold ISD interpretation. Moreover, as demonstrated by Matsumoto (1989), the flux of far infrared radiation is dominated by the component correlated with the atomic hydrogen column density. The spectrum of the HI correlated component indicates the dust temperature of 17-20 °K, depending on the optical property of dust, which is uniform within $\pm 3\%$ (Lange $et$ $al.$, 1989). The ISD component gives a contribution of about 6% to the submillimeter excess and a fluctuation less than 0.2% at a wavelength of 480mm. Its contributions to the flux and fluctuations in the microwave range are much smaller than the measurement errors and upper limits available to date.

The contribution of IPD is considered to be dominant at wavelengths shorter than 100mm since the temperature of IPD is supposed to be higher. However, the spatial distributions of IPD emission in different wavelength ranges are not well known yet. The contribution of larger solid bodies including comets is yet to be clarified, though a modest estimate gives an insignificant contribution.

The ISD emission not associated with HI is yet to be investigated. If the dust to gas ratio in the galactic halo is comparable to or higher than that in HI clouds, the contribution of dust in the halo may not always be negligible. The dust temperature may well depend on the region in which dust is located, since the energy flux of radiation or hot electron responsible for heating dust depends on the region concerned.

The observational data available to date are still too poor to subtract the contributions of these local components with sufficient reliability. Several experiments in preparation with use of sounding rockets and orbiting telescopes such as COBE, ISO and IRTS will be able to separate the local components from the truly primordial background, since a wide region will be surveyed at different wavelengths.

## 3. POSSIBLE ORIGINS OF THE SUBMILLIMETER EXCESS

The excess in the submillimeter range over the Planckian spectrum of 2.74 °K as obtained from microwave and CN measurements has stimulated theoretical attempts which are more or less speculative. A model usually contains many parameters to fit observed data and might therefore be regarded as having too much freedom to be acceptable as a physically sound model. However, there are essentially two parameters, the amount of input energy and the epoch of energy release. These two are often expressed by combinations of parameters, and additional parameters are sometimes contained for making the model realistic.

Each model is assessed on the basis of its plausibility and whether or not the values of parameters are acceptable. Another criterion is the consequences of a model that may result in the background radiation in other wavelength ranges.

The inverse Compton scattering model was proposed almost twenty years ago and contains only two parameters, one the $y$-parameters which is related to the energy density of electrons integrated up to an early epoch and the other the thermodynamic temperature $T_\theta$ of radiation without Comptonization. The values of these two parameters are obtained by fitting the submillimeter spectrum to be $y=0.028\pm0.004$ and $T_\theta=2.75\pm0.03$ °K, which give the Rayleigh-Jeans temperature $T_{RJ}=2.60\pm0.04$ °K (Hayakawa et al., 1987). The value of $T_{RJ}$ appears to be lower than observed values but not inconsistent therewith if systematic errors are properly taken into account (Hayakawa, 1989).

The $y$-parameter is proportional to the product of the electron temperature and density. It has been suggested that such a plasma would explain the X-ray background, since its spectrum could be represented by a thermal bremsstrahlung spectrum with an electron temperature of about $4\ 10^8$ °K. However, such a high temperature requires a relativistic treatment which gives too high a flux in the 260mm band. Moreover, a high thermal energy density of electrons is not easily understandable. These points suggest that the X-ray background has to be explained by other sources, and that there would be energy sources to supply a vast amount of energy to the intergalactic medium. The latter problem will be discussed in what follows, leaving the former for the next section.

In order to attain the electron temperature high enough for inverse Compton heating, electrons have to be reheated to a temperature higher than the thermodynamic temperature of radiation. If electrons in the intergalactic gas are reheated after recombination to attain the $y$-parameter as obtained above, the electron temperature has to be so high as to require a relativistic treatment, since the fraction of gas mass is small. If reheating takes place at an earlier epoch, the electron temperature required is not too high compared with the radiation temperature because of a high matter density, and Compton cooling by radiation suppresses an increase of the electron temperature, thus permitting a nonrelativistic treatment. A large amount of energy can be extracted from dark matter whose energy density exceeds that of ordinary baryonic matter.

At the conference Kawasaki and Signore respectively discussed two reheating sources associated with dark matter, heavy neutrinos which decay into photons and superconducting cosmic strings which emit electromagnetic radiation. They result in several interesting effects not only on the submillimeter spectrum but also on other observable quantities, as will be briefly mentioned below.

The decaying particle model by Fukugita and Kawasaki (1989) is characterized by the product of the particle mass density and the branching ratio to the photon decay and by the lifetime of the particle, which respectively give the energy input and the epoch of electron heating. A small branching ratio permits them to assume a large mass, 10-1000keV, so that these particles do not contribute to the radiation loss of evolved stars. A lifetime of $10^9$-$10^{11}$ s results in a reheating epoch of $z=10^4$-$10^5$ when the electron density is high enough for frequent Compton scattering. The electron temperature is limited to a low enough value by cooling due to ambient radiation. These energy exchange processes give a $y$-value as required.

Although the energy exchange takes place efficiently, some photons do not completely loose their energies and fall in the near infrared range. The near infrared flux they calculated is weaker by an order of magnitude than an upper limit observed by Matsumoto et al., (1988b). However, a set of parameters they assumed are not unique. In fact, Wang and Field (1989) have shown that the near infrared flux could be comparable to the upper limit for a different set of parameters.

The superconducting string model originally proposed for explaining large scale structures and high energy radiation (Ostriker et al., 1986) was pointed out by Ostriker and Thompson (1987) to be responsible for the submillimeter excess. An electric current induced in a superconducting loop due to the passage of a magnetic field emits electromagnetic radiation as the loop moves. The radiation drives a surrounding plasma to form an expanding shock. Dissipation of the expanding motion of matter by ambient photons and matter as well as the decay of the loop heat electrons. Both thermal motion and bulk motion contribute to the $y$-parameter.

The expanding bubble is filled with a hot plasma of a temperature exceeding 10keV and therefore responsible for the S-Z effect and X-ray emission. They would be associated with large scale structures represented by superclusters and voids. The primordial magnetic field required for producing a current may give rise to observable effects such as the Faraday rotation for extragalactic radio sources and the polarization of submillimeter radiation emitted from aligned dust grains (de Bernardis et al., 1989).

The bulk motion contributes to the $y$-parameter in the expanding bubble model. It may be driven also by a pressure wave generated by an acoustic oscillation of a plasma shortly before recombination. At the conference Daly discussed this mechanism to be responsible for the submillimeter excess without invoking speculative particles which form dark matter. Since the result depends on the spectrum of density fluctuations, the model predicts a low amplitude of flux fluctuations.

A model different from the above is based on the pregalactic dust which reprocesses radiation emitted from pregalactic stars and was presented by Carr at the conference referring to an extensive work by Bond et al., (1989). Although the model contains many parameters, there are essentially two, the energy absorbed by a dust grain and the optical depth of dust, which are responsible for the submillimeter flux (Hayakawa et al., 1987). The rate of energy generation is not sufficient, if the luminosity function of pregalactic stars is similar to that in the Galaxy (Lacev and Field, 1988). The energy source requires very massive stars more abundant than expected from a conventional initial mass function. On the other hand, the overproduction of metallic element should be suppressed by reducing the fraction of the stars with masses below 200 $M_\odot$. Thus the model implies a large abundance of very massive stars which eventually form black holes.

The dust model predicts a number of observable effects which more or less depend on detail of the model. Radiation from pregalactic stars is redshifted to optical and near infrared ranges. Although its flux is reduced by the absorption by dust, if the optical depth needed for the submillimeter flux is taken into account, a flux left unabsorbed may be observable in the near infrared range. If the dust consist mainly of silicate, its emission features at about 10mm and 20mm are redshifted and may be observable in the far infrared range. Fluctuations of the submillimeter component may well be different from those of the microwave component, because they are originated from different sources. Different fluctuations in the submillimeter ranges are common to most models, in view of that the superconducting string model forms hot bubbles as already discussed, and the decaying particle model may also produce inhomogeneities if dark matter is clumped by gravitational attraction. The dust model also implies the S-Z effect, since a hot plasma can be produced by the explosive heating associated with very massive stars (Yoshioka and Ikeuchi, 1987).

## 4. X-RAY BACKGROUND

The present status of our understanding of cosmic X-ray background (CXB) was reviewed by Setti (1989). Here I will add some new pieces of information obtained by Ginga and discuss their relevance to cosmic microwave background.

Ginga is suited to observe the diffuse component of X-rays owing to the large area counters with a total effective area of $4000 \text{cm}^2$ and a throughput of $2.4 \text{cm}^2 \text{sr}$. The former allows us to extend the energy spectra of AGNs beyond 10 keV, thus finding a number of objects with spectra flatter than those obtained below 10 keV by past observations. The latter makes it possible to push down the minimum detectable surface brightness to a level of $10^{-9}$ erg $\text{cm}^{-2}$ $\text{s}^{-1}$ $\text{sr}^{-1}$, so that X-ray emission from a cluster of galaxies is found to be extended farther than previously thought and that from a supercluster is indicated. For example, the X-ray diameter of the Virgo cluster is observed to be about $12°$ and the X-ray luminosity thereof is about $1.6 \ 10^{43}$ erg $\text{s}^{-1}$, about 1.5 times larger than thus far adopted (Takano et al., 1989).

Sky regions along the Coma/A1367 supercluster were raster scanned, as shown in Fig.1. The surface brightness profiles of 2-10 keV X-rays across the

222

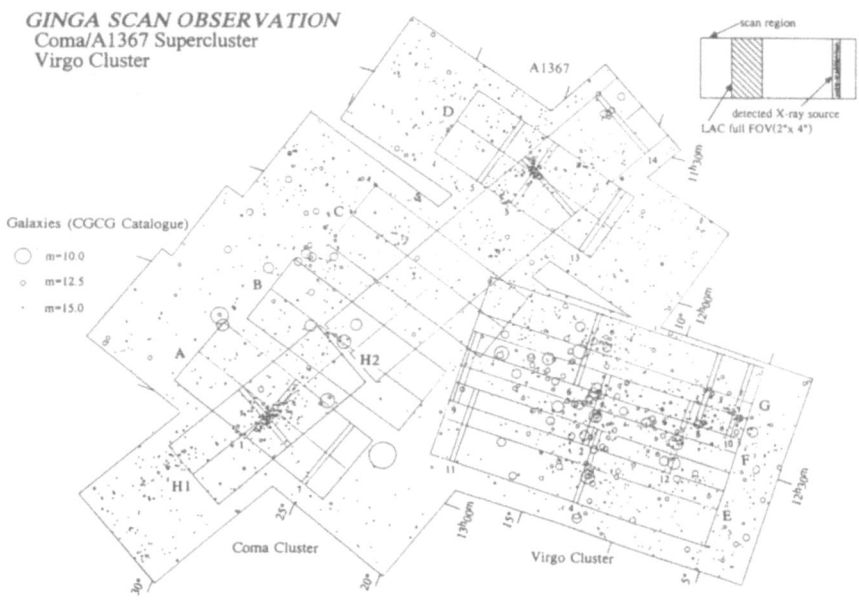

*Figure 1.* The scanned regions superimposed on the optical sky map of galaxies. The brightness of each galaxy in the CGCG catalogue is indicated by a circle whose radius represents a magnitude. Regions A, B, C and D are scanned perpendicular to the line connecting the Coma cluster and A1367, while H1 and H2 are scanned along the line. Regions E, F and G contain the Virgo cluster.

supercluster are shown for four regions in Fig.2. In A and D the Coma cluster and A1367 show prominent peaks, respectively, along with some possible sources yet to be identified. There are several regions, in which the counting rate is the lowest. The surface brightness therein is about 7% lower than that of CXB currently assumed. The distribution of counting rate in the effective beam size of $3.8\ 10^{-4}$ sr observed in B and C is shown in Fig.4. The distribution seems to consist of two peaks. The peak at a higher counting rate could be attributed to emission from the supercluster.

X-ray emission from superclusters has been debated since the first announcement based on a Uhuru observation, because results thus far obtained have been statistically marginal. If we take one-sixth of the data, giving the S/N ratio comparable to that obtained by HEAO/A2, the surface brightness distribution is consistent with that expected from random fluctuations. The whole data give the distribution significantly wider than the Gaussian one with Poisson noise. If, however, the CXB is due to a superposition of discrete sources, the distribution is no longer Gaussian but shows a tail extended to high surface brightness.

Assuming the number-flux distribution of $S^{-2.5}$ as expected for the Euclidian geometry, we generate the simulated distribution of sources for $S > 10^{-13}$ erg cm$^{-2}$ s$^{-1}$. The flux of each source is converted to the counting rate by taking into account the angular response function of the counter, and the contributions of individual sources in a field of view are summed up to give a simulated counting rate. The results of respective steps are shown in three panels of Fig.3. The counting rate distribution shows several peaks which are due to nearby sources, as one can see from the comparison between the first and third panels. The average counting rate of about 4 counts s$^{-1}$ can be increased to the observed value of about 14 counts s$^{-1}$ by reducing the minimum value of S without appreciably changing the shape of distribution. The simulated distribution of counting rate is shown by a solid curve in Fig.4. This shows a high counting rate tail which is due to bright sources. If sources of counting rates higher than 0.7 counts s$^{-1}$ (flux>2.5 $10^{-12}$ erg cm$^{-2}$ s$^{-1}$) are excluded, the simulated distribution is narrowed so as to be consistent with the distribution observed in the lowest counting rate region. Such sources may well be detected by a high resolution device such as the one aboard Einstein, Rosat and ASTRO-D.

If, on the other hand, the surface brightness distribution in regions B and C consists of two populations, although its statistical significance remains to

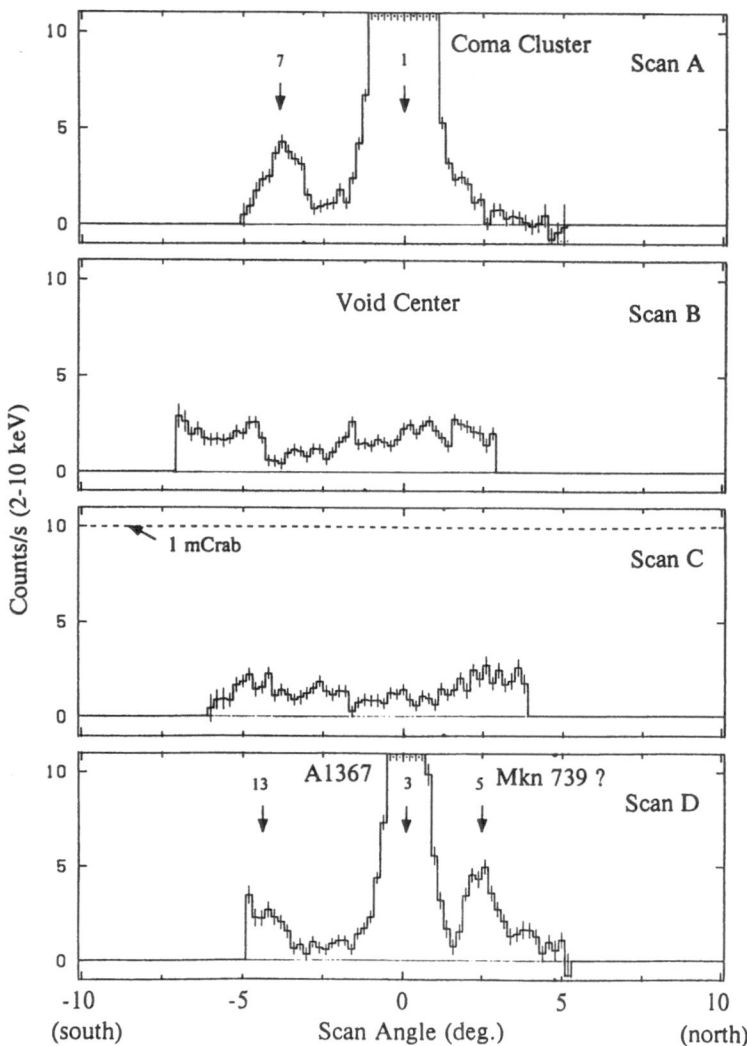

*Figure 2.* X-ray profiles in four regions across the Coma/A1367 supercluster. The ordinate represents the X-ray counting rate after subtracting 13.0 counts s$^{-1}$. The arrows numbered indicate discrete sources. The void center in B does neither show an enhancement nor a depression.

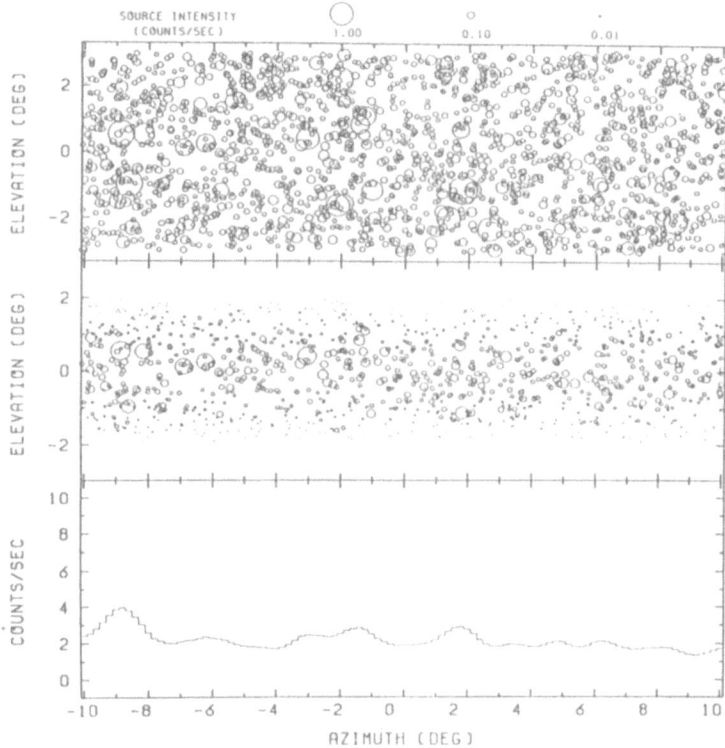

*Figure 3.* The distribution of X-ray intensity generated for the randomly distributed sources with the number-flux distribution
$dN/dS = 2.2 \ 10^{-5}[S(erg \ cm^{-2} \ s^{-1})^{-2.5} \ sr^{-1}]$,
for -11>log S>-13. The upper panel shows the distribution of sources in the region of 20°x6°, in which the value of S is indicated by the circle size. The middle panel shows the distribution to be measured with the Ginga counters by scanning along the line of zero elevation angle. The lower panel represents the counting rate profile to be measured. Note that the contribution of nearby sources remains to be significant even in a deep survey.

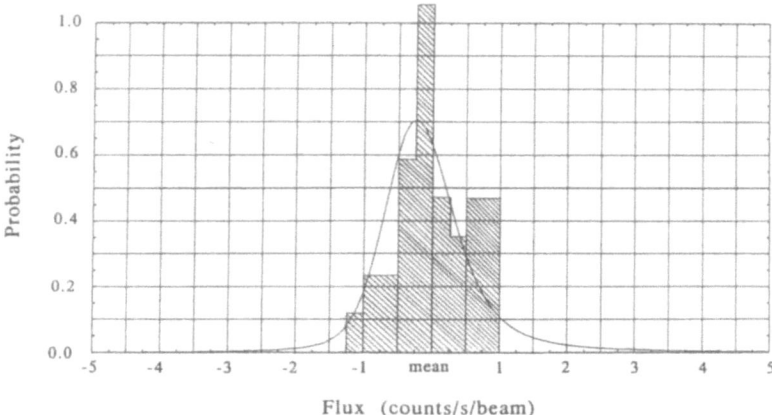

**Figure 4.** The distribution of counting rate observed in regions B and C in comparison with that expected from the dN/dS as given in Fig.3.

be investigated, the lowest counting rate is 13 counts s$^{-1}$ which is compared with 14 counts s$^{-1}$ for the currently accepted CXB rate. If the difference of 1 count s$^{-1}$ is attributed to the Coma/A1367 supercluster, the emission measure $(EM)$ is about 1.5 10$^{-8}$ cm$^{-6}$ Mpc which is compared to about 1.5 10$^{-7}$ cm$^{-6}$ Mpc for the Virgo cluster.

Since the electron column density is approximately given by $EM^{1/2}P^{1/2}$, where P is the thickness of a hot plasma, the S-Z effect due to a supercluster is comparable to that due to a cluster of galaxies, on account of that the value of P of the former is about ten times that of the latter. This would suggest a flux fluctuation of the order of 10$^{-5}$ on the scale of several degrees.

A superposition of superclusters may contribute to CXB, which is comparable to that of clusters of galaxies. Since X-ray emission is supposed to be due to thermal bremsstrahlung of temperatures not exceeding 10 keV, however, it is unlikely that superclusters can account for major part of CXB. The possibility that active galactic nuclei mainly contribute to CXB is discussed by Hayakawa and Piro at the conference, referring to the spectra up to 20 keV and their time variation.

The author thanks K. Koyama, A.E. Lange, T. Matsumoto, S. Sato, and Y. Tawara for making unpublished results available. He also thanks G. De Zotti for his excellent summary of a discussion meeting at the conference and M. Kawada for the preparation of figures.

REFERENCES

Bond, J.R.,*et al.*, 1989, preprint.
de Bernardis, P., Masi, S., Melchiorri, F., and Moreno, G., 1989, *Astrophys. J. (Letters)* **340** L45.
Fukugita, M., and M. Kawasaki, 1989, preprint.
Hayakawa, S., 1989, *Proc. 5th Marcel Grossmann Meeting.*
Hayakawa, S., et al., 1987 *Publ. Astron. Soc. Japan* **39** 941.
Lange, A.E.,*et al.*, 1989, preprint.
Lacey, C.G., and Field, G.B., 1988, *Astrophys. J. (Letters)* **330** L1.
Matsumoto, T., 1989, in this Proceedings.
Matsumoto, T., et al., 1988a, *Astrophys. J.* **329** 567.
Matsumoto, T.,*et al.*, 1988b, *Astrophys. J.* **332** 575.
Ostriker, J.P., and Thompson, C., 1987, *Astrophys. J. (Letters)* **323** L97.

Ostriker, J.P. *et al.*, 1986, *Phys. Letters,* **B180** 231.

Richards, P.L., 1989, in this Proceedings.

Setti, G.C., 1989, in this Proceedings.

Takano, S. *et al.*, 1989, *Nature* **340** 289.

Wang, B. and Field G.B., 1989 *Astrophys. J. (Letters)* **345** in press

Yoshioka, S. and Ikeuchi S., 1987, *Astrophys. J. (Letters)* **323** L7.

# COSMIC BACKGROUND RADIATION
# THE NEXT 25 YEARS

George F. Smoot
Space Sciences Laboratory and Lawrence Berkeley Laboratory
University of California, Berkeley CA 94720

## INTRODUCTION

We are currently celebrating the 25th anniversary of the discovery of the discovery of the Cosmic Background Radiation. As we look back over the history of the field, it is interesting to speculate on where the field will be in another 25 years and what we might do to see that it advances efficiently and effectively. It is well known that it is difficult to forecast the future, even the immediate future, much less 25 years; however, I am going to try basing my prediction by extrapolating trends and what we know to be physical principles.

TRENDS:
Immediately following the discovery of the CBR by Penzias and Wilson, it was recognized that the CBR was a unique and important tool for investigating the early and large scale universe. The understanding of its importance has increased over the years. The CBR traces the geometry of the universe and the evolution of matter and energy in the early universe. The CBR should contain evidence of the fossil progenitors of galaxies and evidence of

*N. Mandolesi and N. Vittorio (eds.), The Cosmic Microwave Background: 25 Years Later, 229–239.*
© 1990 *Kluwer Academic Publishers.*

the growth of structure in the universe. In addition, the CBR ties into particle physics. Its isotropy provides evidence for dark matter in that $\Delta T/T < 10^{-4}$ is too isotropic for galaxies and clusters of galaxies to condense under self-gravity of observed matter in the available time. Likewise the CMB isotropy over causally disconnected regions - the horizon problem - is difficult to explain. Both issues can be resolved by new physics at high energies - inflation and dark matter - and the spectrum and isotropy of the CMB and other cosmic backgrounds provide tests of these explanations.

25 years of effort and development in the field have lead to certain features that indicate trends. Features of the field that are easy to observe are:

1. CBR research uses complicated and specialized equipment (e.g. telescopes, gondolas, satellites) and much of this equipment is necessarily expensive to build and operate.

2. CBR research uses sophisticated techniques both in the experimental observations and in the data analysis. Also these techniques are becoming standardized as they have been around for a while and have been commented on by peers.

3. There is much theoretical development in prediction and interpretation. CBR related theory is a large and active area. We joke that one could develop an expert system for new theories and theoretical interpretation of new CBR observations - e.g.

When a new observations are published, then check against predictions from:

a) cold dark matter theories

b) cosmic strings (including superconducting) theories

c) decaying particles (and vacuum) theories

d) late phase transitions theories

e) cosmic dust theories

using the proper formulas and write a paper with the blanks filled in with the proper new constraints and theory parameters.

4. There is now a growth in group size, collaborations, and longer term programs.

There trends signal a maturing field. No longer can a new person come in and very quickly join the forefront; instead, there is now lore to learn or technology to develop.

## MEASURING AND INTERPRETING THE CMB

For the Cosmic Microwave Background radiation everything else is foreground. That is to say, all currently observable objects in the universe lie between us as observers and interpreters and the source of the CMB. Not only are these objects in the foreground but they are also generally emitters at a level that is now becoming significant ($10^{-5}$ to $10^{-6}$). This means that inevitably that CMB observation/interpretation is becoming a branch of astronomy requiring maps made at multi-wavelengths.

Figures 1 and 2 show examples of the expected galactic emission, radio source count confusion, extragalactic contamination and distortions and a map. These indicate to me that when observations reach the $\Delta T/T \leq 10^{-6}$ level, there will be structure observed in the sky maps. Thus the field will have to have models and data to allow the separation of foreground emission from the cosmic background.

Even when we do manage to find a way to separate the foreground from the cosmic background, we will (hopefully) have discovered regions of structure (anisotropies) and will be in the business of studying these structures - e.g. individual shapes and morphologies. At that point we might get to produce a classification scheme like that for galaxies, the frequency of occurrences and distribution and so forth. This will hopefully be a rich and rewarding field and as you can tell much like extragalactic astronomy or x-ray astronomy.

STATUS of THEORY in 25 years

It is much too difficult to predict the theoretical developments of the next 25 years. We can predict that many of the theories that hold center stage today will be no longer relevant or widely known - viz. the current question by new theorists "What is the Steady State Theory?". We can expect similar questions about theories such as T.O.E., inflation, cold dark matter, cosmic strings. As a close parallel to the Steady State Theory we have the new, updated, and revised Inflation theory which neatly explains the large scale isotropy, lack of monopoles, etc. but has little experimental evidence to

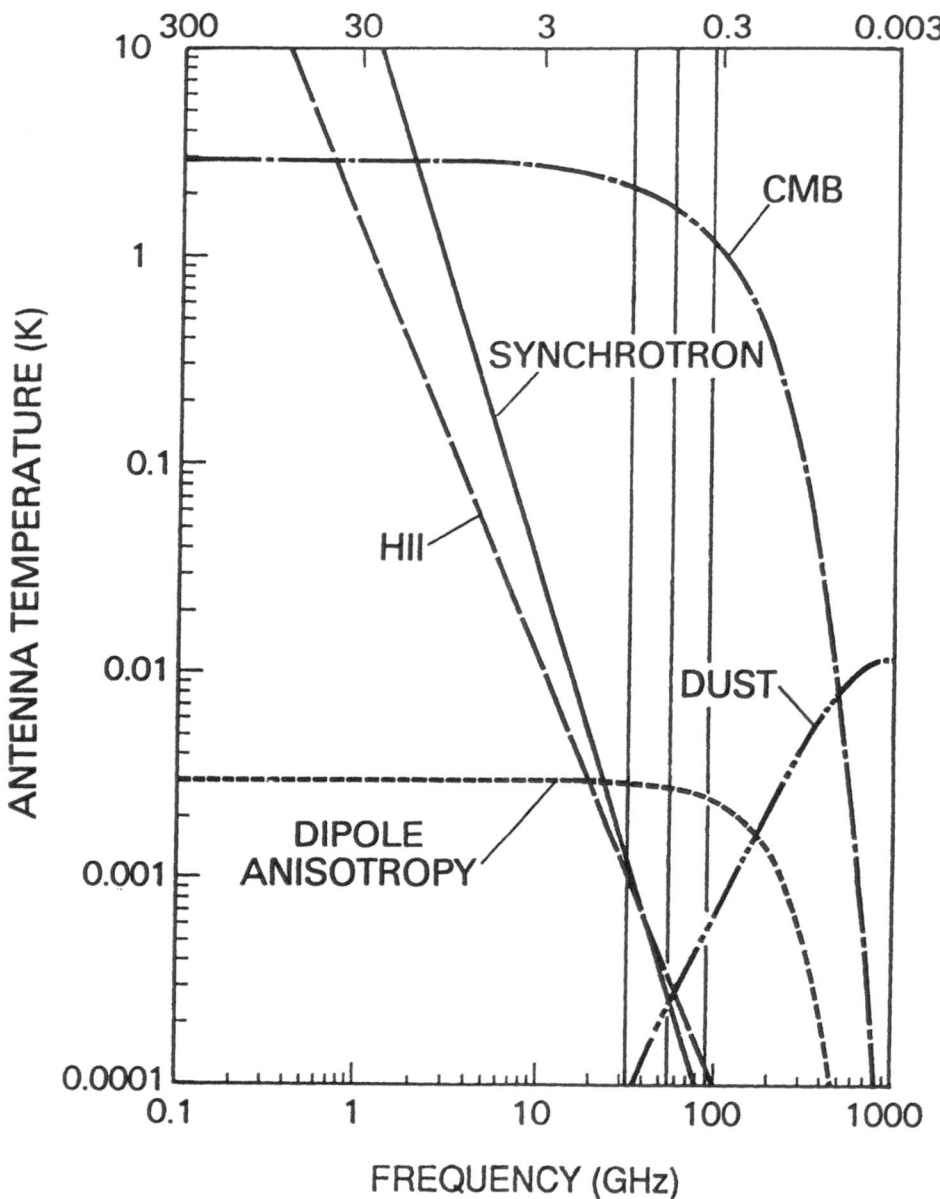

Figure 1. Estimated galactic emission versus frequency. Also shown for reference are a CMB signal and dipole anisotropy amplitude assuming a Planckian spectrum. The three vertical lines are the three frequencies chosen for the COBE anisotropy measurement and are an example of how frequencies might be chosen in a multiwavelength experiment.

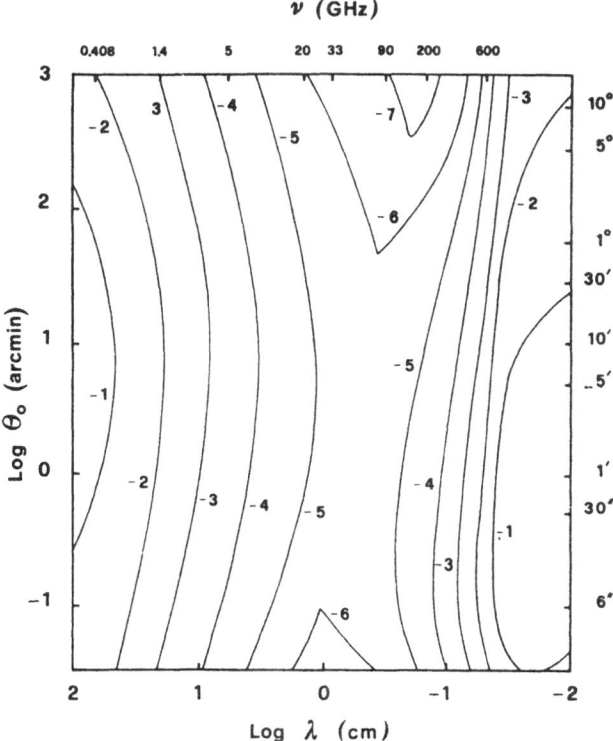

Figure 2. Estimated radio source and extragalactic source confusion limits as a function of frequency and angular scale. Taken from L. Toffolatti et al. in Large Scale Structure and Motions in the Universe, 445, 446, ed. by M.Mezzetti et al. 1989.

solidify it place. A measurement of $\Omega=0.1$ or of non-flatness could easily leave it in the library for elegant, beautiful, unused, and forgotten theories. In fact many of the theories and models that we will be testing in 25 years will be the products of people now in primary school (grades 1 through 6).

Thus researchers should design experiments as free as possible from model bias, as some of the models will no longer be around when the experiment is completed. The field as a whole needs to emphasize experiments that measure the fundamental properties of the CBR and probe the avaliable parameter space. Theories and models are good as examples but do not trust them implicitly but only as they are buttressed by observation.

Guiding Theoretical Ideas of CMB properties

Now having warned everyone not to trust to theory for experimental design, I review current thinking about the overall CMB properties as a rough guide. At the surface of last scattering there are two intrinsic scales affecting the CMB properties - the thickness of the surface of last scattering and the region of causal connectedness. The thickness of region of last scattering is estimated to encompass a range of redshift of about 80 at a redshift of about 1000. This corresponds to a length scale of about $4h^{-1}\Omega^{-1/2}$ Mpc and an angular extend of about $8\Omega^{1/2}$ arcmin. Structures of a size smaller than this will have their features washed out by the near isotropy of Thomson scattering as the CMB photons make their way out to us. This is not to say that one should not look for structure on a smaller scale but that it is likely that any structure on a smaller scale is due to interactions more recent than at a redshift of 1000. It is still worth checking that this picture is correct but as we shall see below making such fine scale maps will be a major undertaking.

The other angular scale of interest is the horizon scale at the surface of last scattering. In a universe without an inflationary epoch, the physical size of a causally connected region is about $3ctd \neq 200h^{-1}$ Mpc or about 2 degrees. In an inflationary universe the entire observable space can be causally connected.

CMB Measurement Goals:

If we want to make measurements on a sensitivity level of $\Delta T/T < 10^{-6}$, we then realize we must make sky maps with spectral sensitivity around the peak CMB signal. At minimum we will want 4 or more frequencies with very well defined bandwidths.

Using a quantum limited receiver, a 1% bandwidth, and a nominal chopping or comparison observational scheme, it takes 3 hours of observation to reach the $10^{-6}$ level. Going to a larger bandwidth helps linearly for blip limited detectors like bolometers and with the square root of the bandwidth for coherent receivers. I estimate that a 10% bandwidth will be the ultimate limit if in 25 years we want to make multiwavelength full sky maps and that it will be extremely difficult to achieve such a wide bandwidth, quantum-limited device while still having good off- beam rejection.

The scales at the surface of last scattering gives us a natural angular resolution of 10 arcminutes on which to look for possible structures from the early universe. This implies a total to 1,500.00 fields of view (pixels) or about 500 years of observation. And we want to map selected areas with higher angular resolution. Even if we stretch to the larger bandwidths, measuring at 4 or more frequencies still requires a total observation time of around 500 years.

Clearly, we are going to have to collaborate or build instruments with arrays of detectors and spectral response.

## EXPERIMENT DESIGNS:

The constraints and factors discussed above point out that several features will be needed for an experimental program in place 25 years from now:

1. Instruments will feature receiver arrays. My predictions are that these arrays will take two primary forms:

a) bolometer focal plane arrays with multichroic splitters: a 20 by 25 element array working at quantum efficiency could achieve the necessary sensitivity to survey the full sky at the desired level and resolution in one year. In practice it is likely to take much longer but be within the scope of a graduate student's lifetime.

b) coherent receiver arrays that synthesize the equivalent of dishes with focal plane arrays: there are serious fundamental problems with trying to put coherent receivers in a close-packed focal plane array with low sidelobes. However it is necessary to use an array of receivers in order to get the observation time to high multiples of the wall clock time. Thus we will need to create a focal plane array by using aperture synthesis and utilize them either directly observing the sky or as feeds for a large reflector.

2. Certain regions and objects will be studied at many wavelengths and even more importantly on various angular scales. Thus the specialized instruments described above will be charged with surveying the whole sky; while, others with higher resolution - OVRO and VLA types, will study certain structures in more detail. Likewise one can expect certain structures will be of enough interest to have other instruments with nominal properties like our survey instruments studying them to greater sensitivity or independently. Here we will be involved in the morphology of bumps and lumps and the independent study and verification of structures.

3. The experiments will be operated from carefully selected sites. We know of certain problems that will not go away but have to be avoided by careful selection of site and experimental design:

a) There are places where the atmosphere is not a problem. The South Pole is a good possibility for the next generation or two of experiments however, in the long run getting above the atmosphere is the only solution.

b) man-made RFI is a serious problem and a growing one at low frequencies and it is only a matter of time until the whole microwave and mm-wave region is contaminated. Near earth orbit is now already a problem with RF at the 1 to 10 volts/meter level common in the centimeter range. There are two sites that offer good promise for CMB astronomy for the foreseeable future, these are the Lagrange point L2 and the Lunar far side (either backside over the polar horizon).

c) Objects in the beam and sidelobes. Both L2 and the lunar far side offer good opportunities to keep extraneous radiation out of the beam by careful antenna and shield designs. Very much care must be used but it is much less severe than for earth-based experiments. Removing the galaxy's emissions from the beam will have a long wait until we mount an out of the galaxy mission.

Clearly the sites suggested and the instruments outlined are major undertakings needing large resources. But then, just how many things are known to $10^{-6}$? This is equivalent to mapping the surface of Venus and Mars to a precision of 20 feet. The people in the field will need to develop a consensus and coordination to be effective. For large projects to get started and sustain themselves, they need the support of the field. Everyone must believe them necessary, good concepts, well thought out, and well managed. Likewise the databases of experience, astronomical and galactic sources, interpretation - the infrastructure of the field is needed to support the overall undertaking.

## PROPOSALS FOR NEW OPERATIONS AND PREPARATION FOR THE NEXT 25 YEARS

I. Hold workshops specifically on experimental designs

  (a) define atmosphere, galactic, etc. limits for various parameter regimes and experiments

  (b) develop designs and technologies such as nearly filled arrays

  (c) develop the design of next generations experiments


II. Maintain and Improve Communication and Coordination System

  1. Exchange addresses including Electronic-mail

  2. Set up on-line system for CMB work

    (a) Names Addresses, and E-mail

    (b) On-line list of articles - title, author, abstracts, keys

    (c) Bulletin board with proposed & work in progress

    (d) Menus for adding names and articles and bulletin board

    (e) Collect hard & electronic copies of articles/theses, etc. for CMB library with visitor/guest facilities

    (f) Data summaries

      (1) CMB data - maps and tables of results

      (2) galactic models

      (3) radio source counts

      (4) software - i.e. experimental/theoretical programs with facility for comments and testing

As an example, I put forward the need for a study of good future sites - e.g. L2 and the moon far side base as well as developing the design of the hardware and techniques for doing the experiment. We could arrange a workshop on experiments designed for the short term at the South Pole and then longer term careful design of complementing experiments at lunar and L2-like sites.

## SITE DISCUSSION

Both the Lagrange point L2 and the lunar far side (or lunar orbiting platform) offer good advantages for serious long term CBR astronomy. The moon provides shielding from the earth. Lagrange point L2 is located on the sun earth line opposite from the sun and at about $10^6$ km from the earth. L2 is a good site because it is far enough from the earth-moon-sun system, which are thus all to one side so that the observing system can be designed to have little contamination from these sources and man-made interference. We can anticipate that most transmitters will be on the earth or in near earth orbit for the next half century and that care can be taken to avoid serious contamination of L2 or the lunar far side. If that cannot be achieved, we have to envision sending equipment to more remote locations or unusual orbits. Both L2 and the moon offer the advantage of relatively low cost of expendables for station keeping. Maintaining equipment at comparable of greater distances or low RFI environments will require substantially greater expenditure of fuels or more sophisticated control system.

For full sky surveys, L2 offers the advantages of providing clear view access to the whole sky in that the earth-moon-sun system is in a small part of the sky and the orbit sweeps around to the opposite side of the sun in six months. L2's disadvantages compared to a lunar base is that it is difficult to:

1) create large (high angular resolution) systems

2) change observing strategies or up-grade the system, do repairs, refurbish

3) to do more sophisticated systematic error checks after the data are being analyzed and the observers have learned from the data - e.g. change shields, frequencies, polarization

Thus instruments at L2 are likely to be used for full sky surveys at moderate angular resolution; while on the lunar sites one would set up observatories in concert with the other astronomical observatories. These facilities would be used to check selected regions of the full sky survey to see if it was correct, calibrated well, and free of serious systematic errors and to perform deeper studies or studies with higher angular resolution and detailed work on regions and structures of special interest. The lunar far side is likely to be the best location for long wavelength measurements of the spectrum of the CBR. While L2 may well turn out to be a superior location for shorter wavelength observation of the cosmic background radiation. We need careful studies and workshops to consider the lunar sites and the facilities likely to be available.

# Future Programs on Background Astronomy

Bianca Melchiorri
Istituto di Fisica dell'Atmosfera del CNR
Pzle L. Sturzo 31, 00144 Rome, Italy

Francesco Melchiorri
Department of Physics,University of Rome La Sapienza
Pzle Aldo Moro 2, 00185 Rome Italy

The future prospects of Background Astronomy are discussed. Three goals appear to be of major relevance: i) the measurement of the Extragalactic Background Spectrum within $1 \leq \lambda \leq 3000 \mu m$ with an absolute accuracy of one part over $10^3$; ii) the measurement of Extragalactic Background Dipole Anisotropy in the $300 \leq \lambda \leq 800 \mu m$ range; iii) the detection and study of spatial distribution of extragalactic diffuse radiations at angular scales between 1 arcminute and several degrees. The possible strategies to reach the above goals are discussed and compared with the experiments already proposed

## 1. INTRODUCTION : A NEW ASTRONOMY

For a long time, since Galilei, astronomers have played a single game: *"we have answered almost all the questions, the few remaining problems could be easily solved if we have a telescope like that we used, but slightly larger"*. This led them from the 5 cm objective lens of Galilei up to the 10 meters mirrors of the new generation of telescopes. Very seldom the doubt arose that this way was the only one available in astronomy.

The discovery of the 2.7 background radiation in 1965 has changed this view. The study of the last scattering surface cannot be carried out by means of the conventional astronomical tools. There are no "objects", like galaxies, to be analized and as we increase the resolution of our telescope more details are expected to be seen. Therefore, instead of maps, we speak in terms of *anisotropies*. In the case of a Gaussian distribution it can be shown that the telescopic appearance of the anisotropies would exactly imitate that of an infinite number of non-resolved sources. What we want to stress here is the following: an enterely new system

241

*N. Mandolesi and N. Vittorio (eds.), The Cosmic Microwave Background: 25 Years Later*, 241–254.
© 1990 *Kluwer Academic Publishers.*

of notations (fine-scale, intermediate, large-scale anisotropies) has been introduced and , as we will see later on, a different experimental approach has been adopted. In brief, a new astronomy is borne: that is the *Background Astronomy*.

As long as the Background Astronomy was limited to the study of the 2.7 K radiation its general character was not clearly recognized. Instead of speaking of a new astronomy in which the background radiation is studied, astronomers were tempted to consider the early universe as a single object, like a large diffuse cloud,in order to force this new field into the standard scenario of conventional astronomy. After all, the confusion limit is reached at the last scattering surface, which has nothing to do with galaxies and quasars.

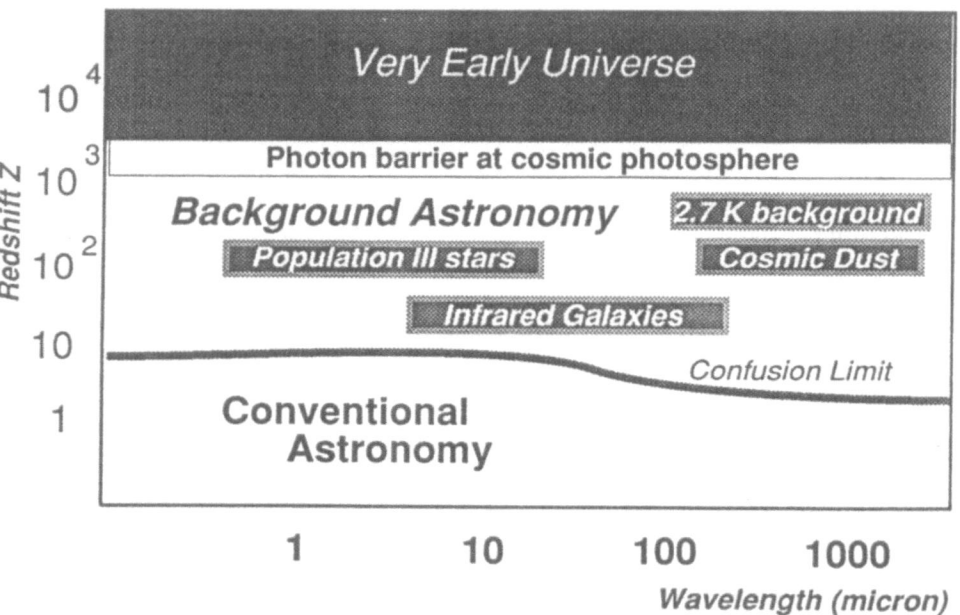

Figure 1: schematic view of the two astronomies: conventional astronomy studies single, well resolved sources. Beyond the confusion limit , background astronomy explores diffuse radiations

We are now at the threshold of the discovery of a new (peraphs two) cosmic backgrounds: a near infrared background has been claimed by Nagoya group and a far infrared background by Nagoya-Berkeley groups. If these observations will be confirmed, we will be led to fill up the region $10 \leq Z \leq 1000$ with matter which emits radiation and the naive Big Bang cosmology has to be modified in order to include this possibility. The theoretical and experimental tools developed for the study of the 2.7 K radiation will be soon applied to the analysis of these new backgrounds and we will be finally initiate officially the new discipline of Background Astronomy. This is illustrated in figure 1, where the position of the demarcation line between conventional and background astronomy is obviously quite arbitrary. Let us briefly outline the main link points to the problem.

## 2. THE BACKGROUND ASTRONOMY

The first question to be studied in the presence of a background radiation is its spectrum, or better still the spectral brightness $I_\nu$. The units adopted in literature are:

$$[Watt.\mu m^{-1}.cm^{-2}.sr^{-1}] = \frac{3 \times 10^{10}}{\lambda^2}[Watt.Hz^{-1}.m^{-2}.sr^{-1}] \qquad (1$$

$$[Jansky(Jy).sr^{-1}] = 10^{-26} \times [Watt.Hz^{-1}.m^{-2}.sr^{-1}] \qquad (2$$

$$T_{antenna} = \frac{\int I_\nu \tau(\nu)d\nu}{\int 8.33 \times 10^{-13}\nu^2\tau(\nu)d\nu} \circ K \qquad (3$$

In the first equation the wavelength $\lambda$ is measured in $\mu m$. In the last equation the spectral brightness in the integral $I_\nu$ is measured in $Watt\ cm^{-2}sr^{-1}/cm^{-1}$. This equation is often employed to compare the radio results (expressed in terms of antenna temperature) with those obtained by means of a bolometric system having a bandwidth of operation defined by the spectral transmittance $\tau(\nu)$ of the filters. In Figure 2 we have plotted the far infrared and radio data containing the most recent informations on 2.7 K radiation and the Nagoya-Berkeley excess. The data are listed also in Table I.

Let us now analyze other possible properties of the extragalactic backgrounds:
a) The polarization

One of the most exciting property of the galactic dust is its ability of polarizing the radiation of the stars. It is believed that this effect is due to the existence of galactic magnetic fields which align the rotating grains. If one apply the same model to the extragalactic dust, it is easy to reach the conclusion that the observed ir excesses must be polarized, unless the cosmic magnetic field is much smaller than the often quoted value ( or upper limit) of $10^{-9}$ Gauss. So far, only very rough computations of the effect are available (see de Bernardis et al. 1989), but they strongly suggest that the polarization state would be an important signature in order to discriminate among various extragalactic backgrounds, as well as to study the cosmic magnetic fields.

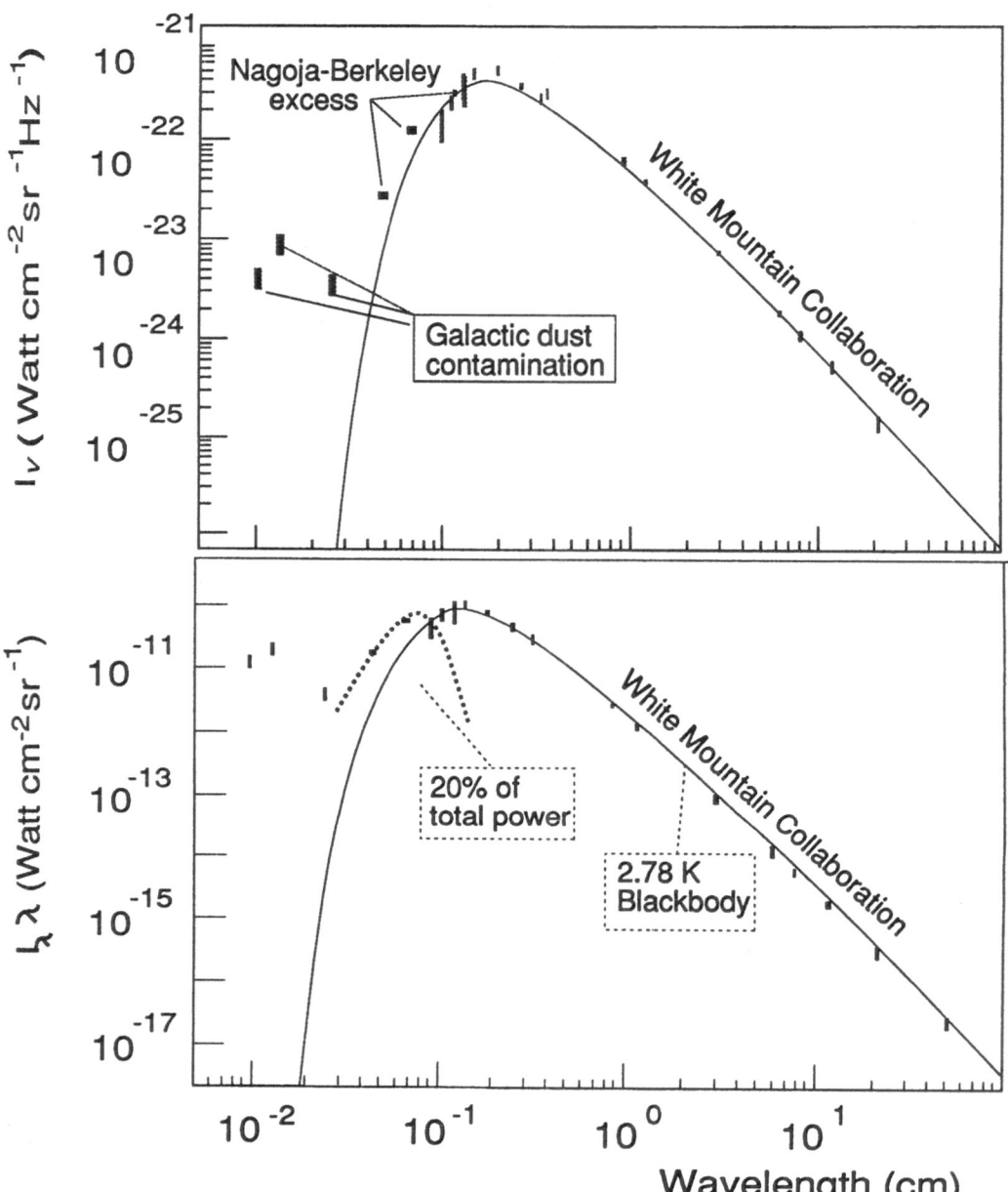

Figure 2:Spectral Brightness and total Brightness of diffuse background radiation in the millimetric and radio region.

One should note that the mechanism of polarization should be re-analyzed, in view of these new possibilities. Peraphs photon scattering is the main mechanism of rotation in cosmic dust and this mechanism is difficult to interrupt: it could even generate angular velocity so high to make the grains very effective emitters in the infrared, or to destroy them. Therefore, theoretical work is needed on this subject, both on the mechanism of alignement and its dependence on the redshift in the cosmological scenario. The experimental upper limits available date back to 1976 and are of the order of 0.01 ( Coletti, Melchiorri, Natale, 1976). Much better results could be obtained today in an appropriately designed experiment: is this a field which could reserve great surprises as soon as the polarization sensitivity will be improved up to $10^{-5}$, as already obtained in the radio region.

## Table I : Recent Observations of Extragalactic Background Spectrum

| Wavelength cm | $T_{CBR}$ K | $\lambda I_\lambda$ $Watt\ cm^{-2}\ sr^{-1}$ | $I_\nu$ $Watt\ cm^{-2}\ sr^{-1}\ Hz^{-1}$ | Reference |
|---|---|---|---|---|
| 50.0 | $2.45 \pm 0.70$ | $1.61 \pm 0.46\ 10^{-17}$ | $2.69 \pm 0.77\ 10^{-26}$ | Sironi et al. (1987) |
| 21.2 | $2.28 \pm 0.39$ | $1.95 \pm 0.34\ 10^{-16}$ | $1.38 \pm 0.24\ 10^{-25}$ | Levin et al. (1988) |
| 12.0 | $2.79 \pm 0.15$ | $1.31 \pm 0.07\ 10^{-15}$ | $5.24 \pm 0.29\ 10^{-25}$ | Sironi and Bonelli (1986) |
| 8.1 | $2.58 \pm 0.13$ | $3.88 \pm 0.19\ 10^{-15}$ | $1.05 \pm 0.05\ 10^{-24}$ | DeAmici et al. (1988) |
| 6.3 | $2.70 \pm 0.07$ | $8.57 \pm 0.24\ 10^{-15}$ | $1.80 \pm 0.05\ 10^{-24}$ | Mandolesi et al. (1986) |
| 3.0 | $2.61 \pm 0.06$ | $7.29 \pm 0.18\ 10^{-14}$ | $7.30 \pm 0.18\ 10^{-24}$ | Kogut et al. (1988) |
| 1.2 | $2.783 \pm 0.025$ | $1.07 \pm 0.01\ 10^{-12}$ | $4.27 \pm 0.05\ 10^{-23}$ | Johnson and Wilkinson (1986) |
| 0.909 | $2.81 \pm 0.12$ | $2.31 \pm 0.13\ 10^{-12}$ | $7.00 \pm 0.39\ 10^{-23}$ | Smoot et al. (1985) |
| 0.351 | $2.80 \pm 0.16$ | $2.36 \pm 0.26\ 10^{-11}$ | $2.77 \pm 0.30\ 10^{-22}$ | Peterson, Richards and Timusk (1985) |
| 0.333 | $2.60 \pm 0.10$ | $2.27 \pm 0.18\ 10^{-11}$ | $2.52 \pm 0.20\ 10^{-22}$ | Smoot et al. (1987) |
| 0.264 | $2.796\ ^{+0.019}_{-0.041}$ | $4.07\ ^{+0.09}_{-0.14}\ 10^{-11}$ | $3.59\ ^{+0.06}_{-0.12}\ 10^{-22}$ | Crane et al. (1988) |
| 0.264 | $2.70 \pm 0.04$ | $3.76 \pm 0.12\ 10^{-11}$ | $3.31 \pm 0.11\ 10^{-22}$ | Meyer and Jura (1985) |
| 0.198 | $2.95\ ^{+0.11}_{-0.12}$ | $7.22\ ^{+0.73}_{-0.79}\ 10^{-11}$ | $4.77\ ^{+0.48}_{-0.52}\ 10^{-22}$ | Peterson, Richards and Timusk (1985) |
| 0.148 | $2.92 \pm 0.10$ | $9.23 \pm 1.09\ 10^{-11}$ | $4.56 \pm 0.54\ 10^{-22}$ | Peterson, Richards and Timusk (1985) |
| 0.132 | $2.75\ ^{+0.24}_{-0.29}$ | $7.60\ ^{+2.93}_{-2.84}\ 10^{-11}$ | $3.35\ ^{+1.29}_{-1.27}\ 10^{-22}$ | Crane et al. (1986) |
| 0.132 | $2.76 \pm 0.20$ | $7.72 \pm 2.25\ 10^{-11}$ | $3.40 \pm 0.99\ 10^{-22}$ | Meyer and Jura (1985) |
| 0.116 | $2.799 \pm 0.018$ | $7.93 \pm 0.23\ 10^{-11}$ | $3.07 \pm 0.09\ 10^{-22}$ | Matsumoto et al. (1988) |
| 0.114 | $2.65\ ^{+0.09}_{-0.10}$ | $6.08\ ^{+1.05}_{-1.03}\ 10^{-11}$ | $2.31\ ^{+0.40}_{-0.39}\ 10^{-22}$ | Peterson, Richards and Timusk (1985) |
| 0.100 | $2.55\ ^{+0.14}_{-0.18}$ | $4.24\ ^{+1.44}_{-1.60}\ 10^{-11}$ | $1.42\ ^{+0.48}_{-0.50}\ 10^{-22}$ | Peterson, Richards and Timusk (1985) |
| 0.071 | $2.963 \pm 0.017$ | $5.03 \pm 0.21\ 10^{-11}$ | $1.19 \pm 0.05\ 10^{-22}$ | Matsumoto et al. (1988) |
| 0.048 | $3.150 \pm 0.026$ | $1.66 \pm 0.13\ 10^{-11}$ | $2.65 \pm 0.21\ 10^{-23}$ | Matsumoto et al. (1988) |
| 0.0262 | - | $3.92\ ^{+0.81}_{-0.74}\ 10^{-12}$ | $3.43\ ^{+0.53}_{-0.67}\ 10^{-24}$ | Matsumoto et al. (1988) |
| 0.0137 | - | $2.00\ ^{+0.37}_{-0.60}\ 10^{-11}$ | $9.14\ ^{+1.69}_{-2.78}\ 10^{-24}$ | Matsumoto et al. (1988) |
| 0.0102 | - | $1.21\ ^{+0.22}_{-0.28}\ 10^{-11}$ | $4.12\ ^{+0.75}_{-0.96}\ 10^{-24}$ | Matsumoto et al. (1988) |

b) Quantum Noise

It is well known that a similar degree of information is contained in measurements of spectral brightness or radiation temporal fluctuations. Several groups have proposed in the past to measure the quantum noise of the 2.7 K radiation (see, for instance, G. Smoot, 1982), but only one group has attempted this difficult task (G. Dall'Oglio et al. 1982). The anticipated cosmic scenario is given in figure 3. De Bernardis and Masi (1982) have shown that two measurements at appropriate frequencies could measure the emissivity of the source, on the basis of Bose-Einstein statistic for photons. From figure 3 one should note that the significant improvement in the performance of bolometers and radio receivers would make possible the observation of the cosmic noise by means of a single detector , without any chopping mechanism. It is therefore conceivable that in the near future the branch of "Noise Astronomy" will be exploited. In this case also a certain amount of theoretical work is needed, so as to define better how the photon noise couples with the instrumental noise as well as the behaviour of photon noise in special cases, like 2.7K photons scattered by hot gas.

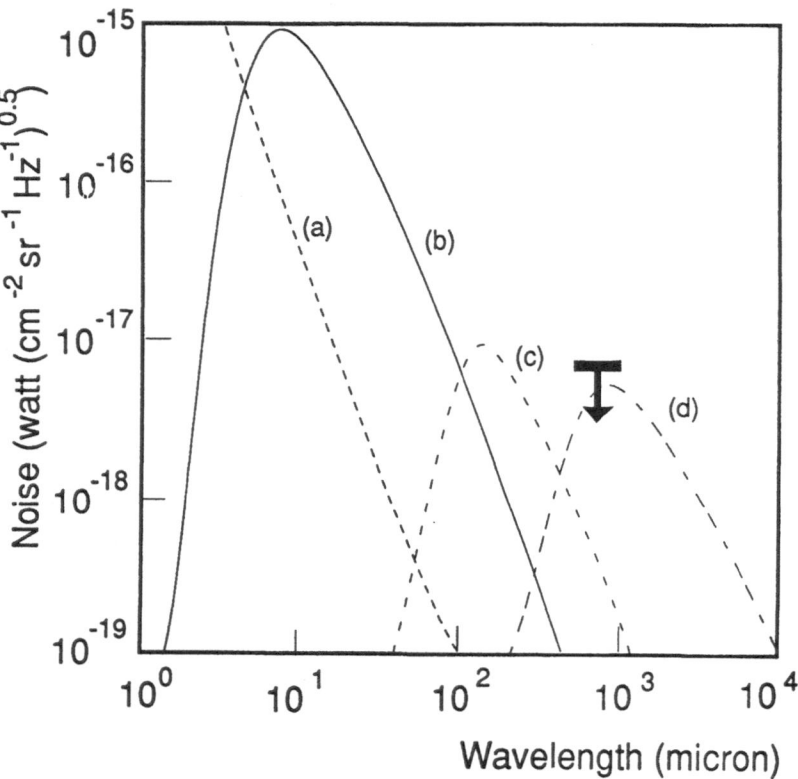

Figure 3:Expected noise scenario in the 1-10⁴ micron region: a)- noise of Zodiacal scattered light; b)- noise of zodiacal emitted radiation; c)- noise of galactic dust emission (T= 12 K, at galactic poles; d)- noise of 2.7K blackbody radiation. The only upper limit available is that obtained by Dall'Oglio et al. (1982)

c) Anisotropies

This is one of the most promising field of research. Let us first recall that every cosmic background is characterized by a "kynematic anisotropy", due to our motion with respect to the distant matter. In terms of temperature a pure blackbody spectrum shows dipole anisotropy $T = T_0(1 + \beta \times \cos \Theta)$ where $\beta = v/c$.

Since we are interested in a generic spectrum , not necessary planckian, we should analize the Brightness anisotropy. As the Brigthness is not a linear function of the temperature, multipole moments appear even in a blackbody spectrum.

In a second order of approximation we have (de Bernardis et al. 1989)

$$I_{oss} \simeq I(\nu) \left[ 1 + (3 - \alpha)\beta \cos \theta + \frac{1}{2}(3 - \alpha)\beta^2 + \frac{\beta^2}{2} \cos^2 \theta (3 - \alpha)(2 - \alpha) \right] \quad (4$$

Where $\alpha$ is the spectral index of the radiation at the frequency of observation. The term in $\beta$ is the Dipole Anisotropy, while the terms in $\beta^2$ are the transverse Doppler effect (not depending on the angle of observation $\theta$) and the "quadrupole" term. In figure 4 we have plotted the available data for dipole terms in the millimetric and submillimetric region, togheter with the expected anisotropies for some extragalactic backgrounds. The data are also collected in Table II. It is important to note that the quadrupole term vanishes in the radio region of the 2.7 K blackbody, being $\alpha = 2$ in the Rayleigh-Jeans region of the spectrum. This fact is a quite obvious consequence of the linear dependence of the Brightness on the Temperature in the Rayleigh-Jeans part of the spectrum, being the temperature anisotropy a pure dipole. A significant contribute to the Dipole and Quadrupole anisotropy is expected by a non-uniform distribution of the sources responsible for the observed background. (Fabbri et al. 1987). In such a case the dipole anisotropy approaches the anisotropies in the counts, which can be as high as several percents. Therefore, the Dipole anisotropy appears to be one of the major cosmological tools of future Background Astronomy.

Both Relic II and COBE are designed to implement this field of research. They are not optimized for submillimetric searches, however. Epifani, Guarini and Melchiorri (1989) have studied this problem: the first important point is that the Dipole Anisotropy of the "excess" $\Delta I_E$ must be greater than the present uncertainty if the amplitude of the Dipole Anisotropy of the 2.78 K alone $\Delta I_{BB}$(of the order of 10%), i.e., $\Delta I_E \geq 0.2 \times \Delta I_{BB}$ in order to disentangle it from the data: this condition is shown in figure 5.

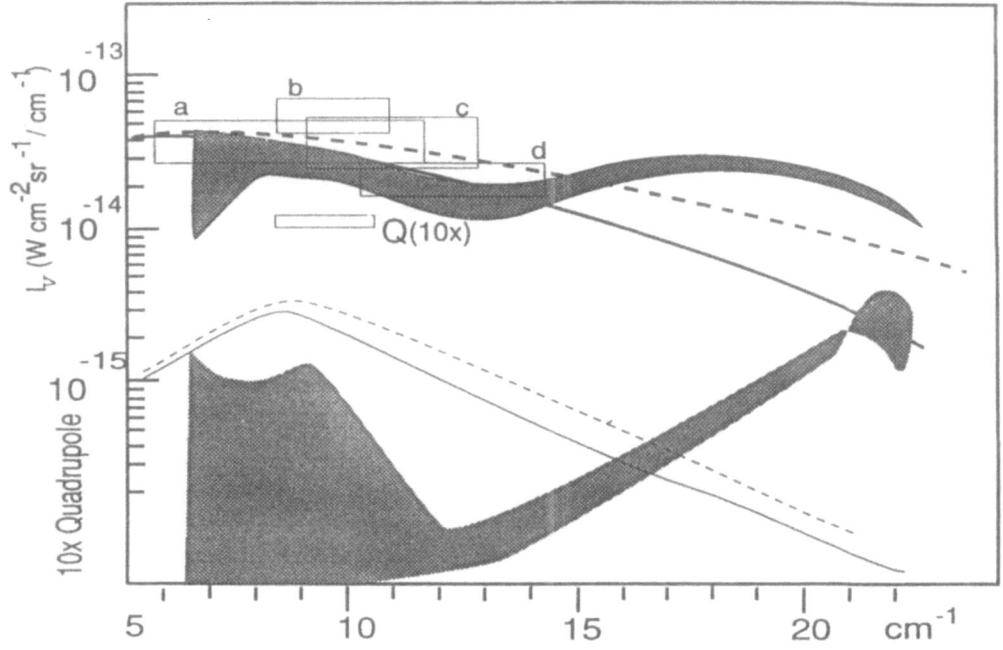

Figure 4:Dipole anisotropy in the case of Gush measurements (gray area),
pure Planckian spectrum (full line) Nagoya-Berkeley excess (dotted
line). The observations are: a- MIT group, b and c- Rome group,
d-Berkeley group. For detailed values see Table II

Secondly, $\Delta I_E$ must be larger than the instrumental noise $3\sigma$ and of the
spurious dipole anisotropy due to galactic dust contamination. Epifani, Guarini
and Melchiorri (1989) have computed the integration time required to fullfill these
requirements: it depends on the central wavelength of operation $\Lambda_0$ as well as the
bandwidth $\Delta\lambda$, as shown in figure 6.

It is interesting to note that none of the experiments listed in Table II appear
to be optimized. New experiments are therefore required in this field.

At smaller angular scales the situation is less clear. In general , one would
expect an increase in the anisotropies as the redshift of emission of the background
becomes shorter. There are however some counter examples: for instance, if the
background is produced by no-barionic decaying particles there is no reason to
believe that their distribution will follow that of the barionic matter. A significant
amount of anisotropy in the sub-millimetric region is expected in the Mock-Gravity
models of Hogan and Bond (1987, see also this volume). In table III we have
collected the data available . The upper limits are in some cases low enough to
put serious constraints to the Mock Gravity theory, as shown in figure 7. More
theoretical work is required in this field. At present there is the suggestion to
explore small angular scales, but due to the lack of a satisfactory theory, all possible
scales experiments will be useful, if the sensitivity is pushed to one part per $10^4$ or
less to the background intensity.

## Table II

Measured and expected values for the dipole anisotropy of CBR+EBR background

| | $\frac{\Delta T}{T}$ | $\Delta I_{obs}{}^a$ | $\Delta I_{cbr}{}^a$ | $\Delta I_{kin}{}^a$ | $\Delta I_{grav}{}^{b\,a}$ | $\Delta I_{gush}{}^a$ | $\lambda_0{}^c$ | $\Delta\lambda^c$ |
|---|---|---|---|---|---|---|---|---|
| *Weiss* (1980) | $(1.0\pm0.3)\cdot10^{-3}$ | $(3.6\pm1.0)\cdot10^{-14}$ | $4.3\cdot10^{-14}$ | $4.3\cdot10^{-14}$ | $\geq 4.4\cdot10^{-14}$ | $(1.0\pm0.7)\cdot10^{-13}$ | 1800 | 720 |
| *Fabbri et al.* (1977) | $(1.1^{+0.5}_{-0.2})\cdot10^{-3}$ | $(3.3^{+1.5}_{-0.6})\cdot10^{-13}$ | $3.6\cdot10^{-13}$ | $4.0\cdot10^{-13}$ | $\geq 4.7\cdot10^{-13}$ | $(2.6\pm0.9)\cdot10^{-13}$ | 1200 | 940 |
| *deBernardis et al.*[d] (1984) | $<1.6\cdot10^{-3}$ | $<1.3\cdot10^{-13}$ | $9.6\cdot10^{-14}$ | $1.1\cdot10^{-13}$ | $\geq 1.3\cdot10^{-13}$ | $(7.3\pm2.4)\cdot10^{-14}$ | 1250 | 1100 |
| *deBernardis et al.* (1984) | $<3.2\cdot10^{-3}$ | $<2.5\cdot10^{-13}$ | " | " | " | " | " | " |
| *Halpern et al.*[e] (1988) | $(1.7\pm0.5)\cdot10^{-3}$ | $(3\pm1)\cdot10^{-13}$ | $2.6\cdot10^{-13}$ | $2.8\cdot10^{-13}$ | $\geq 3.2\cdot10^{-13}$ | $(2.0\pm0.7)\cdot10^{-13}$ | 1170 | 770 |

[a]  $\Delta I_{obs}$, $\Delta I_{cbr}$, $\Delta I_{kin}$, $\Delta I_{grav}$ and $\Delta I_{gush}$ are in units of $W\,cm^{-2}sr^{-1}$

[b]  The equality holds for $\Omega = 1$ ( Fabbri et al., 1988 )

[c]  $\lambda_0$ and $\Delta\lambda$ are in $\mu m$ and the operating spectral region extends from $\lambda_0 - \frac{\Delta\lambda}{2}$ to $\lambda_0 + \frac{\Delta\lambda}{2}$

[d]  uncorrected values

[e]  The reported $\Delta I$ values have been calculated by setting the peak value of the spectral transmission equal to 1 as it is not specified by the authors

## Table III

Measured " small scale " anisotropies for CBR+EBR background

| | $\theta$ | $\sigma$ | $\frac{\Delta T}{T}$ * | $\frac{\Delta I}{I}$ | $\lambda_0(\mu m)$ | $\Delta\lambda(\mu m)$ |
|---|---|---|---|---|---|---|
| *Melchiorri et al.* (1981) | $6^0$ | $2.2^0$ | $3.0 \cdot 10^{-5}$ | $1.1 \cdot 10^{-4}$ | 1130 | 500 |
| *Pajot et al.* (1988) | $6^0$ | $12.7'$ | $1.6 \cdot 10^{-4}$ | $7.3 \cdot 10^{-4}$ | 890 | 280 |
| *Fabbri et al.* (1978) | $2^0$ | $1.5^0$ | $3.0 \cdot 10^{-4}$ | $1.1 \cdot 10^{-4}$ | 1060 | 470 |
| *Caderni et al.* (1977) | $25'$ | $7'$ | $1.2 \cdot 10^{-4}$ | $4.8 \cdot 10^{-4}$ | 1190 | 250 |
| *Meyer et al.* (1983) | $5'$ | $2.1'$ | $4.0 \cdot 10^{-4}$ | $1.6 \cdot 10^{-3}$ | 1650 | 330 |

*     For CBR only , assuming T=2.7K , with T and $\Delta T$ thermodynamic values

has shown a cosec b dependence of the counts (where b = galactic latitude). This situation is rather similar to that found in disentangling the cosmic background radiation from the emission of the earth's atmosphere. Our atmosphere also follows a secant law and measurements at various tipping angles thus discriminating between the extraterretrial signals and atmospheric contribution. Therefore, the same technique has been applied to the study of extragalactic backgrounds. Since the dust emission is well correlated with the HI emission, sometimes the correction has been made by using HI maps instead of dust maps. Using the cosec b method Rowan Robinson claimed to have discovered an extragalactic background at a level of 1-3 MJ/sr around 100 microns: he employed the data provided by IRAS satellite. De Bernardis et al (1988) have carried out a detailed analysis of the cosec b dependence of IRAS data at various wavelengths: this analysis interests our framework since it could tell us the limits of this technique. If we neglect all the $100\mu m$ data closer than 30 degrees to the galactic plane we obtain a quite poor fit with a residual of an amplitude comparable with that indicated by Rowan Robinson: however, the difference between cosecant law and constant offset is not so significant for $b \geq 30°$. If we include the data up to $b \simeq 10°$ the fit is improved and the residual is zero, within the experimental uncertainties ($\pm 0.5 Jy/sr$). No significant improvement is obtained by adding the dependence on the cosec $\beta$, where $\beta$ is the distance from the ecliptic plane. The same analysis, when repeated at $60\mu m$ reveals a significant fit improvement by including the ecliptic dust contribution. We may conclude that the excess claimed by Rowan Robinson is an artifact of data selection and that the limit of the technique is around 0.5 Jy/sr, about ten times greater than the largest expected extragalactic flux at 100 $\mu m$. Some improvement could be perhaps obtained by studying the longitude dependence of dust emission and by correlation for HI regions . F. X. Desert et al. (1988) have produced a map of the "strong" ir cirrus which notably show the deviation from a pure cosecant law.

We believe that the above technique can be used as a first step to produce a "cosecant law free" map at a level of about 0.2 Jy/sr. A further improvement requires the use of the following methods.

b) Multispectral analysis

This technique is again an heritage of the atmospheric noise studies. Let us assume, for the sake of simplicity, that we observe the sky at two frequencies $\nu_1$ and $\nu_2$ and we receive two signals $S_{\nu_1}$ and $S_{\nu_2}$, if the galactic background and the extragalactic background are characterized by two spectral indexes $\alpha_d$ and $\alpha_x$, by observing different sky regions we get the relation

$$S_{\nu_2} = S_{\nu_1} \frac{\nu_2^{\alpha_d}}{\nu_1^{\alpha_d}} + I_x(\nu_1) \left[ \frac{\nu_2^{\alpha_x}}{\nu_1^{\alpha_x}} - \frac{\nu_2^{\alpha_d}}{\nu_1^{\alpha_d}} \right] \tag{5}$$

Therefore, by plotting this relation one can extract the unknown extragalactic background $I_x(\nu_1)$. Let us briefly analyze the strength of this method. We will measure the signals $S_i$ with an uncertainty $\Delta S$ due to the instrumental noise: how will this error propagate to $I_x$ ? The answer clearly depends on the difference $k = \alpha_x - \alpha_d$. If k=0 we cannot disentangle the extragalactic background, because the spectral behaviour is the same for both backgrounds. The method works only in limited regions of frequencies: since the exact behaviour of the extragalactic background is unknown, one is forced to consider the method as a very rough one. A null result cannot be used as an upper limit: it could be due to a strong extragalactic background with k=0.

c) Dipole anisotropy

This is a very promising technique, since all the extragalactic backgrounds must have this signature. It has been proposed by Ceccarelli et al (1983), and applied to the balloon data available at that time. Again in order to evaluate the limits of the method one can apply it directly to IRAS data. The errors are due to the sky roughness and not to the detector noise . This means that to improve the results one has to correct for local cirrus. Masi(1987) has estimated a limit for the minimum detectable dipole anisotropy at 100 microns comparable with that expected by several extragalactic backgrounds. The technique has been recently applied to the search for the submillimetric excess (de Bernardis et al. 1989) . Although the available observations do not allow to confirm the excess they are very close to this goal.

d) Sunyaev Zeldovich Effect

This effect is well known in the case of 2.7 K background. It can be extended to other backgrounds, as first suggested by Fabbri and Melchiorri (1979). No observations are available up to know and the required sensitivity can be obtained only by satellites experiments. The only experiment which can afford this task at present is the soviet telescope Aelita, which flight is planned for 1997. We have to consider this possibility as a long term program, since the SZ effect has still to be confirmed in the case of the 2.7 K background.

## 4. FUTURE PROGRAMS

Bearing in mind the previous listed sections, what will be done in the next 5 years and what is planned for long term programs? We have attempted to answer to these questions by summarizing in figure 6 some of the experiments and the plans discussed in the meeting.

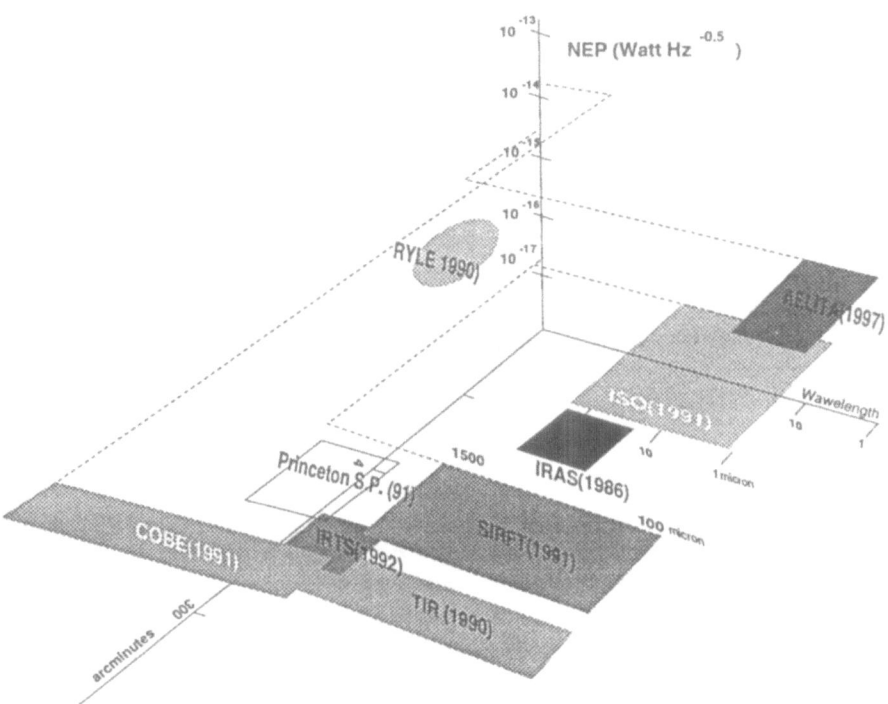

Figure 8: map of the satellite experiments from which cosmological informations can be obtained: the data refer to the anticipated fligth year

Let us stress the areas which seems not fully covered:

a) Small scale anisotropies in the submillimetric region: no satellite experiments are planned, outside Aelita. We believe that balloons experiments and
ground based (Antartica) observations could finally detect such anisotropies, but
satellite programs are needed to have a statistically significant map

b) Radio observations at long wavelength ($\lambda \geq 20cm$) : very few groups
are working in this field. Again satellite experiments could help in carrying out
observations far from local radio contaminations

c) Polarization observations: no programs at all have been discussed in the
conference . This point risks to be neglected for a long time

d) Large scale anisotropies in the submillimetric region: several balloon flights
are planned, but a serious sky coverage can be obtained only through satellite
experiments: it is not clear what COBE could do in this field. Search for quadrupole
- like anisotropies is relevant due to the distortion found by Fabbri et al (1980) and
being not present in the radio maps has not been disputed so far. A millimetric
map of the sky at a level of 0.1 mK is needed.

The few points we have listed and the large number of experiments planned
show the potentiality of the Background Astronomy.

# REFERENCES

Bond,J.R. , Carr B.J., Hogan C.J., 1986, *Ap.J.*,306,428.

Caderni N., De Cosmo V., Fabbri R., Melchiorri B., Melchiorri F., Natale V., 1977, *Phys Rev.* ,D21 ,2424.

Ceccarelli C, Dall'Oglio G., Melchiorri B., Melchiorri F., 1983 ,*Ap.J. Letters* ,275 ,L39.

Coletti, A., Melchiorri F., Natale V., 1975, *in Far Infrared Astronomy*, **Pergamon Press** ,125.

Dall'Oglio G., de Bernardis P., Masi S., Melchiorri B., Melchiorri F., 1982 , *IAU Symposium 104* ,**G.O. Abell, G. Chincarini eds**,135.

de Bernardis P., Masi S., Melchiorri F., Vannoni R, Aiello S., 1988 , *Ap. J.*,326 ,941.

de Bernardis P., Masi S., Melchiorri F., Moreno G., 1989,*Ap.J. Letters* ,340 ,L45.

de Bernardis P, Masi S., Melchiorri B., Melchiorri F., 1989b ,*Ap. J.* ,**August** ,in press.

Desert F.,X., Bazell D., Boulanger F., 1988, *Ap. J.* ,334 ,815.

Epifani M., Guarini G., Melchiorri F., 1989 ,*preprint*, to Ap.J. ,January.

Fabbri R., Melchiorri F., 1979, *Astron. Astrophys* ,285 ,89.

Fabbri R., Guidi I., Melchiorri F., Natale V., 1980, *Phys Rev. Letters* ,44 ,1563.

Fabbri R., Melchiorri B., Melchiorri F., Natale V., Caderni N., Shivanandan K., 1980 , *Phys Rev* ,D21 ,2095.

Fabbri R., Andreani P., Nisini B., 1987 ,*Ap.J.* ,315 ,12.

Gush H.P. ,*Phys. Rev. Letters* ,47 ,745.

Halpern M., Benford R., Meyer S., Muelhner D., Weiss R., 1988 ,*Ap. J.* ,332 ,596.

Lubin P.M., Smoot G. F. , 1981 ,*Ap.J.* ,245 ,1.

Matsumoto T., Hayakawa S., Matsuo M., Murakami H., Sato S., Lange A.E., Richards P.L., 1988 ,*Ap.J.* , 329 ,567.

Pajot F., Gispert R., Lamarre J. M., Peyturoux R., Pomerantz M.A., Puget J.L., Serra G., Maurel C., Pfeiffer R., Renault J.C., 1988 ,*preprint* , to **be published** ,Astronomy and Astrophysics.

Smoot G.F., Bensaudoun M., Bersanelli M., de Amici G., Kogut A., Levin S., Witebsky C., 1987 ,*Ap. J. Letters* ,317 ,L45.

Weiss R., 1980 ,*Ann.Rev.Astron. Astrophys* ,18 ,489.

Smoot G.F., 1982, *IAU Symposium 104* ,**G.O. Abell , G . Chincarini eds** ,153.

# CONCLUDING REMARKS–OBSERVATIONS

R. B. Partridge
Haverford College
Haverford, PA 19041, U.S.A.

This has been the first international conference devoted solely to the cosmic microwave background radiation (CBR). That alone would be a cause to celebrate but the conference here in L'Aquila has also been beautifully organized, well attended and very productive. The presence here of so many younger men and women is a particular pleasure, and a sure sign of the health of our field.

I have been asked to summarize the meeting with an emphasis on the observational side. It strikes me as next to impossible and also wrong-headed to try to summarize the present state of CBR observations without some mention of the past. The progress we have made - and the problems we have encountered - may prove useful guides to future work. A historical approach is even more appropriate since we are celebrating this year the 25th anniversary of the discovery of the CBR by Penzias and Wilson (1965) and Dicke and his colleagues (Dicke, Peebles, Roll and Wilkinson, 1965).

## 1. A ZEROTH ORDER HISTORY

Dave Wilkinson has already presented the tortuous history of the final discovery of the CBR in late 1964 (see also accounts by Alpher, 1989;

*N. Mandolesi and N. Vittorio (eds.), The Cosmic Microwave Background: 25 Years Later, 255–271.*

Weinberg, 1977; and less formally Ferris, 1977). I would like to continue the history from the mid-1960's to the present. Much of what I report here will closely parallel material in a book I am writing on the CBR (1989). One (crude) measure of the progress of a field is the number of papers appearing in it. Most fields begin with one or a handful of seminal papers, grow explosively, then level off. Our field had a slightly different history, as revealed in fig. 1:-an initial burst of activity, a period of diminished interest (or at least fewer results), and a second exponential burst still going on today.

Figure 1 lumps together (with some bias, omissions and judgment calls) both observational and theoretical papers. I would like to suggest, however, that the gross features of this zeroth order history of the CBR result largely from the evolution of observational techniques. Let me try to support that claim by looking more carefully at the evolution of the key observational results.

## 2. FIRST ORDER HISTORY

First, what are the key parameters of the CBR? There are seven (or possibly eight) main observable properties.

1.) $T_0$, the present temperature of the CBR. Of course a single value of $T_0$ may be used to characterize the CBR only if it has a truly thermal spectrum. Therefore let us add another possible observable:

$1_{1/2}$.) The nature and amplitude of spectral perturbations.

2.) $T_1$ , the amplitude of the dipole moment in the CBR intensity, produced in large part by the motion of the Earth.

3.) $T_2$ , the amplitude of (or upper limit on) the quadrupole moment of the CBR. This provides the best constraints on anisotropic cosmological models.

4.) $T_p$, the large scale (linear) polarization of the CBR, produced largely by anisotropy at the epoch of last scattering.

5.) Temperature fluctuations $\Delta T/T$ in the CBR on a scale of $\gtrsim 1°$ , that is, larger than the horizon size at last scattering.

6.) $\Delta T/T$ on a scale of $\sim 0.1°$, the best indicator of many theories of galaxy formation.

7.) And finally $\Delta T/T$ on a scale of $\simeq 0.01°$ (say 10"-30"), not expected unless the Universe was reionized at $z \lesssim 1000$.

Let us bear in mind that all of these parameters can in principle depend

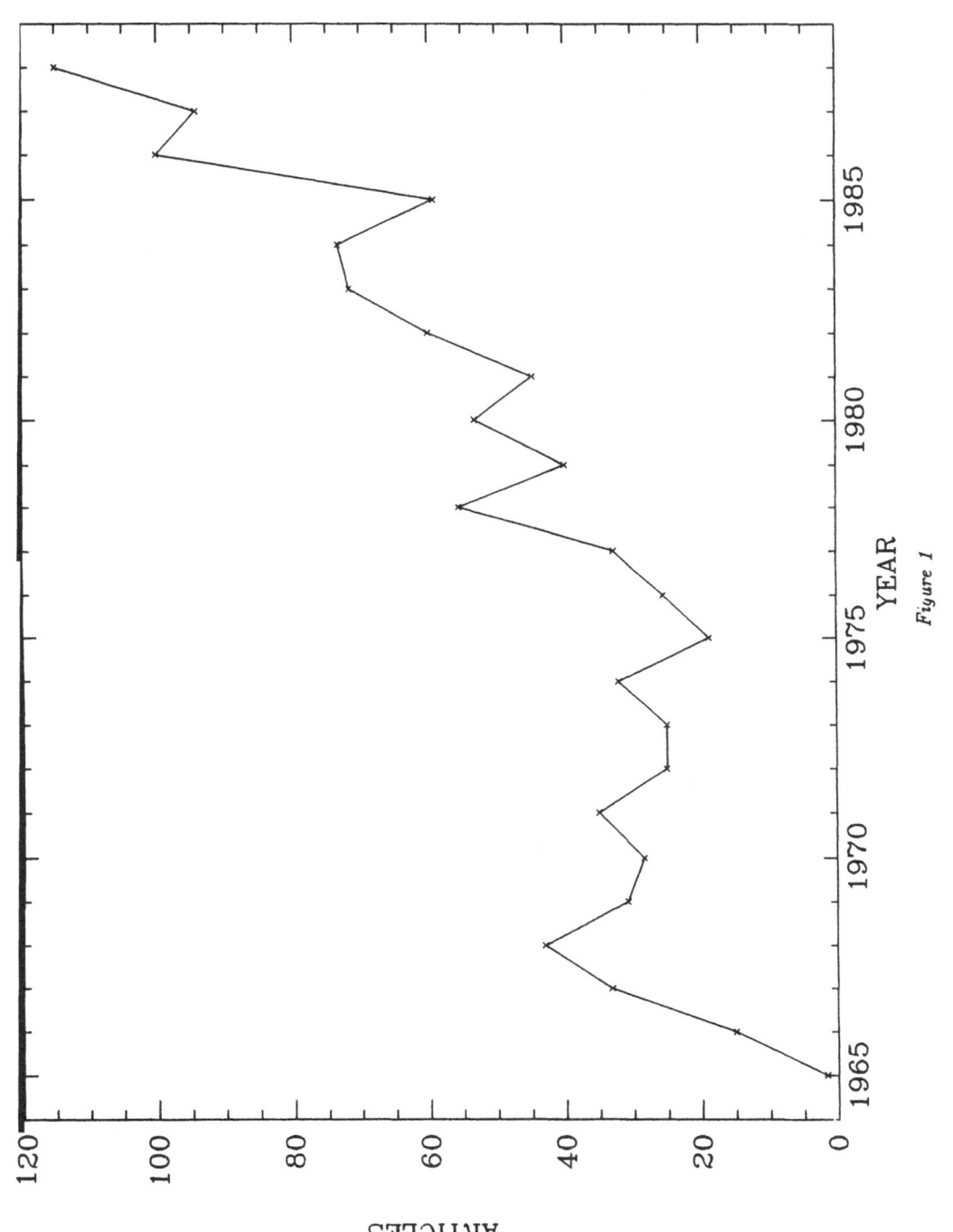

*Figure 1*

on wavelength.

In Table 1, I present a rough first order history of observations in our field. With the exception of the 1964 results - all from Penzias and Wilson's 1965 discovery paper, I have not attempted to attribute the results to individuals or groups. The accuracy in the dates is roughly ±1 yr, i.e., the rough lag time between getting the results, submitting them, having them published and finally having them read by the community.

By 1969, we knew To to ~5% accuracy (e.g., Thaddeus and Clauser, 1966; Wilkinson, 1967; Stokes et al, 1967). On scales from ~1' to 180°, limits on possible CBR anisotropies were below a few ·$10^{-3}$. In other words, much of the observational cream had been skimmed. I suggest that this rapid progress occurred for two reasons. First, conventional radio telescopes (e.g., Conklin and Bracewell, 1967; Epstein, 1967) were used to map the CBR, once the radio astronomers knew the background was there. Second, using readily available microwave components, several groups were able to fabricate quickly small radio telescopes designed specifically to measure the spectrum and large-scale isotropy of the CBR (for an isotropic background, the angular resolution of conventional radio telescopes is unnecessary). Groups from Berkeley, Cambridge, MIT, Princeton and Stanford participated in this early work (see Partridge, 1989, for details). In sum, no new technology was required (though some new observational techniques were).

In the next five years, to ~1974, there was no appreciable improvement in any of the seven CBR parameters. In part, I would argue, effort slackened because some of the principals in the field became interested in some of the other discoveries and problems of high-energy astrophysics and cosmology. After all, the decade 1963-1973 was a bonanza for cosmologists-quasars, pulsars, gravity waves, nucleosynthesis, problems of galaxy formation, and faint radio source counts were all in the air. Some of us switched our allegiance to other fields permanently; some for a period of years.

What kept the field alive during these years of the "dark ages" was in part activity on the theoretical side (outside the scope of this summary), and in part those researchers designing new technology to bring to bear on the CBR. In this regard, I think particularly of those who introduced bolometric detectors to the field, Francesco Melchiorri, Paul Richards and Ray Weiss, and their many coworkers. It was also in the early 1970's that CBR researchers began to employ balloons and high-flying aircraft to get above much of the Earth's atmosphere (e.g., Muehlner and Weiss, 1970;

Table 1: *A first order history of observational progress in CBR studies.*

| | 1964 | 1969 | 1974 | 1979 | 1984 | 1989 |
|---|---|---|---|---|---|---|
| $T_0$ | $3.5\pm1.0$ K | ~5%accuracy; CN results; Wien curvature; | - | - | ~2% accuracy | submillimeter excess |
| $T_1/T_0$ | $\lesssim 0.2$ | $\lesssim 0.002$; hints of dipole | - | dipole detected | dipole amplitude & direction to ~3% | many consistent measurements at various $\lambda$ |
| $T_2/T_0$ | - | $\lesssim 0.011$ | - | $3\ 10^{-4}$ claimed | $\lesssim 3\ 10^{-5}$ | $2\ 10^{-5}$ |
| $T_p/T_0$ | $\lesssim 0.2$ | - | - | $<10^{-3}$ | $\leq 8\ 10^{-5}$ | - |
| $\Delta T/T$:- | | | | | | |
| 1° scale | - | $\lesssim 3\text{-}5\ 10^{-4}$ | - | $\leq 10^{-4}$ | $\lesssim 3\text{-}5\ 10^{-5}$ | $\lesssim 10^{-5}$* |
| 0.1° scale | - | $\lesssim 3\ 10^{-3}$ | - | $\leq 10^{-4}$ | $\lesssim 3\text{-}5\ 10^{-5}$ | $\lesssim 1.7\ 10^{-5}$ |
| 0.01° scale | - | - | - | $\lesssim .05$ | $\lesssim 10^{-4}$ | $\leq 3\text{-}5\ 10^{-5}$* |

*My hunch for $\Delta T/T$ by the end of the year.

Henry, 1971; Smoot, Gorenstein and Muller, 1977).

By 1979, new areas were opening up. Strong limits on large-scale polarization $T_p$ had been established (Nanos, 1979; Lubin and Smoot, 1979) and the first attempt to use aperture synthesis to study the CBR was underway (Martin, Partridge and Rood, 1980).

In the early 1980's a number of new facilities (e.g., the VLA), new technologies (e.g., bolometers), new observing techniques (e.g., the precision cold load of the "White Mountain" collaboration) and new experimental groups (e.g., Cal Tech and Nagoya) were beginning to have a major impact on the field. The experimental pace quickened, and progress was made on all seven of the parameters listed in Table 1 above. That burst of activity has continued unabated until today (with the possible exception of last night when we were all having too good a time at our banquet).

Where do we stand now, and what have we learned from each other here in L'Aquila? I will try to answer those questions below, not in depth (see recent reviews by Wilkinson, 1986; Partridge, 1988; Smoot, 1989; and Partridge 1989), but with special emphasis on the results reported at this conference.

## $T_0$: THE PRESENT STATUS

1.) $\lambda \leq 1$ mm

The most interesting spectral observations are those presented here by Matsumoto and Richards at wavelengths $\leq 1$ mm. These observations suggest either a distortion of the CBR or an addition to it with an energy density 10-20% of the CBR energy. The submillimeter excess has been measured with high precision (that is, the statistical errors are small). As Matsumoto and Richards indicate here, the real question is systematic error. In my view there is no obvious systematic problem which can explain the Berkeley-Nagoya measurements (Matsumoto et al, 1988). Nor, for that matter, is there any widely accepted theoretical model which can explain these measurements (see the papers by Carr, Daly, Danese and Hogan here, and Hayakawa's summary remarks). Quite apart from questions like where the energy comes from (e.g., Lacey and Field, 1988), the spectrum and the small-scale isotropy of the CBR present problems for any model (see Bond, Carr and Hogan, 1989; DeZotti, 1990; Partridge, 1990 and papers here).

Since both COBE and a second Berkeley-Nagoya rocket flight should be producing results by the time this volume appears, I will not belabor the importance of these measurements or the interesting problems they have raised for us in the field.

2.) CN measurements at $\lambda$= 2.64 and 1.32 mm

Interstellar CN provides us with a way to determine To that is completely independent of ground-based or rocket-borne techniques. Hence the sources of systematic error are completely different. As a consequence, the agreement between the CN results reported here by Crane and Meyer and the radiometric and bolometric measurements is especially gratifying. In particular, I note that the CN measurements at 1.32 mm by both Crane et al and Meyer et al now produce values slightly above 2.74 K, the characteristic value of $T_0$ at $\lambda \geq 3$ mm. We have here a tantalizing confirmation of the Berkeley-Nagoya results. It is also worth noting - as Kaiser has reminded us here - that there is good reason to hope that the systematic and statistical errors in the CN observations will be reduced in the next few years. From the beginning CN results have played an important role in determining the CBR spectrum, and it appears they will continue to.

3.) $0.3 \leq \lambda \leq 10$ cm

Several years ago (e.g., Smoot et al, 1985), there seemed to be general agreement that $T_0$ at centimeter wavelengths was 2.74 K with an uncertainty of 1-2%. The situation is now less clear. The newest Berkeley results reported here by De Amici, Bersanelli and Levin tend to lie below 2.74 K, and in particular are in poor agreement with the high altitude 1.2 cm measurement of Johnson and Wilkinson (1987), which gives $T_0 = 2.783\pm0.025$ K.

In some of the new Berkeley results, while $T_0$ is a bit low, measured values of the Earth's atmosphere are a bit high compared to earlier results of the White Mountain collaboration (Smoot et al, 1985) and to atmospheric models (Danese and Partridge, 1989).

4.) $\lambda > 10$ cm

Long-wavelength measurements provide a crucial test of some models of the spectral distortion in the CBR (see, e.g., Danese and DeZotti, 1977). As Sironi and his colleagues have pointed out here and elsewhere, such observations are also very difficult because of foreground emission, especially from the Galaxy. Better galactic maps would help, especially if their absolute flux scale is well determined. It would also be nice to have a 30-60 cm point measured with a cold load calibrator attached directly to the antenna (as pioneered by Wilkinson, 1967).

## $T_1$: PRESENT STATUS OF THE DIPOLE

The one unmistakable anisotropy detected in the CBR has now been measured to an accuracy of $\sim$5% or better (Strukov and Boughn in this volume). Boughn reports an amplitude of 3.36±0.03 mK in thermodynamic temperature at $\lambda$=1.5 cm (where the error is statistical only); Strukov reports 3.25±0.07 mK (which I believe is a weighted average of the Soviet Relict measurement with $T_1$=3.16±0.12 mK, Strukov et al, 1987 and some earlier work). The directions agree well.

More important, there is no evidence that the dipole amplitude depends on wavelength, provided that is that the underlying CBR spectrum is approximately blackbody (see Lubin and Villela, 1985; Halpern et al, 1988 and Masi here). Not so good is the agreement between the CBR dipole and mass or velocity dipoles derived from samples of objects at non-cosmological distances (e.g., optical or IRAS galaxies). The discrepancies between the CBR and other dipoles can tell us much about large-scale velocity flows in the Universe (Gorski, here).

Finally, it is worth noting that measuring $T_1$ at a range of wavelengths is a major goal of COBE.

## $T_2$: PRESENT UPPER LIMITS ON THE QUADRUPOLE MOMENT

The Relict team (Klypin et al, 1987; Strukov and Skulachev here) have shown that $T_2$ <0.1 mK; a slightly model-dependent upper limit of 0.06 mK also can be set, comparable in sensitivity to the best small-scale anisotropy limits discussed below.

The sensitivity of the MIT-Princeton-Haverford balloon observations (Boughn in this volume) is a bit higher than the Relict results, but this work has a larger galactic contamination because of the longer wavelength employed. The balloon observations have not yet been fully analyzed; it will be interesting to see how tight a constraint on $T_2$ they will eventually provide.

Both these experiments produce full sky maps of the CBR. Unlike earlier two-antenna, differential measurements (e.g., Fixsen et al, 1983; Lubin et al, 1983), these two experiments employed a single high resolution horn.[1] Thus

---

[1]The reference in the MIT-Princeton-Haverford experiment was a cold load; in the

the resulting maps required less mathematical deconvolution than differential measurements (or none). I suspect this makes them more directly useful in the search for high-multipole, non-Gaussian features of the kind discussed here by Martinez, Sanz and Stebbins.

## $T_p$: LARGE-SCALE POLARIZATION

Here is an area where there has not been much progress since Lubin's careful work in 1983. Since COBE's microwave receivers are polarized, we can hope to sort out $T_p$ on large angular scales from the COBE results. Likewise I suspect that improved limits on $T_p$ on large angular scales could be obtained now from a careful analysis of the data making up the beautiful 1.5 cm map of the sky shown here by Boughn–a good problem for a student.

## $\Delta T/T$ ON $\sim 1°$ SCALES

Of the seven parameters I listed above, I believe this is the one most actively pursued in the past (and next) few years. Theoretical interest in CBR fluctuations at or slightly above the horizon scale at decoupling is large (Bond and Stebbins here). In addition, a variety of new experimental approaches are being tried, some of which we have heard about here.

What is the present, mid-1989, situation? Davies et al (1987) report a detection of fluctuations in the microwave sky at a level $3\text{-}4\cdot10^{-5}$ at $\lambda = 3$ cm and $\theta = 8°$. As Lasenby explains here, most of the anisotropy signal comes from a single "bump", a bump which shows up more strongly in narrower beam observations, suggesting that it is a real feature on the sky, and less than $8°$ in scale. The Jodrell-Cambridge group is currently conducting 6 cm observations to try to determine if the bump at $\sim 15^h$, $+40°$ is Galactic in origin; the results are not yet definitive, and further work is planned.

A parenthetical remark here. Most of us - most observers, at least - think this effect will either go away or be shown to be local in origin. That is my hunch. Why do I feel this way? In part, I fear, because I have become accustomed to null results and upper limits. One of these days, we must detect real fluctuations in the CBR. Perhaps Davies, Lasenby and their colleagues already have. What will it take to convince us? Francesco Mel-

---

Soviet experiment a low resolution sky horn aligned along the satellite spin axis.

chiorri supplies some answers in his remarks here. I would emphasize the importance of multiple-frequency observations ( $\Delta T/T$ should be independent of frequency if the fluctuations are in the CBR), and the importance of thinking positively. Neither the amplitude nor the angular scale of CBR fluctuations will necessarily be exactly as predicted by theorists, so we should follow the advice of Heraclitus: "If you do not expect it, you will not find the unexpected, for it is hard to find and difficult."

Now back to observations as opposed to exhortations. Two observational programs at angular scales comparable to that of Davies et al, but at higher frequency, are now producing upper limits on $\Delta T/T$ at a few times $10^{-5}$. The first is the 3 mm bolometer experiment described here by Lubin - a very promising technique as shown by both Lubin's talk and the assessment of Dall'Oglio. The second experiment, by Timbie and Wilkinson, makes use of a two-element, phase-switched interferometer. Interferometric techniques sidestep some of the systematic errors involved in direct beam-switched observations, but (of course) introduce others including cross talk between the antennas and, in the case of the two-element interferometer, a strange beam pattern. Nevertheless, I believe interferometry can and should be pushed further on degree scales as well as arcsecond scales (see below). An array of small telescopes with typical separations of $\sim 100\lambda$ and below would be ideal. What I'm describing, of course, is the Very Small Array proposed by the radio group at Cambridge University (see reports in this volume). May it be funded!

## $\Delta T/T$ ON $\sim 0.1°$ SCALES

These scales are well suited to conventional radio telescopes, and considerable effort has been poured into such observations over the past twenty years (e.g., Conklin and Bracewell, 1967; Boynton and Partridge, 1973; Uson and Wilkinson, 1984; Readhead et al, 1989). The upper limits on $\Delta T/T$ on scales of several arcminutes are the best we now have at any scale. The newest entrant in this field is the Owens Valley (Cal Tech) group whose results at 2 cm are described here by Lawrence and by Meyers: $\Delta T/T < 1.7 \cdot 10^{-5}$ in seven independent 1.8' spots separated by 7.1'; and $\Delta T/T \lesssim 3 \cdot 10^{-5}$ in a circumpolar circle of 96 such spots.

Extraordinary care has gone into the control of systematic errors in this work. I suspect Readhead, Lawrence and their colleagues have pushed ground-based observations with conventional radio telescopes (carefully modified in this case) about as far as they can go. Perhaps such measurements can break the $10^{-5}$ barrier, but I think we are rapidly reaching the limits of conventional radio astronomical techniques. Perhaps for this reason–and also because theoretical enthusiasm is shifting to larger scales - only Cal Tech is currently pursuing observations on $\sim 0.1°$ scales.

## $\Delta T/T$ ON $\sim 0.01°$ SCALES

As I have noted, small-scale fluctuations will be smeared out by the finite thickness of the last scattering surface. Calculations by Vittorio and Silk (1984) and by Bond and Efstathiou (1984) show that the autocorrelation function of fluctuations imprinted at last scattering drops rapidly as falls below a few arcmin. Thus in the late 70's there wasn't much enthusiasm among theorists for observations on arcsecond scales. In spite (or perhaps because) of these views, it seemed wrong to me not to try to push limits on $\Delta T/T$ to smaller angular scales if it could be done. Progress since this early work with Martin and Rood (1980) has been rapid, essentially since the VLA, an interferometric array of 27 telescopes, became available in 1980. The first crude limits on $\Delta T/T$ on scales of tens of arcseconds have been pushed down by two orders of magnitude; this work at the VLA is nicely reviewed here by Fomalont. The two groups currently using the VLA have joined forces to make a major effort to push the VLA to its limit using the newly-available 4 cm receivers. We expect to be able to see fluctuations on 10"-50" scales, or to set limits on them, at levels of $\Delta T/T \sim$ few $\cdot 10^{-5}$, i.e., at levels comparable to the best work at arcminute scales.

This technique, too, is reaching natural limits. Several precious days of VLA time will be required to reach our target sensitivity at $\lambda = 4$ cm. As Fomalont indicated, fluctuations in the microwave sky caused by weak, subliminal radio sources already dominate possible cosmological fluctuations, at least at the wavelengths available at the VLA.

To push limits further, we may have to turn back to conventional, filled-aperture, radio telescopes, but use them at high frequencies to keep the beam size small. An interesting limit on $\Delta T/T$ at $\lambda = 1.3$ mm has recently been established by Kreysa and Chini (1989) using a bolometric detector on a large antenna.[2] Their limit of $\Delta T/T \leq 2.6 \cdot 10^{-4}$ emerged as a by-product of observations of faint QSO: no doubt an experiment specifically aimed at detecting CBR fluctuations could reach higher sensitivity. In this sense the plans presented here by Church are encouraging.

## SOME COMMON THEMES

I have just subjected you to a possibly biased review of the observations presented at this conference and of the general status of observational work in the field. To me, some common threads emerge.

1.) The first is that our field, while it does employ specialized techniques, is far from isolated from other fields of physics, astrophysics and cosmology. The links between the CBR and the X-ray background and the near IR background have been explored here by Setti, and by Matsumoto, respectively. Links between CBR observations of small-scale anisotropy and radio source counts (and even the clustering properties of radio sources) are under intense and productive study by Danese and DeZotti and their colleagues in Padua.

One of the hot topics in astronomy these days is large-scale structure, including large-scale velocity flows. The benchmark in that field is the CBR dipole: see the talks by Gorski and Sanz here.

The existence of the CBR implies a high energy cutoff for cosmic rays of intergalactic origin, a point mentioned only by Sciama here and one that has rather faded from sight since the early work of Hoyle (1965), Felten (1965), Greisen (1966), Jelley (1966) and Gould and Schréder (1966) in the mid '60's.

To do nucleosynthesis calculations for the Big Bang you need to know $T_o$ (and also to know how isotropic the early expansion was). In turn, a comparison of the observed and the predicted abundances of light nuclei has provided us with a crucial cosmological datum, the baryon density of the Universe. These same observations can be used to set limits on the number of neutrino families (our astronomical limits would be tighter still if our laboratory colleagues knew the free neutron half-life to higher precision). I find it ironic that one of the major goals announced for the LEP accelerator at CERN is also to set limits on the number of neutrino families. It will be nice to have the astrophysical limit confirmed, but I don't think it is unfair to congratulate ourselves on having done it first (and cheaper!).

---

[2]Their beam size was $\sim 11$"; the beam switch angle 30".

Finally, of course, the rich and explosively growing cosmology-particle physics connection has the CBR as one of its strands. As I have noted optimistically elsewhere, if we can ever measure the angular spectrum of CBR temperature fluctuations, we will have a direct link to predictions of the inflationary paradigm (but see Mattarese's cautionary remarks here).

2.) Next I would like to stress the importance of observing techniques as well as instruments. In various sessions, we've heard about new detectors and new equipment. But old equipment used in new ways can also move the field forward. Among the clearest examples are the use of double beam switching (originally urged on us by Boynton) and the use of an additional detector specifically to monitor atmospheric emission while data in nearby atmospheric windows are being taken. I could also mention the use of large diameter, overmoded cold loads as absolute calibrators for measurements of $T_0$.

Techniques also include means of analyzing the observations once you have them. In the CBR community - but not so far in print - there has been a good deal of discussion of how best to analyze $\Delta T/T$ searches (past treatments of this issue include papers by Boynton and me, 1973, and by Lasenby and Davies, 1983; see also Readhead et al, 1989). I know that fancy statistics are the last refuge of desperate observers, but when you consider the huge effort it cost to improve limits on $\Delta T/T$ on 0.1° scales by a factor of 2-3 (Uson and Wilkinson, 1984 to Readhead et al, 1989), a justified statistical analysis that could do the same would be a boon. I think more work - more published work in particular - on statistics would be quite valuable.

Already showing its value is another "technique" - proper correction for fluctuations introduced by weak radio sources. Such correction may be essential as we push $\Delta T/T$ searches below $10^{-5}$ and to a wider range of angular scales (see Danese's and Fomalont's talks). Deeper high frequency radio source counts - a major effort at the VLA - may be needed before we can sort out the CBR signals from more local ones.

3.) This last consideration leads to my next point: in making searches for CBR fluctuations, we need to cover a larger area in the two-dimensional frequency-scale parameter space. As it happens, most of the better observational limits now lie near a diagonal line stretching from low frequency at low $\theta$ to higher frequency at larger $\theta$. There are exceptions, of course, like the 10 GHz observations of Davies et al (1987) with $\theta=8°$. True CBR fluctuations are expected to have no frequency dependence (unless we count the conversion from antenna temperature to thermodynamic temperature); other "cosmological" fluctuations may (e.g., Hogan, 1980; Bond, Carr and Hogan, 1989). Let us not fail to "expect the unexpected."

4.) We are beginning to encounter some quite fundamental limitation as we seek to improve our knowledge of each of the seven parameters I have listed.

That emission from the Earth's atmosphere is a nuisance or can even dominate error budgets has long been recognized. We have tried to measure it and subtract it; to rise above it using balloons and satellites; and to freeze it (and our students) by going to the arctic and antarctic.

Galactic emission is harder to avoid - and is the fundamental problem for long wavelength spectral measurements. Patchy galactic emission may also be causing trouble in some $\Delta T/T$ searches (e.g., the work of Davies et al, 1987; see also Dall'Oglio's talk here).

Finally, there is confusion by weak radio sources, frequently referred to above. To quote Heraclitus again, "Nature loves to conceal herself."

5.) Last, let me mention a common theme I had not properly appreciated until this meeting: we seem to be entering a new phase of instrument design. The field is beginning to make a new leap forward, propelled by new, more sophisticated apparatus.

In the 1960's, many of our early advances were made with specially designed but cheap equipment (the radiometer Roll and Wilkinson [1966] specially designed to search for the primeval fireball being one example). In the late 70's and early 80's, a larger fraction of CBR observations were made with available, conventional facilities (the 140-foot telescope or the VLA of the U. S. National Radio Astronomy Observatory). One consequence is that we inherited other people's systematics! Now, in the late '80's, we seem to be moving simultaneously in two general directions. One is forward (or outward) into space, following in the footsteps of the Soviet Relict mission. The other in a sense is backwards-back to the design and construction of equipment specially designed to attack a single property of the CBR. Here, I might mention Phil Lubin's elegant balloon-borne telescope; the hope to construct arrays of bolometers to map the CBR to <0.01 mK on a range of angular scales (Peterson et al here); and my sentimental favorite, a Very Small Array to limit $\Delta T/T$ interferometrically. These experiments, however, are no longer cheap. The CBR is becoming big science as well as fun science. When we next meet, as I hope we will in 5 or 10 years, we are certain to have some interesting, and perhaps even startling, new results to discuss.

REFERENCES

Alpher, R. A. 1989, in preparation; see also Alpher, R. A., Gamow, G., and Herman, R. C. 1967, *Proc. Nat. Acad. Sci.*, **58**, 2179.

Bond, J. R., Carr, B. J., and Hogan, C. J. 1989, submitted to *Astrophys. J.*

Bond, J. R. and Efstathiou, G. 1984, *Astrophys. J. (Letters)*, **285**, L45.

Boynton, P. E., and Partridge, R. B. 1973, *Astrophys. J.*, **181**, 243.

Conklin, E. K., and Bracewell, R. N. 1967, *Phys. Rev. Letters*, **18**, 614.

Danese, L. and De Zotti, G. 1977, *Riv. Nuovo Cimento*, **7**, 277.

Danese, L., and Partridge, R. B. 1989, *Astrophys. J.*, **342**, 604.

Davies, R. D., Lasenby, A. N., Watson, R. A., Daintree, E. J., Hopkins, J., Beckman, J., Sanchez-Almeida, J., and Rebolo, R., 1987, *Nature*, **326**, 462.

DeZotti, G. 1990, in *IAU Symposium 139*, Galactic and Extragalactic Background Radiation, Kluwer Intl. Publ.

Dicke, R. H., Peebles, P. J. E., Roll, P. G., and Wilkinson, D. T., 1965, *Astrophys. J.*, **142**, 414.

Epstein, E. E. 1967, *Astrophys. J. (Letters)*, **148**, L157.

Felten, J. E. 1965, *Phys. Rev. Letters*, **15**, 1003.

Ferris, T. 1977, *The Red Limit*, Wm. Morrow and Co., New York.

Fixsen, D. J., Cheng, E. S., and Wilkinson, D. T. 1983, *Phys. Rev. Letters*, **50**, 620.

Gould, R. J., and Schréder, G. P. 1966, *Phys. Rev. Letters*, **16**, 252.

Greisen, K. 1966, *Phys. Rev. Letters*, **16**, 748.

Halpern, M., Benford, R., Meyer, S., Muehlner, D. and Weiss, R. 1988, *Astrophys. J.*, **332**, 596.

Henry, P. S. 1971, *Nature*, **231**, 561.

Hogan, C. J. 1980, *Month. Not. Roy. Astr. Soc.*, **192**, 891.

Hoyle, F. 1965, *Phys. Rev. Letters*, **15**, 131.

Jelley, J. V. 1966, *Phys. Rev. Letters*, **16**, 479.

Johnson, D. G., and Wilkinson, D. T. 1987, *Astrophys. J. (Letters)*, **313**, L1.

Klypin, A. A., Sazhin, M. V., Strukov, I. A., and Skulachev, D. P. 1987, *Soviet Astron. Letters*, **13**, 104.

Kreysa, E., and Chini, R. 1989, in *Proc. of the Third ESO-CERN Symposium on Astronomy, Cosmology and Fundamental Particles*, in press.

Lacey, C. G., and Field, G. B. 1988, *Astrophys. J. (Letters)*, **330**, L1.

Lasenby, A. N. and Davies, R. D. 1983, *Month. Not. Roy. Astr. Soc.*, **203**, 1137.

Lubin, P., Melese, P., and Smoot, G. 1983, *Astrophys. J. (Letters)*, **273**, L51.

Lubin, P. M. and Smoot, G. F. 1979, *Phys. Rev. Letters*, **42**, 129.

Lubin, P. M., and Villela, T. 1985, in *The Cosmic Background Radiation and Fundamental Physics*, ed. F. Melchiorri, Editrice Compositori, Bologna.

Martin, H. M., Partridge, R. B., and Rood, R. T. 1980. *Astrophys. J. (Letters)*, **240**, L79.

Matsumoto, T., Hayakawa, S., Matsuo, H., Murakami, H., Sato, S., Lange, A. E., and Richards, P. L. 1988, *Astrophys. J.*, **329**, 567.

Muehlner, D., and Weiss, R. 1970, *Phys. Rev. Letters*, **24**, 742.

Nanos, G. P. 1979, *Astrophys. J.*, **232**, 341.

Partridge, R. B. 1988, *Reports on Progress in Physics*, **51**, 647.

Partridge, R. B. 1989, *3 K: The Cosmic Microwave Background Radiation*, in preparation for Cambridge University Press.

Partridge, R. B. 1990, in *IAU Symposium 139*, Galactic and Extragalactic Background Radiation, Kluwer Intl. Publ.

Penzias, A. A., and Wilson, R. W. 1965, *Astrophys. J.*, **142**, 419.

Readhead, A. C. S., Lawrence, C. R., Myers, S. T., Sargent, W. L. W. Hardebeck, H. E., and Moffet, A. T. 1989, submitted to *Astrophys. J.*.

Roll, P. G. and Wilkinson, D. T. 1966, *Phys. Rev. Letters*, **16**, 405.

Smoot, G. F. 1989, in *The Berkeley Workshop on Particle Astrophysics*.

Smoot, G. F., De Amici, G., Friedman, S., Witebsky, C., Sironi, G., Bonelli, G., Mandolesi, N., Cortiglioni, S., Morigi, G., Partridge, R. B., Danese, L., and De Zotti, G. 1985, *Astrophys. J. (Letters)*, **291**, L23.

Smoot, G. F., Gorenstein, M. V., and Muller, R. A. 1977, *Phys. Rev. Letters*, **39**, 898.

Stokes, R. A., Partridge, R. B., and Wilkinson, D. T. 1967, *Physical*

*Review Letters*, **19**, 1199.

Strukov, I. A., Skulachev, D. P., Boyarskii, M. N., and Tkachev, A. N. 1987, *Soviet Astron. Letters*, **13**, 65.

Thaddeus, P., and Clauser, J. F. 1966, *Phys. Rev. Letters*, **16**, 819.

Uson, J. M., and Wilkinson, D. T. 1984, *Astrophys. J. (Letters)*, **277**, L1.

Vittorio, N., and Silk, J. 1984, *Astrophys. J. (Letters)*, **285**, L39.

Weinberg, S. 1977, *The First Three Minutes*, Basic Books, New York.

Wilkinson, D. T. 1967, *Phys. Rev. Letters*, **19**, 1195.

Wilkinson, D. T. 1986, *Science*, **232**, 1517.

# CONCLUDING REMARKS-THEORY

P. J. E. Peebles
Joseph Henry Laboratories
Princeton University

A summary may attempt to weave threads of the debate into a well ordered tapestry, as Bruce Partridge has so elegantly done for the experimental side. I shall instead play the role of agitator, emphasizing points of controversy for the purpose of stimulating action on the theoretical front.

To begin with a positive note, I should like to echo Dennis Sciama's point, that the discovery of the microwave cosmic background radiation (the CBR) marked a central landmark in the development of physical cosmology, to be compared to Hubble's discovery of the law of general recession of the nebulae and Lemaître's interpretation of Hubble effect as the expansion of the universe. Once the CBR was shown to be close to isotropic with a spectrum close to blackbody, it was generally accepted as almost tangible evidence that the universe really did expand from a dense state, for no one has been able to see how these properties of the CBR could have been produced in the universe as it is now. That encouraged people to consider in detail the physics of an expanding relativistic universe, and it gave an invaluable datum, the entropy per baryon number. There followed an enormous growth of research, such as physics at the relatively modest redshifts at which galaxies formed and ranging back all the way to the Planck epoch. The results have occupied us at length at this conference and many others.

N. Mandolesi and N. Vittorio (eds.), The Cosmic Microwave Background: 25 Years Later, 273–277.
© 1990 Kluwer Academic Publishers.

Before the discovery of the CBR, there was a legitimate difference of opinion on whether the most reasonable world model is the Big Bang cosmology of Friedmann, Lemaître and Gamow, or the Steady State Theory of Bondi, Gold and Hoyle, or something else. It is worth reflecting on some similarities between the role of the Steady State theory in cosmology of the early 1960s, before the CBR, and the role of the inflationary scenario at the present time. The inflation of inflation has a longer pedigree than the continuous creation hypothesis of the Steady State theory, but there is a fundamental similarity: both were conceived by bold strokes of imagination, rather than drawn from clues from Nature, for the purpose of solving perceived problems. To my mind the great promise of inflation is that it offers a way by which we might derive initial conditions for classical cosmology out of the physics of the inflation epoch, without having to learn how to deal with the relativistic singularity. Steady State deals with the singularity in another way, by eliminating evolution (which leads to de Sitter line element, just as for the inflation epoch). The Steady State theory also was thought to have the advantage of resolving the time- scale problem that is now seen to have been the result of an overestimate of Hubble's constant.

The successes of these theories in resolving the problems they addressed is not evidence that either is on the right track, because they were invented for the purpose of solving the problems. Such successes are not even necessary, for the problems need not be real, as we see in the time-scale puzzle of the 1950s. Instead, one must look for new predictions. In the case of the Steady State theory, an interesting line of attack was the radio source counts as a function of flux density, because the Steady State theory makes an unambiguous prediction. The theory has been abandoned, but its legacy, such as source counts, still is very much with us, as we have heard at this conference.

Inflation begat the biased cold dark matter (BCDM) theory of galaxy formation, and BCDM begat an intense interest in the search for anisotropy in the CBR. Matarrese is right in emphasizing that BCDM is not equivalent to inflation: within inflation the density parameter need not be unity if one is willing to invoke an effective cosmological constant to make space curvature negligibly small; the spectrum of density fluctuations need not to be scale-invariant, and the primeval density fluctuations need not be Gaussian. However, BCDM does seem to be the simplest possible outcome of inflation, and if BCDM proved successful it surely would enhance the credibility of the inflation scenario. For that reason, it would be of considerable interest to know whether BCDM is on the right track. Thus the talk I have heard at this conference of the design of experiments to maximize sensitivity to the anisotropy predictions of BCDM is good science. However, with all due respect to Dick Bond and the others who have worked to excellent effect to explore the possibilities offered by this theory, I would suggest that caution is in order.

The evidence for inflation is virtually non-existent (again, apart from the fact that it resolves the problems it was designed to solve, and, of course, apart from our desperation for any sensible-looking way to deal with the early universe!). There is evidence that BCDM is on the right track, including its success in dealing with the small-scale structure of the galaxy distribution, but there is also some clouds on the horizon. Examples are the predicted late epoch of assembly of galaxies, that seems difficult to reconcile with observations of quasars and an ionized intergalactic medium at redshifts $z \sim 4$, and of galaxies with apparently old star populations at $z \sim 1$; the small predicted fluctuations in the mass distribution on larger scales, that make it difficult to account for observed lager-scale structures such as the Perseus-Pisces supercluster; and, as discussed at this meeting, the Nagoya-Berkeley submillimiter excess. These clouds do not show that this theory is wrong, but rather that we do not yet have a sound guide to weather this or any of the other currently popular theories has a reasonably chance of being right. Thus, to be blunt, if you are contemplating a long-term experiment to test BCDM, you should consider the possibility that you will end up working on a theory that has gone out of fashion. That is why I felt a little more comfortable with the philosophy behind Partridge's remark, that he decided to work on anisotropy measurements at the angular scales sampled by VLA because that regime was not throughly explored, at least in part because early theoretical discussions indicated there would be nothing of interest there. Now, of course, there is considerable interest in small-scale anisotropy as a test of scattering by the plasma in young galaxies.

How will the anisotropy searches will turn out, what will be the dominant source of the anisotropy? In sense we already know. At relatively long wavelengths, Birkinshaw showed us that the Sunyaev-Zel'dovich effect undoubtedly exists and Rephaeli emphasized that it is easy to check that the integrated effect of known clusters is $\delta T/T \sim 1 \times 10^{-5}$ with coherence length $\sim 1$ arcmin, which is close to the observational bounds. Depending on the cosmological parameters, scattering by plasma in young galaxies could suppress the primeval signal and insert anisotropy due to plasma motions at $\delta T/T \sim 1 \times 10^{-4}$ to $10^{-5}$. Bond put it well: he is going to have to 'dirty up his maps' with such foreground astronomical noise to assess the chances of picking up signals from decoupling or from cosmic strings.

Anisotropy measurements on larger angular scales, $\gtrsim 3°$, have the advantage that, in conventional cosmology, scattering by plasma has little effect, and the broad angular average suppresses fluctuations from radio sources and the Sunyaev-Zel'dovich effect. Present observational bound already are interesting: if galaxies trace mass on scales $10< hr < 30$ Mpc, and the mass autocorrelation function vanishes at larger separations, then the predicted quadrupole moment $a_2$ of the CBR is about at the RELICT bound described by Strukov. If $a_2$ were detected at about the present bound, it would be a fascinating clue to large-scale structure and a serious problem for theories that produce clustering of galaxies by local mass rearrangement, of the sort discussed by Hogan.

The message to theorist on the spectrum of the CBR is that we must pay serious attention to the Nagoya-Berkeley submillimiter excess. If real, the excess almost certainly signals interesting activity at high redshifts, whether reradiation of starlight by dust, which Carr showed would happen at $z \sim 30$ or Compton-Thomson scattering at still higher redshifts, as discussed by Danese, or mixing of very small scale temperature inhomogeneities at $z \sim 10^4$, as Daly considered.

Thermal radiation from dust is the most attractive of the above three models for the submillimiter excess, because we know the effect has to be present at some level, but there is a serious problem with isotropy. In any reasonable variant of the dust scenario I can think of, the space distribution of the submillimiter source is the same as that of the dust; and the dust surely is where the galaxies are. (Whether giant or dwarf galaxies dominate the submillimiter luminosity is a second order consideration; the space distributions are similar.) The orders of magnitude for this picture are simple: the present galaxy clustering length $r_0 \sim 5h^{-1}$Mpc at the Hubble length $c/H$ gives fluctuations $\delta i/i \sim (Hr_0/c)^{1/2}$, which works out to about 5% on the angular scale $Hr_0/c \sim 6$ arcmin. This seems hard to hide. But there certainly is no crisis: as noted above, we have other theories.

Finally, I should like to point out that we are celebrating two anniversaries, that of the discovery of the CBR, and, as the main topic of this meeting, the detection and interpretation of deviations of the CBR from homogeneous sea of blackbody radiation. The latter anniversary may be a negative number, but if so the impression I bring away from this meeting is that the odds are we do not have long to wait: reasonable-looking theoretical estimates are being pressed on several sides by very active experimental efforts. The anniversary may be positive. Possibly we will remember this meeting as the one where we discussed significant signals, perhaps in the spectrum measurements, perhaps in one of the anisotropy experiments; and those who were arguing for what turned out to be the right explanation will be justified in remembering that too.

This work was supported in part by the US NSF.